AQA

AIM HIGH
FOR THE TOP GRADES

D0244042

GCSE

Chemistry

Ripon Grammar School

Book No _233_

Name	Form	Date Issued
James Hamilton	4B	12.9.12
Imogen Martin	4C	15/9/14

Richard Grime

with **Rob Wensley**

Series editor
Nigel English

Longman
Part of Pearson

Longman is an imprint of Pearson Education Limited, Edinburgh Gate, Harlow, Essex, CM20 2JE.

www.pearsonschoolsandfecolleges.co.uk

Text © Pearson Education Limited 2011
Edited by Stephen Nicholls
Typeset by Tech-Set Ltd, Gateshead
Original illustrations © Pearson Education Ltd 2011
Illustrated by Tech-Set Ltd, Geoff Ward, Tek-Art
Cover design by Wooden Ark
Cover photo: A layer of bubbles on water. The bubbles are maintained by washing-up liquid, which is made up of molecules with a hydrophilic head and a hydrophobic tail. © Science Photo Library Ltd: Dr Jeremy Burgess.

The rights of Richard Grime, Nigel Saunders and Martin Stirrup to be identified as authors of this work have been asserted by them in accordance with the Copyright, Designs and Patents Act 1988.

First published 2011

15 14 13 12 11
10 9 8 7 6 5 4 3 2

British Library Cataloguing in Publication Data
A catalogue record for this book is available from the British Library

ISBN 978 1 408253 79 3

Acknowledgements

The authors and publisher would like to thank the following individuals and organisations for permission to reproduce photographs:

(Key: b - bottom; c - centre; l - left; r - right; t - top)

viii Imagestate Media: John Foxx Collection. **x** PhotoDisc. **1** www.imagesource.com. **2–3** Corbis: Peter Ginter / Science Faction. **4** Pearson Education Ltd: Trevor Clifford (cr, bl); Trevor Clifford (cl); Trevor Clifford (br). Shutterstock.com: Vibrant Image Studio (t). **5** Pearson Education Ltd: Trevor Clifford (tr); Trevor Clifford (tl); Trevor Clifford (bl, br). **10** Trevor Clifford (r); Peter Gould (l). **11** Getty Images: Time & Life Pictures (t). Pearson Education Ltd: Naki Kouyioumtzis (b). **12** Digital Vision (b). Shutterstock.com: Andrei Nekrassov (t). **13** Corbis: Jason Hawkes. **14** Pearson Education Ltd: Trevor Clifford (tr). Peter Gould (bl). Shutterstock.com: Buquet Christophe (tl). **15** Getty Images: Eco Images / Universal Images Group (br). Martin Stirrup. Pearson Education Ltd: Trevor Clifford (c). Shutterstock.com: Jiri Vaclavek (l). **16** Martin Stirrup (l). Photolibrary.com: Aflo Foto Agency (r). **17** DK Images: Tim Ridley (b). Shutterstock.com: Niels Quist (t). **18** PhotoDisc: Glen Allison (tr). Rex Features: Alex Segre (b). Shutterstock.com: Vicente Barcelo Varona (tl). **19** Construction Photography: Grant Smith (b). Martin Stirrup (t). **20** Shutterstock.com: Lee Prince (l); Jens Mayer (c); Denis Selivanov (r). **21** Shutterstock.com: jennyt. **22** Photos.com (l). Shutterstock.com: Audrey Snider-Bell (r). **23** Alamy Images: Leslie Garland Picture Library. **24** PhotoDisc: Photolink (l). Shutterstock.com: Roy Palmer (r). **25** Peter Gould. **26** Corbis: Joel Stettenheim (b). PhotoDisc: StockTrek (t). **28** Shutterstock.com: Olaf Speier (r); holligan78 (l). **29** Science Photo Library Ltd (b). Shutterstock.com: PJF (c); MC_PP (t). **30** Pearson Education Ltd: Naki Kouyioumtzis (l). PhotoDisc (r). www.imagesource.com (c). **31** Construction Photography: David Stewart-Smith (t). Martin Stirrup. **34** Peter Gould. **38** Photos.com: Jupiterimages. **43** Shutterstock.com: Zoran Karapancev. **50–51** Getty Images: G. Brad Lewis. **53** Science Photo Library Ltd: Andrew Lambert Photography. **56** B&H Colour Change (r). Shutterstock.com: Andresr (l). **57** iStockphoto: bojan fatur (b). Science Photo Library Ltd: AJ Photo (t). **58** iStockphoto: Michael Utech (bl); Aleksandar Jaksic (r). PhotoDisc: Photolink / Tracy Montana (tl). **60** Peter Gould (br). Science Photo Library Ltd: Roger Job / Reporters (t). Shutterstock.com: Chris Hill (bl). **62** Pearson Education Ltd: Richard Smith. **63** Science Photo Library Ltd: Martin Bond. **65** Pearson Education Ltd: Trevor Clifford (r); Trevor Clifford (l). **66** Science Photo Library Ltd: Martyn F. Chillmaid (r). Shutterstock.com: Dusan Zidar (l). **67** Pearson Education Ltd: Jules Selmes. **68** Digital Vision: Getty Images (b). Shutterstock.com: Daniel Rajszczak (t). **77** iStockphoto: Lars Nilsson. **79** Shutterstock.com: Monica Johansen. **80** Shutterstock.com: iofoto (t); Bronskov (b). **92–93** PhotoDisc: Photolink. **94** Science Photo Library Ltd: Sheila Terry. **95** Pearson Education Ltd: Trevor Clifford (bl, br). Peter Gould. **98** Pearson Education Ltd: Trevor Clifford (t, b). ix PhotoDisc: Photolink / F Schussler. **102** Shutterstock.com: mirounga. **103** CERN Geneva (r). Shutterstock.com: jordache (l). **104** Science Photo Library Ltd: George Steinmetz. **105** Pearson Education Ltd: Trevor Clifford. **106** Shutterstock.com: Blaz Kure. **108** Alamy Images: Steve Hamblin (t). Shutterstock.

com: Craig Jewell (b). **109** Shutterstock.com: Kevin Britland. **110** Shutterstock.com: Elnur. **111** Science Photo Library Ltd: Pascal Goetgheluck. **112** Pearson Education Ltd: Mark Bassett. **114** Science Photo Library Ltd: Susumu Nishinaga (t); Eye of Science (b). **120** Pearson Education Ltd: Stefan Grippon. **124** Pearson Education Ltd: MindStudio (t). Peter Gould. **127** Shutterstock.com: Antonio S. **128** Shutterstock.com: Yenyu Shih. **136–137** Alamy Images: David Wall. **138** Martin Stirrup. **140** iStockphoto: redmonkey8 (t). Martin Stirrup. **142** Peter Gould. **144** Peter Gould. Shutterstock.com: Losevsky Pavel (t). **147** Corbis: Document General Motors / Reuter R (t). Shutterstock.com: Sergey Peterman (b). **148** Peter Gould. Science Photo Library Ltd: Charles D. Winters (c); Andrew Lambert Photography (b). **149** Creatas (l). Pearson Education Ltd: Debbie Rowe (r). **150** Alamy Images: Art Directors & Trip (tl). Martin Stirrup. **151** Photolibrary.com: Stockbyte / George Doyle. **152** Martin Stirrup. Peter Gould. **153** Heat in a Click (t). iStockphoto: Daniel Laflor (b). **157** Pearson Education Ltd: Trevor Clifford. **159** Pearson Education Ltd: Trevor Clifford (r); Trevor Clifford (l). **160** Alamy Images: ScotStock. **162** Peter Gould. **164** FLPA Images of Nature: Nigel Cattlin (b). Pearson Education Ltd: Trevor Clifford (t). **167** Shutterstock.com: terekhov igor. **168** Science Photo Library Ltd: Sheila Terry (b). Shutterstock.com: Ingvar Bjork (t). **170** Science Photo Library Ltd: Martin Bond. **173** PhotoDisc: Photolink. **174** Heat in a Click. **184–185** FLPA Images of Nature: Chris Mattison. **186** Alamy Images: Interfoto. **190** Photolibrary.com. **191** Pearson Education Ltd: Trevor Clifford (t). Science Photo Library Ltd: Charles D. Winters (b). **192** Pearson Education Ltd: Trevor Clifford (b). Shutterstock.com: Jens Stolt (t). **193** Shutterstock.com: travis manley (r); Gina Sanders (l). **194** Alamy Images: sciencephotos. **199** Pearson Education Ltd: Trevor Clifford (r, l). **200** Shutterstock.com: douglas knight (r); Dariusz Majgier (l). **201** Shutterstock.com: Jodie Johnson. **202** Martin Stirrup. Shutterstock.com: Stephen Aaron Rees (l). **203** Shutterstock.com: Shell114. **204** PhotoDisc: StockTrek. **206** Pearson Education Ltd: Trevor Clifford. **207** Pearson Education Ltd: Trevor Clifford. **208** iStockphoto: Judi Ashlock (t). Photos.com. **210** Shutterstock.com: maksimum. **212** Getty Images: AFP (t). Shutterstock.com: Mona Makela (b). **213** iStockphoto: Gene Chutka. **220–221** Science Photo Library Ltd: Jim Varney. **222** Pearson Education Ltd (l). **223** Pearson Education Ltd. **224** Pearson Education Ltd: Trevor Clifford (t, b). **225** Pearson Education Ltd: Trevor Clifford. **226** Pearson Education Ltd: Trevor Clifford (l, r). **228** Pearson Education Ltd: Trevor Clifford (l). Photos.com. **229** Pearson Education Ltd: Trevor Clifford. **230** Pearson Education Ltd: Trevor Clifford (t, c, b). **234** Science Photo Library Ltd. **238** Science Photo Library Ltd: Charles D. Winters (b). Shutterstock.com: Tom Nance (t). **239** Pearson Education Ltd: Trevor Clifford. **240** Shutterstock.com: Wutthichai (r); SunnyS (cr); Nikola Bilic (cl); Lucie Lang (l). **241** iStockphoto: Jean Thirion. **242** Shutterstock.com: Mircea Bezergheanu.

All other images © Pearson Education

Every effort has been made to contact copyright holders of material reproduced in this book. Any omissions will be rectified in subsequent printings if notice is given to the publishers.

Introduction

This student book has been written by experienced examiners and teachers who have focused on making learning science interesting and challenging. It has been written to incorporate higher-order thinking skills to motivate high achievers and to give you the level of knowledge and exam practice you will need to help you get the highest grade possible.

This book follows the AQA 2011 GCSE Chemistry specification, the first examinations for which are in November 2011. It is divided into three units, C1, C2 and C3. Within each unit there are two sections, each with its own section opener page. Each section is divided into chapters, which follow the organisation of the AQA specification.

There are lots of opportunities to test your knowledge and skills throughout the book: there are questions on each double-page spread, ISA-style questions, questions to assess your progress and exam-style questions. There is also plenty of practice in the new style of exam question that requires longer answers.

There are several different types of page to help you learn and understand the skills and knowledge you will need for your exam:

- Section openers with learning objectives and a check of prior learning.
- 'Content' pages with lots of challenging questions, Examiner feedback, Science skills, Route to A*, Science in action and Taking it further boxes.
- 'GradeStudio' pages with examiner commentary to help you understand how to move up the grade scale to achieve an A^x.
- 'ISA practice' pages to give you practice with the types of questions you will be asked in your controlled investigative skills assessment.
- Assess yourself question pages to help you check what you have learnt.
- Examination-style questions to provide thorough exam preparation.

This book is supported by other resources produced by Longman:

- an ActiveTeach (electronic copy of the book) with BBC video clips, games, animations and interactive activities
- an Active Learn online student package for independent study, which takes you through exam practice tutorials focusing on the new exam questions requiring longer answers, difficult science concepts and questions requiring some maths to answer them.

In addition there are Teacher Books, Teacher and Technician Packs and Activity Packs, containing activity sheets, skills sheets and checklists.

The next two pages explain the special features that we have included in this book to help you learn and understand the science and to do the very best in your exams. At the back of the book you will also find an index and a glossary.

Contents

How to use this book

These two pages illustrate the main types of pages in the student book and the special features in each of them. (Not shown are the end-of-topic Assess yourself question pages and the Examination-style question pages.)

Section opener pages – an introduction to each section

An introductory paragraph to help put what you will be learning into context. There are two section openers for each unit.

A list of the learning objectives you will have achieved by the end of the section.

Test yourself on what you should have learned previously that will help with your understanding of this section.

Content pages – covering the AQA specification

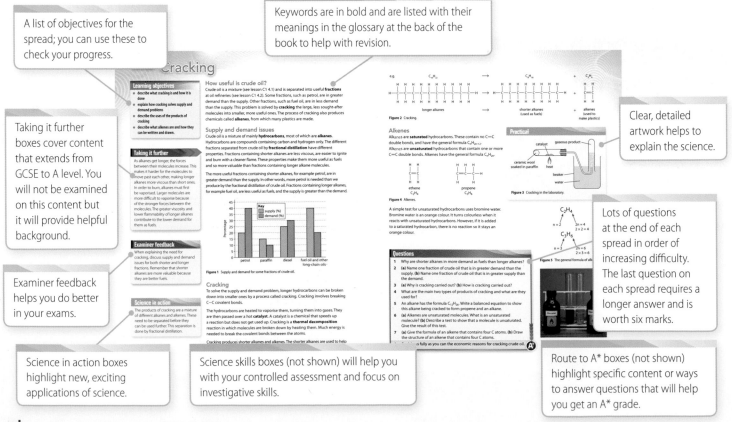

A list of objectives for the spread; you can use these to check your progress.

Keywords are in bold and are listed with their meanings in the glossary at the back of the book to help with revision.

Taking it further boxes cover content that extends from GCSE to A level. You will not be examined on this content but it will provide helpful background.

Clear, detailed artwork helps to explain the science.

Examiner feedback helps you do better in your exams.

Lots of questions at the end of each spread in order of increasing difficulty. The last question on each spread requires a longer answer and is worth six marks.

Science in action boxes highlight new, exciting applications of science.

Science skills boxes (not shown) will help you with your controlled assessment and focus on investigative skills.

Route to A* boxes (not shown) highlight specific content or ways to answer questions that will help you get an A* grade.

ISA practice pages – to help you with your controlled assessment

> This question requires a longer-text answer for which you will also be assessed on your use of English and of specialist terms.

> The questions are similar to the ones you will be asked in your controlled assessment papers.

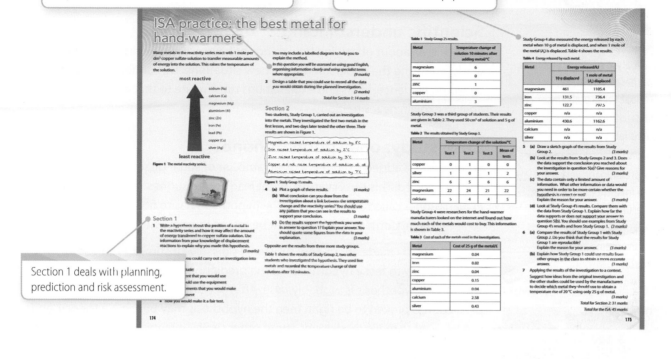

> Section 1 deals with planning, prediction and risk assessment.

GradeStudio pages – helping you achieve an A*

> 'GradeStudio' questions focus on the new exam questions, which require a longer answer.

> Three student answers are given at three different grades, B, A and A*, so you can see how they improve.

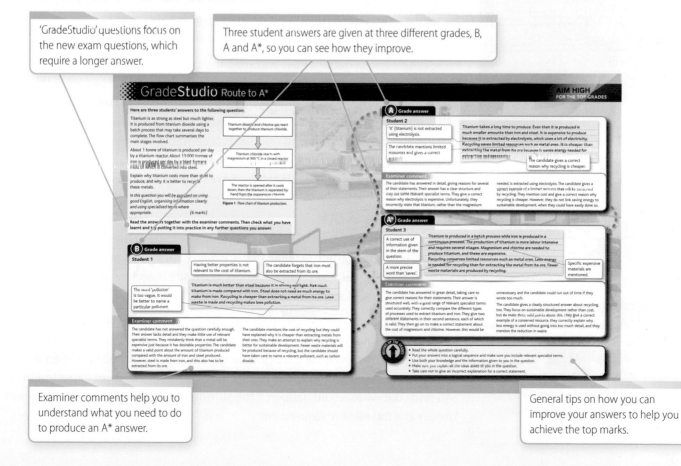

> Examiner comments help you to understand what you need to do to produce an A* answer.

> General tips on how you can improve your answers to help you achieve the top marks.

Researching, planning and carrying out an investigation

Do you take sugar?

Scientific understanding

Science is all about observing how things behave, trying to understand how they work, and using the understanding in new situations. This understanding must be based on evidence. Ideas and explanations must relate to that evidence, and be able to explain new situations.

Case study: one spoon or none?

When sugar is added to coffee, a given amount of sugar dissolves quicker if the coffee is stirred. This **observation** might lead you to suggest an explanation. The explanation is called a **hypothesis.** The hypothesis has to fit all the known facts. To see if the hypothesis is true, it must be tested in a new situation.

A **prediction** is made about how substances dissolve if stirred. Predictions provide a way to test a hypothesis – if the prediction is tested and proved wrong, the hypothesis is probably false. After several investigations in which the prediction is proved to be right, then the hypothesis may become a **theory**, and can be used to make predictions about other similar events.

Testing the prediction

In this investigation, the **independent variable**, the one you change or select, is the amount of stirring. The **dependent variable**, the one that you measure, is the time for all the sugar to dissolve. **Control variables** are other variables you need to keep the same, for example the volume of coffee used, if the test is to be fair.

All scientific investigations have **hazards**, things that can go wrong with the experiment and cause injury to people or objects. The biggest hazard in this investigation is the risk of scalding yourself. To minimise the **risk**, or the chance of it happening, **control measures** are used. A control measure is something that reduces the hazard to a level of risk that is acceptable. The control measures could be to wear gloves, firmly hold the beaker with a clamp, and ensure the kettle is not near the edge of the bench or anywhere it could be knocked over.

Preliminary work

Before you carry out your investigation, you need to be certain the investigation will give you results that are meaningful. **Preliminary work** allows you to find out how much sugar you should use. Too much sugar and it will take too long to dissolve, or some of it will never dissolve.

You also need to decide on the **range** of your independent variable, that is, the amount of stirring. The range should be evenly spaced, using at least six different values if possible, and wide enough to ensure any trend can be identified. If you carry out the investigation without stirring at all, this is a **control experiment**, and finds out what would happen if the stirring didn't take place. It is sometimes known as a **baseline** measurement.

Validity, reliability, precision and accuracy

Preliminary work is always used to find out what values of one or more variables are going to provide meaningful results, never to check that the experiment will work.

It may be that your preliminary work helps you improve the **validity** of your investigation. Will your investigation allow the prediction and as a result the hypothesis to be confirmed or disproved? The results you obtain may suggest that there are more variables that need to be controlled than you originally thought, if you are to adequately test the prediction.

You should repeat your measurements to make sure your results are reliable. You should get nearly the same value each time. The more close the measurements are to each other and the **mean** of the results, the greater the **precision** of the results. This does not mean that your results are **accurate**. For your results to be accurate they must be close to the true value.

To be able to draw a conclusion from a set of results, your data must be **repeatable**. This means that if you repeated the investigation the data obtained would be the same or similar.

If you change the method or use different equipment, or if someone else does the investigation, and the results are still similar, then we say that the results are **reproducible**.

You can check whether the findings are valid by looking for similar evidence from classmates or on the Internet, or by trying a different method to see if you get the same answer to the investigation.

Route to A* (A*)

If you get an anomalous result, think about it – was it an error in measurement, or do you need to amend your hypothesis or your experimental technique?

Plan your experiment in advance: you must decide what range you will use for the independent variable, what controls you need and how you will record your results.

Science in action

In 1994, McDonald's were successfully sued for damages when a customer was scalded by her hot coffee. She suffered burns to 16% of her body, and was in hospital for 8 days. She claimed that McDonald's should have warned her that the coffee was hot. As a control measure, McDonald's now print a caution on every paper cup.

Questions

1 Suggest a hypothesis to explain why sugar dissolves faster in coffee when stirred.

2 Suggest two more control variables affecting an investigation into dissolving.

3 Describe how you would carry out a preliminary experiment to determine a suitable temperature to use for the dissolving sugar investigation.

4 Explain the difference between a control variable and a control experiment.

5 Make a prediction that you can test that would confirm or disprove your hypothesis from question 1.

6 Describe how you would investigate your prediction from question 5. (A*)

Presenting, analysing and evaluating results

Learning objectives

- describe how to record experimental data
- describe how to report and process your experimental data
- evaluate the data collected, identifying errors
- analyse the evidence from your data and other data
- use research to confirm whether the findings are valid.

Examiner feedback

The independent variable, that is, the one you change, is usually recorded in the left-hand column of your table. The right-hand column is for the dependent variable. You may need several columns if you do several trials for each value of the independent variable. Both columns should have the units in the heading, to save writing them out for each measurement.

Recording and displaying results

It is best to collect your results in a table that you have already prepared. Table 1 shows the results of an experiment measuring the percentage of sugar remaining undissolved after 60 seconds when coffee has been stirred for different periods. A **mean** for each value of the independent variable should be calculated. Sometimes you can spot a trend or pattern in a set of results from the table, but it is more likely that you will need to see them as a bar chart, if the independent variable is **categoric**, or as a line graph, for **continuous variables**.

Table 1 Example table for recording results of an investigation.

Time of stirring/s	Sugar undissolved after 60 s (%)				
	Trial 1	Trial 2	Trial 3	Trial 4	Mean
10	63	66	64	67	
20	29	31	32	33	
30	12	16	14	25	
40	11	8	9	11	
50	7	11	9	8	
60	9	7	6	8	

Calculating the mean

If the results are 12, 16, 14 and 25%, you discard 27 as it is not close to the rest, then $\frac{12 + 16 + 14}{3} = 14\%$

Examiner feedback

When calculating a mean, make sure you discard anomalous results and record the mean calculated to the same number of decimal places as the original readings.

Patterns and relationships

When plotting a line graph you should first plot the points and then look for a pattern in the points. Sometimes there are errors. Table 2 lists common errors and how to deal with them. If there are still anomalous results, these will show up as points that will not lie on your best-fit line or curve. Plot the line or curve leaving these out.

Table 2 Common errors in results, and what to do about them.

Error	Description	Solution
Random	Results spread around the true value	Calculate the mean
Systematic	Readings to be spread about a different value, than the true one	A change to the equipment or method is needed
Zero	Systematic error in which the instrument used for measurement is not set at zero	Find the difference to zero, then add it to or subtract it from the results

If a set of scales is not set at zero before you step on it, the measurement will have a zero error.

The best-fit line or curve shows you the relationship between the two variables. Straight lines indicate a linear relationship. In Figure 1, the top graph shows a **positive linear** relationship, while the middle graph shows a **negative linear** relationship. If the line goes through the origin, where the axes meet, then the relationship may be **directly proportional**. This is only the case when the origin of the graph is truly zero on both axes.

Curved lines indicate a more complex relationship, as shown in Figure 2.

When describing a curve or complex relationship, make sure that you say what the starting pattern is, how it changes and when it changes. If comparing two sets of results make sure you describe the similarities and the differences.

graphs showing a linear relationship

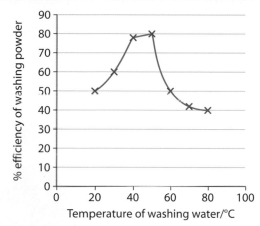

Figure 2 A curve plotted from experimental results.

graph showing directly proportional relationship

Figure 1 Examples of line graphs.

Analysing the evidence

Your conclusion must relate to the investigation. It is important not to go any further than the evidence you have. It may be possible to identify further evidence required to support or provide more detail for your conclusion. Look at Figure 2. This is a graph of how temperature affects the effectiveness of biological washing powder. The **interval**, or gap, between temperatures in this investigation was 10°C. What conclusion can you draw? It seems that the powder becomes more efficient as temperature increases up to 40°C, plateaus to 50°C, then falls back rapidly. Is it really a plateau between 40 and 50°C? You really need more data here, so your conclusion should be: 'it rises to 40°C, and starts to drop at 50°C, but I need more data between 40 and 50°C to know what's happening here. It seems from the graph that it should reach a peak.'

Which temperature should you choose?

Questions

1   Calculate the means for the experimental data in Table 1.
2   Plot a graph of the results from Table 1. Draw in your best-fit line.
3   What type of relationship does this show?
4   Use your graph to predict the percentage of sugar undissolved after 70 seconds.
5   Describe the patterns of the results on your graph.
6   Look at Figure 2. Select two further temperatures at which to carry out the investigation to improve your conclusion. Explain the reasons for each choice.

The Earth provides

Everything we use comes from the Earth in one way or another. A few things can be used directly, such as limestone blocks for building. However, mostly the Earth provides the raw materials that must be turned into useful products by physical or chemical processes.

In this unit you will first learn about the fundamental nature of materials; of atoms and elements, and how they form chemical compounds. It is our understanding of these fundamental ideas that has allowed us to create the products we need. You will then go on to look at three key topics of great importance to the built environment and the way we live in it.

Limestone is a common rock that has been used directly for buildings for thousands of years. Today, millions of tonnes of limestone are dug from quarries every year and chemically processed to make the key building materials cement and concrete. What effect does this have on the environment?

Most metals are found locked up in chemical compounds within the rocks. How can they be extracted, processed and used for buildings, machinery and cables? What are the environmental hazards associated with mining and metal production, and how can we make the most of our finite resources?

Crude oil, an ancient biomass found in the rocks, is the raw material for many useful products. Some of these are important fuels. How can fuels such as petrol be extracted from this complex mixture? What are the advantages and disadvantages of burning these fuels and what alternatives do we have to their use?

Test yourself

1. How are the physical properties of solids, liquids and gases controlled by the way their particles are arranged?

2. What is the difference between an element, a mixture and a compound?

3. What effect does acid rain have on limestone, and how does this compare with the simple acid/base reaction?

4. What is the reactivity series with regard to metals and their reactions?

5. What are fossil fuels and how did they form?

6. How does our use of fossil fuels affect the environment?

Objectives

By the end of this unit you should be able to:

- explain simple chemical processes in terms of the atoms involved and how they form compounds
- describe the chemistry of limestone and limestone products
- evaluate the advantages and disadvantages of our use of limestone and limestone products, including environmental issues
- describe the processes involved in extracting metals from rocks
- explain why some metals are easier and cheaper to extract than others
- analyse data on the properties of different metals and use this to explain their uses
- evaluate the possible ways in which metal extraction may have to develop in the future, as our resources are used up
- explain how useful products may be obtained from crude oil
- evaluate the social, economic and environmental impacts of the uses of fuels.

What are materials made of?

Learning objectives

- explain that all substances are made from atoms
- use the symbols for some common elements
- explain the difference between elements and compounds
- interpret the chemical formula of a compound
- describe the simple structure of an atom.

One 3-litre balloon contains 80 000 000 000 000 000 000 000 atoms or 8×10^{22} atoms.

sulfur, S

potassium, K

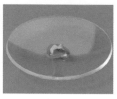

sodium, Na

copper, Cu

Some common elements and their symbols.

Route to A*

It is well worth familiarising yourself with the names, symbols and properties of the common elements. Questions will make a lot more sense on first reading, saving you a lot of time.

Atoms

All materials are made up of **atoms** – the paper and ink on this page, the air you are breathing and every part of your body. Atoms vary in size but they are all incredibly small – about 0.000 000 1 mm wide. That means you would need 10 million of them end to end to make 1 mm.

Elements

A substance made up of only one type of atom is called an **element**. There are about 100 different elements. In most elements, like sulfur or copper, the atoms are joined together. In a few, such as the helium inside the balloons in the photo, the atoms are separate.

Each element has a unique, internationally recognised **chemical symbol** of one or two letters. Most symbols clearly derive from the name of the element in English. However some are less obvious, for historical reasons (see Table 1).

Table 1 Some elements and their symbols.

Single letters (always capital)	Double letters (first letter capital)
H Hydrogen	Al Aluminium
C Carbon	Ca Calcium
N Nitrogen	Cl Chlorine
O Oxygen	He Helium
S Sulfur	Mg Magnesium
I Iodine	Zn Zinc
Some 'oddities'	
Na Sodium – from the Latin name *natrium*	
K Potassium – from the Latin name *kalium*	
Fe Iron – from the Latin name *ferrum* (iron and steel are called ferrous metals)	
Cu Copper – from the Latin name *cuprum* (from Cyprus)	
Pb Lead – from the Latin name *plumbum* (plumbers used to work with lead pipes)	

Compounds

Compounds are substances made from two or more types of atom joined together. This joining happens during a chemical reaction. Many familiar substances, for example water and alcohol, are compounds.

Each compound has a **chemical formula**. The chemical formula for a compound contains two types of information. The symbols in the chemical formula tell you which types of atom are present. FeS, for example, is the formula for iron sulfide, which contains iron atoms and sulfur atoms combined together.

The formula FeS also tells us that there is one iron atom for every sulfur atom in the compound. We say that the **ratio** of iron atoms to sulfur atoms is 1 : 1. Some chemical formulae contain small (subscript) numbers after the symbol. The chemical formula for water, for example, is H_2O. The number tells you that there are two hydrogen atoms for every one oxygen atom. The ratio of hydrogen atoms to oxygen atoms in water is 2 : 1.

You can have the same atoms joined together and get different compounds. CO_2 is carbon dioxide and CO is carbon monoxide. Both are gases, but carbon monoxide, CO, is poisonous. Knowing the chemical formula is important.

Inside the atom

In their earliest ideas about atoms, people thought of atoms simply as spheres. We often draw them like this when we are making simple models of chemical reactions. One hundred years ago, however, scientists discovered that the atoms themselves were made of even smaller particles.

Atoms have a central **nucleus**. This is made up of **protons** and **neutrons**. Outside the nucleus are the **electrons**. The shape of the atoms is formed by the moving 'cloud' of electrons that surrounds the nucleus. Protons, and electrons are electrically charged, while neutrons have no charge (see Table 2).

Atoms of the same element always have the same number of protons in the nucleus. If the number of protons is different it is a different element. When atoms are separate, for example in helium, they have the same number of protons as electrons. Helium atoms have two protons, each with a $+1$ change, and two electrons, each with a -1 charge, so the charges cancel out and the atoms are neutral overall.

During chemical reactions electrons in atoms can be lost, gained or shared. This means that atoms that are joined together may have an overall charge. We call charged atoms **ions**; and you will learn more about them later.

sodium chloride, NaCl   iron(III) oxide, Fe_2O_3

sodium fluoride, NaF   copper(II) sulfate, $CuSO_4$

Some compounds and their formulae.

Table 2 The charge on atomic particles.

Name of particle	Charge
proton	$+1$
neutron	0 (neutral)
electron	-1

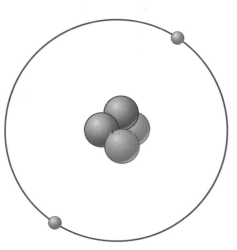

Figure 1 This simple model of a helium atom shows the electrons in orbit around the nucleus. In fact they form a fuzzy cloud, and the scale is very different.

Questions

1   What are the symbols for sodium, sulfur, iron, lead and potassium?

2   Brass is a metal containing copper and zinc atoms. Is brass is an element?

3   Copy and fill in the table to show the ratio of the different atoms in each compound. The top one has been done for you.

Chemical formula	Hydrogen atoms	:	Sulfur atoms	:	Oxygen atoms
SO_3	0	:	1	:	3
SO_2		:		:	
H_2SO_4		:		:	
H_2O		:		:	

4   Copy and fill in the table to show the chemical formula for each compound. The top one has been done for you.

Iron atoms	:	Sulfur atoms	:	Oxygen atoms	Chemical formula
1	:	1	:	4	$FeSO_4$
1	:	0	:	1	
2	:	0	:	3	
1	:	1	:	0	

5   A single neon atom has 10 protons.
    (a) How many electrons will it have? **(b)** What will be its charge overall?

6   A student stirs iron powder and sulfur powder together in a test tube. She then heats the test tube. A chemical reaction takes place and iron sulphide is made. Write an account of her experiment using the words *element, mixture, compound, symbol* and *chemical formula*.

Electrons rule chemistry

Learning objectives

- describe how most materials are made
- describe the formation of ions
- explain why metal/non-metal compounds have ionic bonds
- describe how non-metal compounds are formed and the type of bonding that occurs
- explain what happens to the total mass in chemical reactions.

Electrons control chemical reactions

When a chemical reaction takes place between atoms, electrons may move from one atom to another or be shared between atoms.

When metals and non-metals combine, the metals give up some electrons and the non-metals take them. Because the electrons are negatively charged, the once-neutral atoms become charged particles called **ions**.

- When a metal atom loses an electron it becomes a **positive ion**.
- When a non-metal atom gains an electron it becomes a **negative ion**.

Positive and negative ions are attracted to one another and can form compounds. Sodium and chlorine form a compound in this way, called sodium chloride. This is common salt. The **chemical bonds** between the ions are called an **ionic bonds**.

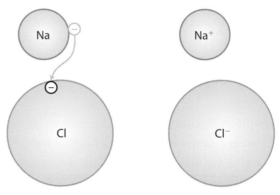

Figure 1 After an electron is transferred from each sodium atom to each chlorine atom, the sodium and chloride ions are arranged in a regular pattern to form a salt crystal.

Share and share alike

When non-metals combine, the atoms share electrons and form **molecules**. One carbon atom and two oxygen atoms can combine in this way to form carbon dioxide: CO_2. Two hydrogen atoms and one oxygen atom form water: H_2O.

Atoms of elements such as oxygen can also share electrons together and make molecules without any other elements. The oxygen in the air forms O_2 molecules from two oxygen atoms.

Strong chemical bonds form when atoms share electrons to make molecules. These are called **covalent bonds**.

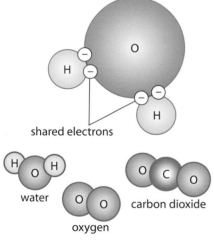

Figure 2 Electrons are shared to make bonds within molecules.

Conserving atoms

When chemicals react, the atoms rearrange themselves. We say they are conserved: that is, no atoms are lost or gained. You always end up with the same number of each type of atom as you started with. You can show this as a **balanced equation**. For example, sodium hydroxide **neutralises** hydrochloric acid to form sodium chloride and water:

$$NaOH + HCl \longrightarrow NaCl + H_2O$$

1Na, 1O, 2H, 1Cl	1Na, 1O, 2H, 1Cl
in reactants	in products

What if it's not that simple?

If an equation does not balance 'first go' you may need more of one reactant or product. To show this, you put a number in front of the formula. For example, $2H_2O$ means two molecules of water: $H_2O + H_2O$.

Balancing more difficult equations

The reaction of hydrogen and oxygen to form water does not balance 'first go'. Here's a step-by-step guide.

Table 1 Balancing equations.

	Step	Reactants		Products	Balanced?
1	Write the word equation	hydrogen + oxygen	⟶	water	
2	Add the formulae	H_2 + O_2	⟶	H_2O	
3	Count the atoms	2H, 2O	⟶	2H, 1O	✗ no
4	Add another water molecule	H_2 + O_2	⟶	$2H_2O$	
5	Recount	2H, 2O	⟶	4H, 2O	✗ no
6	More hydrogen needed	$2H_2$ + O_2	⟶	$2H_2O$	
7	Recount	4H, 2O	⟶	4H, 2O	✓ yes

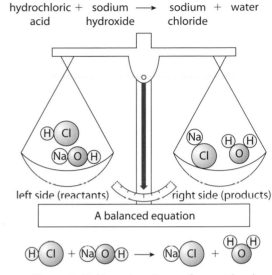

hydrochloric + sodium ⟶ sodium + water
acid   hydroxide   chloride

left side (reactants)   right side (products)

A balanced equation

Figure 3 Making salt and water from acid and alkali: it's just a case of rearrangement.

Conserving mass

As the number of atoms does not change during a chemical reaction, it follows that the mass does not change either. For example, when calcium oxide reacts with carbon dioxide to form calcium carbonate:

$$CaO + CO_2 \longrightarrow CaCO_3$$

1Ca, 1C, 3O   1Ca, 1C, 3O   the atoms add up on

in reactants   in products   each side…

… and by experiment you will find the masses add up too:

56 g of calcium oxide and 44 g of carbon dioxide give 100 g of calcium carbonate

or

11.2 g of calcium oxide and 8.8 g of carbon dioxide give 20 g of calcium carbonate.

This is known as the **conservation of mass**.

Questions

1. Describe what happens in terms of electrons when potassium atoms react with fluorine atoms to form potassium fluoride.

2. What type of chemical bonds would you find in:
   (a) magnesium chloride, $MgCl_2$; **(b)** carbon dioxide, CO_2; **(c)** methane, CH_4; **(d)** copper oxide, Cu_2O?

3. Count up the atoms on each side of this equation to check that it balances:
   $6CO_2 + 6H_2O \longrightarrow C_6H_{12}O_6 + 6O_2$

4. Balance the equation for the reaction of sodium hydroxide (NaOH) and sulfuric acid (H_2SO_4) to form sodium sulfate (Na_2SO_4) plus water (H_2O).

5. Balance the equation $CO + Fe_2O_3 \longrightarrow Fe + CO_2$

6. 100 g of calcium carbonate gives 56 g of calcium oxide when heated. What mass of carbon dioxide is driven off in this reaction if no other products are formed?

7. Treating 56 g of calcium oxide with water makes 74 g of dry calcium hydroxide. What mass of water has reacted with the calcium oxide to make this?

8. Use the balanced equation for the reaction of hydrogen and oxygen to explain how atoms are conserved in a chemical reaction. If 2 g of hydrogen reacts completely with 16 g of oxygen, what mass of water will be formed?

C1 1.3 Electrons and the periodic table

Learning objectives

- describe how electrons are arranged in the energy levels (electron shells) around the first 20 elements
- explain how the different elements may be arranged in a periodic table
- use the position of elements in the periodic table to work out their properties.

Energy levels

The electrons move round the nucleus of an atom forming an 'electron cloud', which gives the atom its shape. The pattern is not haphazard, and the electrons can only fit into certain zones. These zones are called energy levels or **electron shells**.

Atoms have many energy levels but the electrons usually occupy the lowest ones. For the first 20 elements, the pattern is a simple one.

- Level one, next to the nucleus: one or two electrons only
- Level two: up to eight electrons
- Level three: up to eight electrons (for the first 20 elements)
- Level four: any more electrons (for the first 20 elements)

If you know the number of electrons in the atoms, you can work out how they are arranged in the levels. This is called the **electronic structure** of the atom. It can be drawn on a 'flat' version of the atom as in Figure 1, or written as the numbers in each level, in turn.

Filling them up

- Hydrogen's single electron fits in the first level, and helium's two electrons fill up the first level completely.
- At number three, lithium's third electron must start a new level.
- Carbon is number six, so it half-fills the second level.
- Neon has 10 electrons, which fill both levels one and two.
- Sodium has 11 electrons, so the 11th electron starts a third level.
- Elements 12–18 (magnesium to argon) progressively fill this level.
- Calcium, at number 20, is the last element to show this simple pattern; its last two electrons are found in the fourth level.

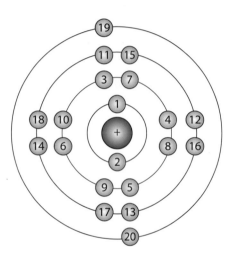

Figure 1 How electrons fill the energy levels of atoms for the first 20 elements. The grey circles represent positions that electrons can fill, numbered in order. You can mark a cross (×) to show the position of an electron, as in Figure 2.

Route to A* ⍟A*

Electrons rule chemistry. A clear understanding of how electrons fit into energy levels, and the significance of the energy levels, is the key to success in chemistry.

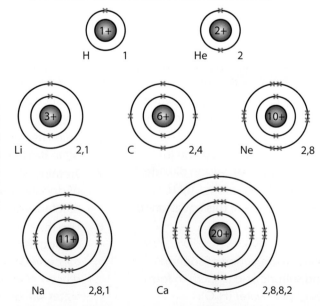

Figure 2 Electrons usually fit into the lowest available energy level.

8

Different properties

Different elements have different properties. However, there is a pattern to this variation that is controlled by the number of electrons in the highest energy level, or outer shell. You can see this in the **periodic table**. Each vertical column, or **group**, shows elements with similar properties. The group number tells you how many electrons there are in the highest energy level.

The periodic table may look complicated but you don't need to learn it by heart. You simply need to know how to use it. If you know the properties of one element in a vertical group, you know that the properties of other elements in that group are likely to be similar. For example magnesium (Mg) is a reactive metal, so you would expect calcium (Ca) to be a reactive metal too, as it has a similar electronic structure.

Metals are on the left of the table, non-metals on the right, divided by a zig-zag line as shown in Figure 3. Many of the metals you are familiar with from everyday life, such as iron (Fe) and copper (Cu), are found in a block that squeezes in between groups 2 and 3.

The elements are listed in order of their **atomic numbers** – hydrogen has atomic number 1, helium 2, lithium 3 and so on. The atomic number is the number of protons in an atom of the element, which is the same as the number of electrons in an uncharged atom of the element. If you add the number of protons and the number of neutrons in the atom's nucleus, you get the mass number.

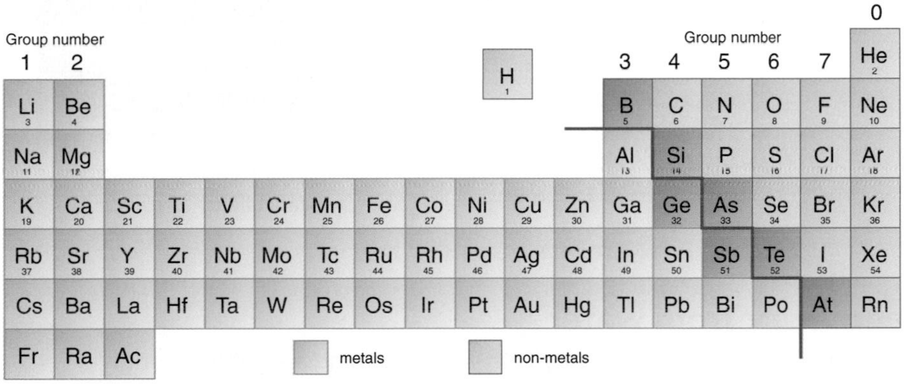

Figure 3 The periodic table of elements.

Examiner feedback

It is worth becoming familiar with the layout of the periodic table. You will use it over and over again when you study chemistry.

Taking it further

The elements closest to the zig-zag line separating metals and non-metals in the periodic table – the symbols shaded in purple in Figure 3 – often have properties that are between those of metals and non-metals. For example, silicon is a semiconductor.

Questions

1. Magnesium has 12 protons in its nucleus. How many electrons does a magnesium atom have? How many electrons will fit in each energy level?
2. Draw the electronic structures of: oxygen (O: eight electrons), aluminium (Al: 13 electrons), chlorine (Cl: 17 electrons) and potassium (K: 19 electrons).
3. Use the periodic table to see how many electrons each of these elements have in their outer shell: **(a)** oxygen (O); **(b)** iodine (I); **(c)** barium (Ba); **(d)** boron (B).
4. From the periodic table alone, which of these elements are metals and which are non-metals? Kr, Mo, Br, Ba, Rh, Ti, P.
5. Look at the periodic table. **(a)** What do iodine (I), bromine (Br), chlorine (Cl) and fluorine (F) all have in common? **(b)** What is different about their electronic configurations?
6. Oxygen is a reactive non-metal. What chemical properties would you expect the element sulfur (S) to show and why?
7. Calcium is a reactive metal. What chemical properties would you expect the element strontium (Sr) to show and why?
8. Sodium (Na) is a very reactive metal; chlorine (Cl) is a very reactive non-metal. Use their positions in the periodic table and their electron configurations to suggest properties for lithium (Li) and fluorine (F).

A*

A closer look at groups 1 and 0

Group 1: the exciting group

Group 1 contains very reactive metals such as lithium (Li), sodium (Na) and potassium (K). These all have just one electron in their outer shell, which can be lost to give a positive ion when they react with another element. This happens easily, which is why group 1 metals are so very reactive. Because of this similar electronic structure they all react in a similar way. Group 1 metals react very strongly with oxygen.

The balanced equation for the lithium reaction is:

$$4Li + O_2 \longrightarrow 2Li_2O$$

You could write the balanced equations for sodium and potassium simply by changing Li to Na or K. The atoms combine in exactly the same way in each case.

Practical

If you dissolve a group 1 metal compound in water, and the resulting solution is heated in a Bunsen burner, a coloured flame is produced. The colour comes from the metal in the solution. For example:

$$4Li + O_2 \longrightarrow 2Li_2O \text{ (scarlet flame)}$$

$$4Na + O_2 \longrightarrow 2Na_2O \text{ (yellow-orange flame)}$$

$$4K + O_2 \longrightarrow 4K_2O \text{ (lilac flame)}$$

These flame colours are so characteristic that they can be used to test for group 1 elements.

Lithium emits a characteristic scarlet light.

Keep away from water

Group 1 metals also react violently with water, forming a soluble, strongly alkaline hydroxide compound and giving off hydrogen gas. So much heat energy is produced in this reaction that the hydrogen gas often catches fire.

$$\text{Sodium} + \text{water} \longrightarrow \text{sodium hydroxide} + \text{hydrogen}$$
$$2Na + 2H_2O \longrightarrow 2NaOH + H_2$$

Again, substitute Li or K for Na and you have the balanced equations for the lithium or potassium reactions.

Sodium in water

Group 0: the dull group?

Group 0 contains the unreactive gases helium (He), neon (Ne) and argon (Ar). Helium has two electrons in its outer level, the other elements in group 0 have eight. This arrangement is very stable and makes the gases unreactive.

The unreactive nature of these gases, called the noble gases, is sometimes exactly what is needed. One hundred years ago, airships were filled with reactive hydrogen, which proved disastrous on many occasions when they exploded.

Modern airships, and 'floaty' party balloons, are filled with helium. Its density is low enough for it to float up through air, but its extreme lack of reactivity makes it perfectly safe.

A bright idea

Old-style light bulbs work by heating a thin filament of tungsten metal to 2000 °C. In air the tungsten would quickly react and burn out. The solution was to put the wire in a glass bulb full of unreactive argon gas.

Neon is also chemically unreactive – but it glows red-orange if electricity is passed through it. This property is used in brightly coloured neon lights. Neon has a very low boiling point, – 246 °C, and liquid neon is used as a very low temperature refrigerant. Krypton (Kr) and argon (Ar) are used in some lasers.

The noble gases were hard to spot

Air contains approximately 20% reactive oxygen and 80% 'unreactive' nitrogen. What was not suspected until just over 100 years ago was that, hidden within the nitrogen, there was about 1% truly unreactive gas…

In 1892, scientists discovered they could make nitrogen react with hot magnesium. However, when this reaction was performed with the nitrogen from the air, about 1% would not react. They realised that they must have discovered something new and called it 'noble gas', after the similarly unreactive 'noble metal' gold. It was later found that most of this 1% noble gas in air is argon, along with smaller amounts of neon. Helium was discovered in the Sun by analysis of sunlight before it was found on Earth.

This could never happen to a helium-filled airship, or balloon.

A neon sign

Questions

1 Rubidium (Rb) is in group 1. How will it react with water?

2 Write the balanced chemical equation for the reaction of sodium with oxygen.

3 Write the word and balanced chemical equations for the reaction of potassium with water.

4 Sodium and potassium are always kept stored in oil. Why do you think that is?

5 Krypton (Kr) is in group 0. Describe its likely chemical properties.

6 Explain why argon is used in filament light bulbs.

7 Explain why the noble gases remained undiscovered for so long.

8 Hydrogen is cheaper to produce industrially than helium. A nitrogen/oxygen mixture comes free in the air. So why do we fill party balloons with helium instead of cheaper hydrogen and fill deep-sea divers' tanks with an expensive helium/oxygen mixture instead of cheaper compressed air?

A*

Earth provides

Many old buildings in Oxford are built from limestone.

Science skills

About 70 million tonnes of limestone were quarried in Britain in 2008. We used:

- 20 million tonnes for road 'chippings'
- 21.5 million tonnes as aggregate for concrete
- 17.5 million tonnes crushed for construction ballast
- 7 million tonnes to make cement
- 2 million tonnes for the chemical industry
- 1.5 million tonnes for agriculture – neutralising soils
- 0.5 million tonnes as straight building stone.

Use the data provided above to draw a pie chart showing the uses of quarried limestone.

Rock of ages

Almost everything we use comes from the Earth in one way or another. Rock is a material that we can sometimes use straight from the ground. Blocks of stone are cut out from quarries. One example is limestone, a common rock with many uses. It has been used in building for thousands of years. Today, most limestone is crushed and used as chippings or changed by chemical processes into other useful products such as **cement**. Limestone is a very important raw material.

The effects of extraction

Digging out rocks leaves big holes – quarries. Quarries are noisy and dirty. Rock is blasted from the quarry face by explosives, spreading dust far and wide. Wildlife habitats can be destroyed as the dust and other pollutants settle over the surrounding countryside. Unfortunately, the best limestone is often found in scenic areas, such as the Peak District or the Yorkshire Dales. Quarries can be ugly places that spoil the natural beauty of the landscape, damaging tourism and upsetting local people. Rock-crushing and sorting machinery rattles away all day and huge lorries rumble down local roads. Most people wouldn't like a quarry next to their home.

However, we need the limestone for making concrete and other important materials. So we have to have quarries somewhere if we are to maintain or improve our standard of living in Britain. Quarries also provide employment for local people, which can be very important in rural areas. They also bring money into the community, which helps the local economy by supporting local shops and services.

This quarry in Yorkshire produced some of the 70 million tonnes of limestone used in Britain in 2008.

A company wishing to open up a quarry might meet local opposition on environmental grounds yet get support from the local council for providing work. They will need to weigh things up carefully to see if a new quarry is economically as well as environmentally viable. How much rock is available and how easily can it be dug out? Are there good transport links to get their product to their customers? What price can they get for that product and is that price likely to rise or fall? Quarries that make money in economic boom years might go bankrupt during an economic recession, when construction projects across the country slow down or stop.

Of course, once the rock has been removed, the soil can be replaced and trees planted. Some old quarries have been successfully turned into nature reserves or country parks. Elsewhere, old quarries have been used to build shopping malls, such as Bluewater in Kent. More imaginatively, an old clay pit in Cornwall has been turned into the amazing Eden Project.

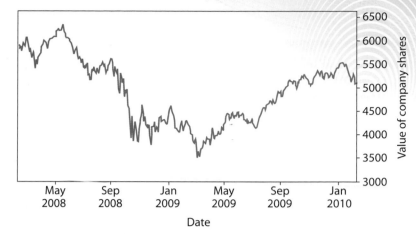

Figure 1 FTSE 100 graph. 2008 ended in a business slump that badly affected construction projects across the country.

Questions

1. Roofs used to be made from slate, a rock quarried in Wales. Why do you think cement tiles are used instead these days?

2. 2008 ended in an economic slump.

   (a) How do you think that will have affected the amount of limestone used in Britain? **(b)** What effect might that have had on the price that limestone could be sold for? **(c)** How might this have affected quarrying companies?

3. The problem with quarrying is that it affects everything around it. Before a new quarry is opened, the owners have to consider a lot of different factors. These include:

   - the cost of extracting the rock
   - the availability of people living nearby to work in the quarry
   - the amount of noise that will result from blasting away the rock
   - the dust and dirt that will be produced
   - the impact of quarrying on the landscape
   - the impact of the quarry on the local wildlife
   - the extra traffic generated
   - the effect on local shops
   - the effect on tourism if the quarry is in a tourist area
   - the need for new roads to transport the stone away
   - the price for which the rock could be sold
   - what could be done with the quarry after all the rock has been removed.

   (a) Sort out the factors in the list into those that affect the environment directly, indirectly or not at all. Put them in a table. **(b)** What effect would a large quarry have on: **(i)** the local shops? **(ii)** tourism? **(c)** Who would make money from having a quarry? Explain your answer.

4. A company wants to open a new quarry in a scenic part of the Yorkshire Dales. Write a balanced report for the local council on the pros and cons of this.

The Eden project near St Austell in Cornwall was built in a disued clay pit.

Route to A*

In your exam, you may well get case-study questions related to problems around the exploitation of the Earth's mineral resources. Make sure that you justify any statements you make and bring in examples to back up your arguments.

Limestone chemistry

The chemical structure of limestone

Limestone is the compound calcium carbonate. Its chemical formula is $CaCO_3$. This means that for every calcium atom there is one carbon atom and three oxygen atoms. All carbonates have one carbon atom and three oxygen atoms arranged like this in a regular arrangement known as a **lattice**. This regular arrangement is reflected in the shape of any crystals that form, as shown for calcite in the photograph.

Calcium carbonate in an irregular, non-crystalline form.

Calcite: the natural crystal form of calcium carbonate.

Limestone reacts with acid

Drop a piece of limestone into acid and it will fizz steadily. The acid reacts with the calcium carbonate to form a salt, water and carbon dioxide. For example, with hydrochloric acid:

calcium carbonate + hydrochloric acid $\longrightarrow$ calcium chloride + water + carbon dioxide

$$CaCO_3 + 2HCl \longrightarrow CaCl_2 + H_2O + CO_2$$

Acid rain

Rainwater is naturally slightly acidic and, over several million years, it will slowly dissolve limestone. This process can open cracks into fissures or even hollow out gigantic cave systems such as those at Cheddar, in Somerset. Pollution can make rain much more acidic. Sulfur dioxide from power stations can dissolve in rain to form sulfuric acid. Nitrogen oxides from car exhausts can dissolve to form nitric acid. These can make **acid rain**, which is strong enough to dissolve limestone over much shorter timescales.

In polluted areas, old limestone buildings have been badly damaged by acid rain and the fine detail of many statues has been lost forever.

However, limestone can be used to fight back against pollution. Lakes and rivers that have been acidified by acid rain can be restored by adding powdered limestone to neutralise the acid. The calcium sulfate produced is harmless and is in any case not very soluble in water. This is safer than adding an alkali, a base disolved in water, because if you add too much limestone it will simply remain undissolved and settle out harmlessly.

neutralisation

calcium carbonate + sulfuric acid ⟶ calcium sulfate + water + carbon dioxide

$$CaCO_3 \quad + \quad H_2SO_4 \quad \longrightarrow \quad CaSO_4 \quad + \quad H_2O \quad + \quad CO_2$$

(acid)   (neutral salt)

A common pattern

Carbonates of other metals such as magnesium, copper, sodium and zinc react in the same way. For example, copper carbonate is a green solid. If you put a piece in sulfuric acid it will fizz and dissolve to form the salt copper sulfate.

copper carbonate + sulfuric acid ⟶ copper sulfate + water + carbon dioxide

$$CuCO_3 \quad + \quad H_2SO_4 \quad \longrightarrow \quad CuSO_4 \quad + \quad H_2O \quad + \quad CO_2$$

Copper carbonate (green) and copper sulfate (blue).

This limestone statue has been badly damaged by acid rain.

Questions

1 Why do crystals of calcium carbonate all have a similar shape?

2 **(a)** What is the name of the salt that is formed when limestone dissolves in hydrochloric acid? **(b)** Explain why you need 2HCl to balance the equation.

3 In old churchyards in polluted areas you will find tombstones made from rocks such as limestone or granite. On some the inscriptions are still easy to read but on others you can barely make out any writing. Which type is likely to have been made from limestone? Explain the reason for your choice.

4 Fish tanks containing a lot of fish can get 'self-polluted' and acidic. You can buy solid white blocks to put in the tank that slowly dissolve over time to overcome this. What are the blocks likely to be made from, and how do they help to protect the fish?

5 Powdered limestone can be used to neutralise acid spills in the laboratory. Complete and balance this equation for the reaction that occurs if spilt nitric acid is neutralised in this way.
$CaCO_3 + HNO_3 \longrightarrow Ca(NO_3)_2 + \underline{\hspace{1cm}} + \underline{\hspace{1cm}}$

6 **(a)** What would you see happen if you put some copper carbonate into sulfuric acid? **(b)** What is the gas given off? **(c)** The solution turns blue. What chemical causes this to happen?

7 Zinc carbonate ($ZnCO_3$) reacts with sulfuric acid to form zinc sulfate ($ZnSO_4$). Write a balanced chemical equation for this reaction.

8 Fish prefer water to be nearly neutral. What would happen if you tried to neutralise acid lakes with alkali but put too much in? Why is this not a problem with limestone?

A helicopter dropping powdered limestone into an acidified lake.

New materials from limestone

Learning objectives

- explain how limestone and other carbonates can be broken down by thermal decomposition
- describe how calcium oxide can be turned into calcium hydroxide
- explain that calcium hydroxide can be used to neutralise acid and improve soils
- describe the test for carbon dioxide.

Thermal decomposition

If limestone is heated strongly, the compound is broken down and the atoms are rearranged.

- The carbon atom takes two oxygen atoms to form carbon dioxide.
- The calcium atom is left with just one oxygen atom; this is calcium oxide.

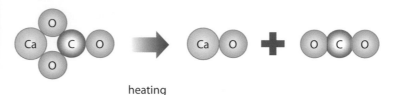

$$\text{calcium carbonate} \xrightarrow{\text{heating}} \text{calcium oxide} + \text{carbon dioxide}$$

Figure 1 Particle model for the quicklime reaction. Note that calcium carbonate and calcium oxide are ionic compounds held together by ionic bonds, whereas carbon dioxide is a molecule with covalent bonds.

As with all chemical reactions, mass is conserved. Yet if you measure the 'before and after' masses for this reaction, the mass will appear to be lower after the reaction. The reason is simple when you think about it: only the calcium oxide remains to be measured. Carbon dioxide is a gas, which escapes into the atmosphere and is lost. So, for example, 100 g of reactant limestone will apparently only give you 56 g of product, but that 'product' is all calcium oxide. The 44 g of carbon dioxide will have escaped unmeasured.

Route to A*

Atoms of different elements have different masses. These are shown in the periodic table, and can be used to work out how much of each product you will get in a reaction. Use a periodic table to calculate the masses of the compounds in Figure 1, and compare your figures with the example in the text.

Science in action

The calcium oxide formed in this way, once called lime, glows with a bright white light when heated. This light was used to illuminate the stage in the theatre before electric lights were invented. This is where the phrase 'in the limelight' comes from.

Limelight may be history in the theatre but campers still use it. Camping gas lights have a 'mantle' coated with calcium oxide around the flame to give out a bright white light.

Campers in the limelight.

Green copper carbonate breaks down to black copper oxide when heated.

Breaking down a compound by heating it like this is called **thermal decomposition**. All metal carbonates will break down in this way. For example, green copper carbonate breaks down to black copper oxide if heated over a Bunsen burner.

$$CuCO_3 \xrightarrow{\text{heating}} CuO + CO_2$$

For some group 1 metal carbonates, such as potassium carbonate, the temperature needed to make this decomposition happen is too high to achieve with just a Bunsen burner.

Reacting calcium oxide

Calcium oxide is a very strong alkali. It reacts with water to form calcium hydroxide. A lot of heat is given out in this reaction. If you spilt calcium oxide on your skin it would react with the water in your body and give you a bad burn.

calcium oxide + water $\longrightarrow$ calcium hydroxide
$$CaO + H_2O \longrightarrow Ca(OH)_2$$

energy given out

Science in action

Farmers have used these reactions for centuries, roasting limestone over charcoal fires in special kilns. They made calcium hydroxide to use as a simple fertiliser to make their soil less acid. Calcium hydroxide also helps to break up the soil so that plants can grow well.

calcium hydroxide + soil acids $\longrightarrow$ calcium salts + water
(a base)   (an acid)   (neutral)

Calcium hydroxide fertiliser was made locally by farmers all over the world.

Testing for carbon dioxide

A solution of calcium hydroxide is called limewater. This can be used to test for carbon dioxide. If you bubble carbon dioxide through limewater, it turns cloudy as a precipitate of white, insoluble calcium carbonate forms. Limewater is alkaline – the man in the photograph should wear eye protection.

calcium hydroxide + carbon dioxide $\longrightarrow$ calcium carbonate + water
$$Ca(OH)_2 + CO_2 \longrightarrow CaCO_3 + H_2O$$

Questions

1   Copy and complete this equation. Make sure it is balanced.
$CaCO_3 \longrightarrow CaO +$ _____

2   Why have farmers made calcium oxide from limestone for hundreds of years?

3   Iron carbonate ($FeCO_3$) breaks down just like calcium carbonate. What new chemicals would you get if you heated this? Write a word equation for the reaction.

4   In the thermal decomposition of copper carbonate, the black solid left behind weighs less than the original copper carbonate. Why?

5   Zinc carbonate ($ZnCO_3$) breaks down in the same way to give zinc oxide (ZnO). Write a balanced equation for this reaction.

6   Calcium hydroxide can neutralise acids such as sulfuric acid. Complete this equation and make it balance.
$Ca(OH)_2 + H_2SO_4 \longrightarrow CaSO_4 +$ _____

7   Describe each stage of the 'limestone cycle' of reactions, giving the balanced equations.

8   Calcium oxide has to be handled very carefully. Explain why, as fully as you can.

The air we breathe out contains carbon dioxide and turns limewater cloudy.

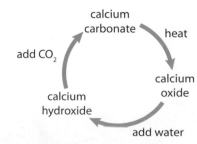

Figure 2 This reaction completes the 'limestone cycle' of reactions.

New rocks from old

Learning objectives

- describe how cement and concrete are made
- evaluate the advantages and disadvantages of concrete as a building material, including the environmental effects of its production
- explain why concrete needs to be reinforced with steel.

Rock or 'artificial rock'?

These statues look similar. One took a sculptor many weeks to make by chiselling the limestone. The other was made by simply pouring concrete into a mould.

Concrete or rock?

Making cement, mortar and concrete

To make **cement**, limestone is heated with clay in a big oven, called a rotary kiln because it keeps turning to mix everything. Thermal decomposition of the limestone occurs to give calcium oxide, which is mixed with other chemicals from the clay. The roasted product is ground to form a light grey powder. This is cement. Cement forms a paste with water that soon sets solid by a series of chemical reactions. The new compounds form tiny crystals that interlock tightly to make a hard, rock-like material.

Cement can be mixed with sand to make **mortar**, which makes a paste with water for binding bricks together.

Cement is usually mixed with sand, **aggregate** – gravel or crushed rock – and water to make **concrete**, which is cheaper and stronger than pure cement. Concrete forms a thick liquid when first mixed, and can be poured into any shape. Slow chemical reactions make the cement in it set after a few hours and eventually it becomes rock-hard. It is used to make roads, bridges and the frameworks and foundations of buildings.

The aggregate, which makes up the bulk of the concrete, can vary depending on what is available locally, from natural sand and gravel, through crushed rocks and quarry waste, to crushed and recycled building rubble left over from demolition. You can even use ash from power stations. Locally sourced materials help to keep costs down.

Cement and concrete's carbon footprint

Making cement from limestone produces carbon dioxide, which adds to Britain's carbon footprint (this contributes to global warming). The amounts produced are relatively small (just 3% of the total CO_2 emissions of the UK), but are still significant. Fortunately, concrete re-absorbs some of this CO_2 as it continues to harden over a year or so.

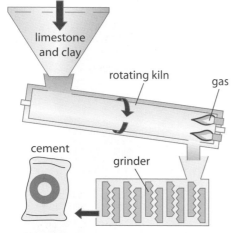

Figure 1 A rotary cement kiln.

limestone and clay

rotating kiln

gas

cement

grinder

The Royal Festival Hall was opened in 1951 as part of the Festival of Britain.

Concrete: beauty or the beast?

The architects who designed the Royal Festival Hall in London wanted to show off their building material. They left the concrete exposed for all to see. Some people like it but many find the buildings ugly, especially now that the gleaming white concrete is streaked and dirty.

The Bahá'í Lotus temple in Delhi is also made of concrete. Concrete is so versatile you can build whatever you like from it.

Getting the best out of concrete

Building with quarried stone is a slow, labour-intensive and expensive process. Each block has to be carefully cut to shape and fitted into place by an expert craftsman.

In comparison, building with concrete is quick, cheap and easy.

- Cement is relatively cheap to produce on an industrial scale.
- Powdered cement is easy to store and transport.
- Aggregate of one sort or another can be found almost anywhere and is relatively cheap.
- Liquid concrete can be easily mixed on-site.
- Liquid concrete can be moulded into any shape: floor, girders, domes, ornaments.
- Once it sets, concrete really is rock-hard.

Concrete has one problem that it shares with quarried limestone: it is very strong if you try to squash it, so it can support the weight of very large buildings such as skyscrapers. However, it is brittle, so if you stretch or bend it, it can crack easily. To overcome this problem, concrete beams or girders are reinforced with steel rods that stop the concrete from stretching and cracking.

The Bahá'í Lotus temple in Delhi, India, which was completed in 1986.

Steel rods are used to strengthen concrete.

Questions

1. Every tonne of limestone that is turned into cement releases 440 kg of carbon dioxide into the atmosphere. Suggest one problem caused by cement manufacture.
2. How does concrete reduce this effect as it slowly hardens?
3. Why is concrete used rather than pure cement?
4. How can the choice of aggregate help to keep the concrete costs down?
5. Explain why it would be hard to build a limestone block skyscraper.
6. Many modern buildings now have a concrete framework but are faced with other more expensive materials such as brick or tiles. Why is this?
7. Old buildings often had wooden beams over doors and windows to support the weight of stone or brick above. The wood was strong and could bend slightly without cracking. Today these beams are more likely to be made of concrete. **(a)** What advantage does wood have over simple concrete for this use? **(b)** How could the concrete beam be changed to overcome this problem? **(c)** What is the disadvantage of using wood if you want your building to last a very long time?
8. Explain as fully as you can why concrete has become such a popular and important building material over the last 100 years.

Examiner feedback

You won't be expected to *know* the properties of other building materials: wood, brick, steel or glass. However, in the examination you might be *given* a table of their properties to compare with concrete.

While concrete is a very useful material there are significant environmental effects associated with its production and use. This can also form the focus of examination questions.

Digging up the ore

Where do metals come from?

Metals are found in chemical compounds in the rocks of the Earth. Some metals are quite common. A field of mud contains tonnes of aluminium and a lot of iron too. The trouble is, the metals are extremely difficult to extract. Fortunately, natural processes sometimes concentrate metals in certain rocks from which it is relatively easy and cheap to extract them. Rocks like this are called **ores**.

The Bingham Canyon copper mine, Utah, USA.

Copper ore. Most copper ores contain a very small percentage of copper.

Iron ore. Most iron ore is used to make steel.

Some ores, like iron ore, are relatively common. Others, like copper ore, are very rare, so when we find a big body of concentrated copper ore we just keep on digging until it has all been dug up. The Bingham Canyon mine in Utah, USA, is now nearly 1 km deep. Open-cast mining, which is just digging a big open hole in the ground, presents the same problem for the environment as limestone quarrying; see lesson C1 2.1. Pollution issues may be even greater, as many ores are toxic. Also, only a fraction of the mined rock is useful, so the rest is left in unsightly **spoil** tips. Mining companies have to allow for environmental clean-up costs when working out the economics of their mine. They may have to landscape unsightly areas and remove toxic materials for safe disposal. On the other hand, mining provides work and trade for the local area. It also provides essential raw materials for industry.

Science skills
A small Welsh gold mine was just about breaking even in 2007. Table 1 shows how the price of gold varied over the next two years.

Table 1 Gold prices (2008–9).

Date	Price of gold / $/oz
03/2008	1000
06/2008	900
09/2008	700
12/2008	890
03/2009	920
06/2009	940
09/2009	1050
12/2009	1200

a Plot a line graph of the gold price over the two years.

b The mining company's costs did not vary much over the period. Explain what would have happened to the finances of the company over this period.

Is it worth it?

Ores contain metal compounds in a concentrated form. However, that doesn't necessarily mean that it is worth the cost and effort to extract it. That will depend on:

- how concentrated the ore is
- how easy it is to get the ore out of the ground (cheaper for open-cast mining than deep underground mining)
- how easy it is to extract the metal from the ore
- what price you can get for the metal you extract
- the long-term environmental clean-up costs.

The economics of the process change over time as the demand for the metal and/or its price changes. A mine that is just breaking even might suddenly start making a fortune if the price of its metal rises – or just as suddenly go bankrupt if the price falls.

Extracting the metal from the ore

There are four stages in extracting a metal from its ore.

1.	The ore is mined.	Mining can be by open-cast digging or may involve tunnelling down to underground deposits, which is more difficult and expensive.
2.	The ore is separated from impurities.	The waste from these impurities, spoil, is left by the mine. Spoil tips may contain poisonous metals, like copper and lead, that dissolve in rainwater and leak into the soil. Most plants that are sown on spoil tips die.
3.	The ore is converted to the metal.	Usually the ore is heated in air or with carbon. In both cases impurities can react with air to form poisonous substances like sulfur dioxide.
4.	The metal formed is then purified.	See lesson C1 3.3 for examples.

Pure metals such as copper or aluminium are turned into useful products, for example electrical wires, saucepans and bikes. When the product has worn out and is no longer useful, it can be recycled. This is particularly important for metals like aluminium that are expensive to make, and for very rare metals. For example, you can get more gold per tonne out of recycled mobile phones than you can out of the richest gold ore.

Recycling metals:

- saves money
- means that reserves of metal in the ground last longer
- avoids waste and pollution
- lessens the effects of mining on people and the environment
- saves energy.

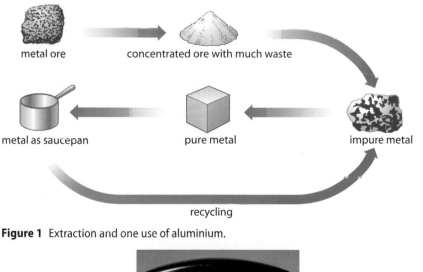

Figure 1 Extraction and one use of aluminium.

Aluminium is expensive to make and definitely worth recycling.

Questions

1. What are the similarities between open-cast mining and quarrying for limestone?
2. How do spoil tips affect the environment?
3. Ordinary garden mud contains a few per cent of aluminium. Why is this not used as aluminium ore?
4. A new, important use has been found for a very rare metal. **(a)** What will be the probable effect on the price of the metal? **(b)** Will this make mining the metal more worthwhile?
5. Why do plants growing on spoil tips die?
6. What are the advantages of recycling?
7. Draw a flow chart to show how a copper kettle can be made from copper ore and then recycled.
8. A typical 100 g mobile phone might contain 16 g of copper and 0.03 g of gold. Discuss the potential advantages and disadvantages of recycling the metals in mobile phones.

Metal from the ore

Gold was well known to the ancient Egyptians.

most reactive

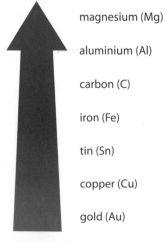

magnesium (Mg)

aluminium (Al)

carbon (C)

iron (Fe)

tin (Sn)

copper (Cu)

gold (Au)

least reactive

Figure 1 The reactivity series of metals.

Which metals are common?

Aluminium and iron are two metals we use a lot. As you can see from Table 1, they are also very common in rocks. Some other important metals, however, such as copper and gold, are really very rare.

Iron is king

Iron is the metal we use most. In 2007, the annual global production of iron peaked at almost 1.5 billion tonnes. That is 20 times as much as all the other metals put together. Production on this scale helps to make iron cheap compared with aluminium, even though aluminium is more common.

Table 1 Metals in rocks.

Metal	Percentage of the rocks of the Earth's crust
aluminium	7
iron	4
magnesium	2
copper	0.0045
tin	0.0002
gold	0.000 000 5

Table 2 Metal prices.

Metal	Price/tonne (January 2010)
aluminium	£3600
copper	£4500
iron	£1600

Reactivity and metals

Gold is a very **unreactive** metal. It does not react with other elements and so is found in rocks as uncombined metal. It may be rare but, if you are very lucky, you could find a nugget of pure gold.

In contrast, aluminium is a very reactive metal. Because of this, the aluminium atoms are tightly combined in compounds with other atoms such as oxygen. Mud contains plenty of aluminium but you cannot easily extract it. You never find aluminium as a pure element in the Earth's crust.

There is plenty of aluminium in this mud. But how could you tell?

Carbon reduction and oxidation

Carbon is more **reactive** than metals such as iron and copper, so it can displace these metals from their compounds, pushing the metal out and taking its place. **Displacement reactions** with carbon are used to get less-reactive metals from their ores.

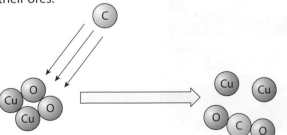

Figure 2 Carbon can displace metals such as copper from their oxide ores.

Many metal ores are oxides. When we react the ore with carbon, the carbon combines with oxygen from the ore to form carbon dioxide. The carbon is oxidised. This is an **oxidation** reaction.

The metal oxide ore has its oxygen taken away. The metal oxide is reduced to the metal only. This process is called **reduction**. Reduction is the chemical opposite of oxidation, so the two reactions always go together. Figure 3 shows an example with copper oxide.

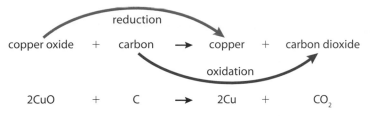

Figure 3 Reduction and oxidation in the copper oxide/carbon reaction.

Many metals, such as iron, tin and lead, as well as copper, have been extracted by this method for thousands of years. Originally the carbon was in the form of charcoal. Today coke, made from coal, is used instead. The oxide ore and coke are heated to very high temperatures in a furnace. As they react, the newly formed metal melts and can be run out through a tap at the bottom.

Why aluminium is so expensive

Aluminium is more reactive than carbon. You can't use the carbon reduction method to get aluminium from its ores. Aluminium ore first has to be heated up and melted. The **molten** ore is then split apart using large amounts of energy in the form of electricity. This process is called **electrolysis**. This is a very costly process that makes aluminium much more expensive than iron.

Figure 4 Molten metal can be run out from the bottom of the furnace.

Questions

1. How much more common is iron than gold?

2. Explain why rare gold was discovered before the much more common aluminium.

3. Tin oxide (SnO_2) reacts with carbon to give tin and carbon dioxide.
   (a) Write this as a word equation, showing the oxidation and reduction arrows. **(b)** Write the equation in symbols.

4. Write a simple word equation for the reduction of iron oxide by carbon.

5. Complete and balance the equation for the reaction of the iron ore magnetite with carbon.
   $$Fe_3O_4 + C \longrightarrow Fe + CO_2$$

6. Why is iron not produced commercially by the electrolysis of molten iron oxide?

7. Aluminium powder and iron oxide react violently to give molten iron.
   (a) Write a word equation for this reaction and explain what is happening.
   (b) Why is this reaction not used to make iron commercially?

8. Aluminium is nearly twice as common as iron in the Earth's crust, yet costs more than twice as much to buy. Use your understanding of the chemistry of the production methods used to explain this.

Developing new methods of extraction

Learning objectives
- describe how iron is extracted from its ore
- explain why copper is easier to get from good-quality ores
- explain why new methods of copper extraction are being developed.

Molten iron pours from a blast furnace.

Wanted by all

Iron is the most widely used metal, so we need to produce plenty of it to supply the world. Over the last 300 years scientists have refined the carbon reduction method to make it very efficient. Iron, however, is not as easy to extract as less-reactive metals, such as copper or lead.

Iron plays hard to get

Carbon reduction reactions need a kick-start of energy to get them going. For copper, the heat from a simple fire will do it. However, for iron, much more energy is needed. You can make coke burn at a very high temperature by providing more oxygen. In a **blast furnace**, air is pumped through the burning coke, raising the temperature to 1600 °C or more, hot enough to reduce the ore and melt the iron.

Iron ore (iron oxide), coke (carbon) and limestone are tipped in at the top of the blast furnace. The limestone is only there to remove impurities. The main reaction is the reduction of iron oxide and the oxidation of carbon.

$$\text{iron oxide} + \text{carbon} \longrightarrow \text{iron} + \text{carbon dioxide}$$

Most iron ore used today is only about 50% iron oxide. This will produce about 350 kg of iron per tonne (1000 kg) of ore. However, the impurities from 1 tonne of ore will react with the limestone to make just under 1 tonne of waste material called **slag**, which is dumped. Slag heaps can be unsightly and dangerous.

Copper's easy – if you can find it

Copper is usually found as copper sulfide crystals scattered through the ore body. In the past, this was converted to copper oxide and then the copper was extracted from the ore by carbon reduction in a furnace. For the best ores, heating alone was enough. However, good-quality, copper-rich ore is getting harder to find. There are just a few major copper mines dotted around the world. Even in these the ore only contains about 1% copper. Separating out the copper by reduction in a furnace is too expensive, so nowadays other methods are used.

Copper sulfide ore.

In one method, acid is sprayed onto the rock. Soluble copper compounds dissolve out of the rock and the **leachate** solution is collected. The copper is then removed from the leachate by electrolysis. This works because the copper forms positive ions in the solution. These positive ions move towards the negative electrode, where they can be collected – see Figure 1.

What is the future?

We use large amounts of copper for water pipes and electrical wiring so the copper mines will soon be exhausted. Many old mines are now reworking their old waste tips to get out more of the copper by **leaching** and electrolysis.

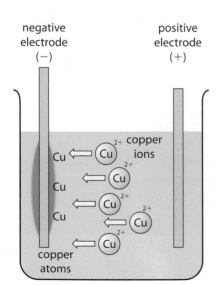

Figure 1 Electrolysis of copper sulfate solution. Positive copper ions move towards the negative electrode.

This process can extract copper from rocks containing just 1% ore. However, scientists are working hard to find ways of extracting smaller and smaller fractions. One new method, called **bioleaching**, involves using special bacteria that 'eat' the copper ions from the rock. They produce a copper-rich leachate. Another method, **phytomining** (see lesson C1 3.4), uses special plants that absorb metals from the soil as they grow.

Even if we find a new source of copper, however, environmentalists have started to object to mining companies ripping great holes in the Earth.

- Mines will inevitably be surrounded by waste heaps.
- Rainwater will naturally leach these, carrying toxic metal salts into surrounding rivers.

A new leaching method that does not affect the environment so much involves drilling down to the ore. Acid is then pumped down to the ore and comes back up to the surface rich in copper for processing: the process is illustrated in Figure 2. You don't need to dig big, ugly holes in the ground at all.

Any old iron will do

In some ore-rich areas the rivers can carry quite high concentrations of dissolved copper salts, or salts of other metals. Iron is more reactive than copper, so if you put scrap iron in these rivers, the iron will displace the copper from solution. The copper can then be collected and refined.

$$\text{copper sulfate} + \text{iron} \xrightarrow{\text{displacement}} \text{iron sulfate} + \text{copper}$$

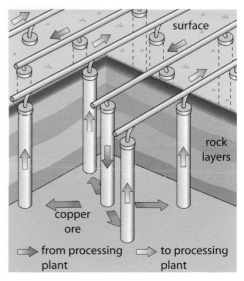

Figure 2 Leaching copper the environmentally friendly way?

Questions

1 Explain how the blast of air in a blast furnace helps to raise the temperature.

2 Limestone reacts with any sand (SiO_2) in iron ore to form calcium silicate. Complete the balanced chemical equation:

$$CaCO_3 + SiO_2 \longrightarrow CaSiO_3 + \underline{\hspace{1cm}}$$

3 **(a)** Write out the simple word equation for the blast furnace and add the oxidation and reduction arrows. **(b)** If the iron oxide is Fe_2O_3, write a balanced equation for this reaction.

4 Explain why copper cannot be extracted from 1% copper ore in a furnace.

5 Explain why copper collects at the negative terminal during electrolysis.

6 Draw a flow chart for the production of copper from low-grade ores by leaching.

7 Complete and balance the equation for the displacement of copper from copper sulfate by iron: $CuSO_4 + \underline{\hspace{1cm}} \longrightarrow FeSO_4 \underline{\hspace{1cm}}$

8 Can you suggest a method for cleaning up rivers polluted with copper salts? Explain how it works. Are there any possible disadvantages to your method?

Practical

If you put an iron nail into copper sulfate solution, it will soon become coated with copper.

Finite resources

Learning objectives

- evaluate the economic, social and environmental effects of mining, extracting and recycling metals
- explain why we need to find ways of recycling metals effectively.

Scarce resources

We all live on the Earth and get everything we need from it. There are more than six billion people on the planet, and we are consuming the Earth's resources at an ever-increasing rate. Some metal ores will become scarce over the coming decades.

Earth from space.

Science skills

a Which metals might run out in your lifetime using today's mining techniques?

b Iron is unlikely to run out in the near future, but steels used for machinery or other high-performance uses are alloyed with metals such as vanadium and manganese that might run out. What problems would society face if we ran out of metals for these important alloys?

c As metals get scarce, the price will rise. How might this help to make more metal available?

Table 1 Summary of common metals and their reserves.

Metal	Uses	Proven reserves will last until...
tin	rust-proofing for steel; also used in bronze and solder	2030
copper	electrical wiring, water pipes; also used in brass and bronze	2030
tungsten	the filament in 'old fashioned' light bulbs; also added to steel to make it very hard and strong	2050
aluminium	cans, saucepans and aeroplanes	2050
nickel	added to steel for acid resistance, and in coinage	2100

Health and safety rules may not be applied as strictly in some countries as in the West. This man is not wearing any eye protection.

Are we exploiting poorer countries?

Some countries are rich in mineral resources, while others have few. Industrial societies are buying up more than their fair share of the global resources. Sometimes this is because the countries of the developed world have used up their own resources. However, often it is because it is cheaper to mine in developing countries because the wages of the miners are so low. This does have some benefits, as it provides jobs for people in the developing countries. However, sometimes the working conditions are not very good.

Zambia is a poor country. Its economy relies on exports of copper from its huge copper mines. Forty years ago Zambia was encouraged to take out huge loans from the World Bank to develop these mines. They said it would raise the standard of living for all Zambians.

d What happened to the price paid for copper in the late 1970s?

e What effect do you think this had on the people and economy of Zambia?

f Do you think the advice from the World Bank was good or not? Explain your answer.

g In the boom of 2007 the price shot up to $8000 per tonne. What do you think happened to the price in the recession that followed in 2008?

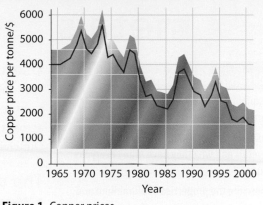

Figure 1 Copper prices.

The economics of recycling

Recycling can help to make our resources last longer and reduce pollution. It also reduces our energy consumption. However, it costs money. One of the problems is that waste material may well contain different metals mixed up. Recycling is only economically viable at the moment for metals that are easy to separate, such as aluminium, iron or steel, or expensive metals such as gold. As technology improves, perhaps we will be able to do this for all metals.

Reclaiming spoilt land

Mines and associated industrial activities can leave great areas of spoilt, contaminated land. How can we reclaim these 'brownfield' sites for housing or recreation? Help is at hand from a surprising source.

Many plants won't grow on soil contaminated with metals, but others do well. The brassicas, a plant family that includes cabbages and sprouts, not only thrive but take up the metals into their leaves. This has led to the development of environmentally friendly **phytomining**. Brassicas are grown on metal-rich soils. As they grow they take any metals out of the soil, cleaning the land. At the end of the season they are harvested and burnt. The metal rich ash left behind can then be collected and metals such as cadmium, nickel and cobalt may be extracted.

Examiner feedback

You need to understand that new methods of extraction need to be economic and this will partly be determined by how scarce the metal becomes and how much of it is needed.

Questions

1 Brassicas grown on metal-rich soils may contain up to 1% nickel by dry mass. **(a)** If a 1 hectare field gives 10 tonnes of dry biomass, how much nickel could be recovered? **(b)** If the price of nickel were £10 000 per tonne, what would this be worth? **(c)** If the growing and extraction costs were £400 per hectare, would this phytomining be economically viable?

2 Alfacture is a large company that produces one million tonnes of aluminium goods every year. It has its own plant for extracting aluminium from aluminium ore but also uses recycled aluminium. The energy used to extract the metal from its ore costs £100 per tonne.

The energy used to recycle aluminium costs just £5 per tonne. The company currently uses 40% recycled aluminium in its products. **(a)** Calculate Alfacture's energy costs for its annual production of aluminium goods. **(b)** The company hopes to increase the amount of recycled aluminium it uses to 50%. How much money a year would it save on energy costs? **(c)** Why is it so important to recycle aluminium, which is a common metal? Explain your answer fully.

3 Suggest what will happen when current copper reserves run out. How might we prevent this situation arising?

Heavyweight and lightweight metals

Learning objectives

- explain why the main properties of metals are so important
- explain that iron and copper are transition metals and why they have many everyday uses
- describe the properties of aluminium and titanium and how these affect their use.

Useful metals

Technology relies on the strength and ease of shaping of metals for everything from aeroplanes to tin cans. Thousands of kilometres of electric cables snake across the country to bring us the power we need for our electrical equipment. Our civilisation is heavily dependent on metals.

The useful properties of metals

- Metals are strong, which makes them good structural materials. We use them to build machines, bridges and the frameworks for large buildings.
- Metals are easy to shape: they can be bent, pressed, drawn and rolled. Car body panels are pressed out of sheet steel.
- Most metals have high melting points. Engines don't melt when they get hot in use.
- Metals also conduct heat and electricity. They are used in electrical wiring and as **heat sinks**, to conduct heat away from microprocessors in computers, to stop them overheating.

The properties of metals make them a useful construction material.

Figure 1 Iron and copper are transition metals.

Everyday heavyweights

Iron and copper are two of the **transition metals** from the central block of the periodic table. The transition block contains many important, everyday metals.

Iron

Iron, usually used in the form of steel (see lesson C1 3.6), is by far the most widely used metal. It can be made hard and tough for machinery, tools, bridges or girders. Or it can be made into sheets that can be rolled, cut and pressed into shape for anything from car bodies to cans. It is extremely versatile. However, it has one big weakness: it rusts.

Copper

Copper is particularly useful as it is such a good electrical **conductor**. It is also soft and bendy, so it is great for electrical wires. It is also a good conductor of heat. The best saucepans often have copper bottoms. Copper is not very reactive: it does not react with water, and so does not corrode like iron and steel. This makes it useful for pipes for plumbing. It also means it can be used for roofing on important buildings, although it is too expensive for general use in this way.

Copper is our electrical conductor of choice.

The lightweight champions

Steel may be fantastic for cars, trains and ships, but its high density makes it useless for aircraft. An aircraft made from steel would weigh twice as much as a modern plane. It would have to use too much energy to get off the ground. Aluminium is not a transition metal. It is not as strong as steel but its density is very much lower. It can be used to make a plane that is both strong enough *and* light enough to fly. Without aluminium there could be no commercial airliners.

Commercial aircraft rely on lightweight yet strong aluminium.

Table 1 Properties of widely used metals.

Metal	Density / g/cm³	Strength	Melting point/°C
pure aluminium	2.7	low–medium	660
steel	7.7	high	1540
titanium	4.5	high	1670

Supersonic fighter jets fly so fast that the wings get hot enough to melt aluminium. A new 'supermetal' was needed that was strong and had a low density but a high melting point. Titanium fitted the bill perfectly. Titanium is as strong as steel but less dense, though denser than aluminium. A bonus is that its melting point is higher than that of steel – high enough to withstand the frictional heating caused by supersonic flight. In fact titanium is such a supermetal we would use it to replace steel all the time if it were not so expensive.

High-performance fighters need even stronger (but more expensive) titanium.

Resisting corrosion

Aluminium and titanium share another useful property: they both resist corrosion well. That is why aluminium foil stays shiny. Titanium resists **corrosion** much better than aluminium and stainless steel. It can be used safely where even stainless steel would corrode away, for example in nuclear power stations, or inside the human body.

Questions

1   Suggest three differences between transition metals and group 1 metals.

2   What 'key property' makes the metal useful in each of these cases?
    (a) Iron girders of bridges **(b)** steel used to make car bodies **(c)** copper water pipes **(d)** copper electricity cables.

3   Iron bridges have to be repainted regularly. What would happen if they were not?

4   Why do sheet copper roofs last longer than corrugated iron roofs?

5   Titanium is denser than aluminium. Explain why it is still an excellent material for aircraft manufacture.

6   What would happen in time if the hip replacement shown on the X-ray photograph were made from steel?

7   Given iron's problem with rusting, why don't we use other metals like copper or aluminium for cars and bridges instead?

8   Compare the properties of steel, aluminium and titanium to explain which metals are used in aeroplanes and why. Some lower-performance supersonic planes use titanium for the nose cones and wing leading edges only. Suggest two reasons for this.

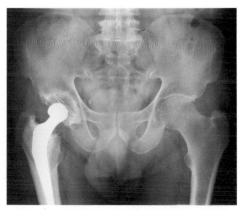

Titanium is used for hip replacement surgery.

Improving metals

Learning objectives

- explain that iron from a blast furnace is impure and has limited uses
- describe how removing these impurities produces pure iron
- describe how the atoms are arranged in pure iron
- describe how and why iron is turned into steel
- explain why the properties of alloys are related to their structures.

Iron straight from the blast furnace

The iron that comes out of a blast furnace is only about 96% pure. It contains impurities like carbon, silicon, sulfur and phosphorus. These impurities make the iron **brittle**, so it is not very useful. To get pure iron these impurities are removed by reacting them with oxygen. Jets of pure oxygen are blasted through molten iron. The oxides formed are then easily separated from the iron.

Like other pure metals, pure iron is soft and easily shaped.

Steel is an alloy

The softness of iron means that it is too soft for many uses. Most iron is converted into steels. Steels are **alloys**. Steel alloys are a mixture of iron with carbon or other metals. The different atoms that are added change the properties of the pure metal, in this case iron. Alloys are harder.

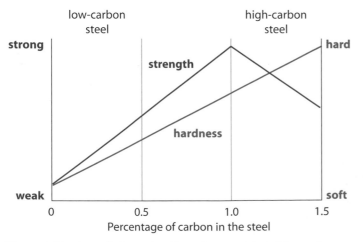

Bronze is an alloy used for statues.

Different types of steel do different jobs.

The best steel for the job

Alloys can be designed to have properties for specific uses. Steels are made by adding carefully calculated amounts of carbon to the pure iron. This affects both the hardness and the strength of the steel, as shown on the graph.

Low-carbon steels are made by adding small amounts (up to 0.25%) of carbon to the molten pure iron. This small amount of carbon makes the iron stronger and a little harder. These steels are easily shaped. They are used for things like wire, nails, cans and car bodies.

As a little more carbon is added, the steel becomes even stronger. However, increased hardness means that the steel can no longer be easily pressed into shape. However, it is good for making hammers.

If slightly more carbon is added (up to 1.5%), **high-carbon steels** are made. These are even harder but the strength drops a little, so they can be brittle. They are used for

Figure 1 How just a little carbon affects the properties of the steel.

cutting tools, drill bits and masonry nails. Masonry nails are used for hammering into bricks or concrete. Using high-carbon steel nails means that they do not bend as low-carbon iron nails would. However, they can snap if struck incorrectly.

If any more carbon is added the steel becomes very hard but very brittle, just like the 'cast iron' that comes straight out of the blast furnace.

Special steels

Sometimes other metals are added to steel to give new alloys with special properties. Nickel is added to make a steel that is both hard and very strong, for machinery. Tungsten is added to make a very hard steel for cutting tools. Adding 15% or so of chromium produces stainless steel, which does not corrode, making it ideal for cutlery, for example.

Other metal alloys

Aluminium is made stronger for aircraft by adding just 4% copper to make an alloy called **duralumin**. Many other metals are strengthened in this way. Even gold has to have a little copper added to harden it. Without it, gold rings would quickly distort or wear away when used. There are three other important alloys of copper.

Table 1 Important alloys of copper.

Name of alloy	Composition	Special property	Uses
brass	70% copper, 30% zinc	harder than pure copper	electrical fittings, screws
bronze	90% copper, 10% tin	harder than pure copper	bells
cupronickel	75% copper, 25% nickel	harder than pure copper	coins

Masonry nails are made from high-carbon steel.

Modern 'silver' coins are really cupronickel.

Questions

1 What effect do impurities have on iron straight from the blast furnace? How are they removed?

2 Explain how adding different-sized atoms to a metal in an alloy helps to make it harder.

3 Which type of steel would be best for **(a)** scissors and chisels? **(b)** making car body panels? Explain your answer.

4 Why should the builder in the photograph be wearing safety goggles?

5 Why are aircraft built from duralumin, not pure aluminium, and why is copper's high density not a problem?

6 Brass is made from copper and zinc. Why are bolts and hinges made from brass and not copper? Explain the difference in properties.

7 Cupronickel also has the added property that it is more resistant to chemical corrosion. Why does this make it useful for coinage?

8 Explain the effect of adding small amounts of carbon to iron and describe how the metal's properties change with the amount of carbon added.

Examiner feedback

You do not have to remember the details of all these alloys. However, you might get a table of alloys and their properties and have to answer questions about their suitability for different jobs.

Route to A*

You may be given unfamiliar data to work with in examination questions. Practise data questions like those here, which use data from these pages, so that you become good at spotting the relevant information and are able to use it to justify your answers.

Assess yourself questions

1 Complete and balance these chemical equations.

(a) $CaCO_3 \longrightarrow$ ____ $+ CO_2$ *(1 mark)*

(b) $CaO +$ ____ $\longrightarrow Ca(OH)_2$ *(1 mark)*

(c) $Ca(OH)_2 + CO_2 \longrightarrow$ ____ $+ H_2O$ *(1 mark)*

(d) $CaCO_3 + HCl \longrightarrow CaCl_2 +$ ____ $+$ ____ *(3 marks)*

2

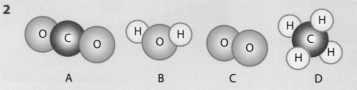

Figure 1 Molecules.

(a) Which of these diagrams in Figure 1 shows a molecule of an element? *(1 mark)*

(b) Which of these diagrams shows a molecule of a compound? *(1 mark)*

(c) (i) Give the chemical formulae for molecules A–D. *(2 marks)*

(ii) Give the chemical names for molecules A–D. *(2 marks)*

(d) Which subatomic particle is shared between atoms in molecules like these, to make the chemical bonds? *(1 mark)*

3 Limestone ($CaCO_3$) breaks down to quicklime (CaO) when heated. Three students performed an experiment to see how much quicklime they could get by heating 10 g of limestone. They set their balance to zero with a crucible in place and measured out exactly 10 g of limestone. They then heated the crucible strongly, reweighing it regularly. They kept heating until the mass stopped going down. They each repeated their experiment three times.

Table 1 The students' results.

| | Quicklime produced from 10 g of limestone/g | | | |
	Expt 1	Expt 2	Expt 3	Mean
Student 1	5.6	5.7	5.5	
Student 2	5.65	5.66	5.64	
Student 3	5.62	5.68	6.92	

(a) Why does the mass as measured on the balance go down during this reaction? *(1 mark)*

(b) Why did they have to keep heating 'until the mass stopped going down'? *(1 mark)*

(c) Calculate the mean result for each of the three students. *(3 marks)*

(d) (i) Which student had been given an older, less precise balance to work with? *(1 mark)*

(ii) How will this have affected their results? *(1 mark)*

(e) Which student appears to have obtained the most **reliable** results? Explain your answer. *(2 marks)*

(f) Which student ran out of time and didn't heat their final piece of limestone long enough? Explain your answer. *(2 marks)*

(g) (i) The theoretical amount of quicklime produced from 10 g of calcium carbonate is 5.6 g. Whose final answer appears to be the most accurate? *(1 mark)*

(ii) A detailed analysis of the limestone used shows that the residue after heating is indeed slightly greater than 5.6 g. Suggest a possible reason for this. *(1 mark)*

4 Describe the three reactions that make up the 'limestone cycle', giving the balanced chemical equations for each stage.

In this question you will be assessed on using good English, organising information clearly and using specialist terms where appropriate. *(6 marks)*

5 Iron is made by heating iron oxide with coal in a blast furnace.

(a) Which element is coal mostly made of? *(1 mark)*

(b) Complete this word equation:

iron oxide + carbon $\longrightarrow$ iron + ____ ____ *(1 mark)*

(c) Which reactant is oxidised in this reaction? *(1 mark)*

(d) Which reactant is reduced in this reaction? *(1 mark)*

(e) Why can't aluminium be produced from aluminium oxide by this reaction? *(1 mark)*

6 Laos is a poor country with underdeveloped heavy industry. Bamboo grows well in its hot climate.

In London, scaffolding is built from steel tubes that are screw-clamped together. In Laos, scaffolding is built from bamboo poles lashed together with natural string.

(a) Which do you think would be stronger, the steel or bamboo? *(1 mark)*

(b) What would you notice if you picked up a steel pole and a bamboo pole? *(1 mark)*

(c) Suggest *two* reasons why bamboo is used in Laos rather than steel. *(1 mark)*

(d) Steel scaffolding poles last longer than bamboo poles. Why is that not a problem in Laos? *(1 mark)*

(e) Broken bamboo poles are simply thrown away. Why is that not an environmental problem? *(1 mark)*

(f) Large bamboo poles cost more in the UK than their steel equivalents. Suggest *two* reasons for this. *(2 marks)*

7 **Table 2** Properties of metals.

Metal	Melting point /°C	Strength (1 = low, 50 = very high)	Cost £/tonne	Density g/cm³
aluminium	660	1 (pure) 5 (alloyed)	3600	2.7
steel	1540	20	1600	7.7
titanium	1670	10	12 000	4.5
tungsten	3400	50	9000	15.3

(a) For each use below, suggest a suitable metal and give a reason (from Table 2).

  (i) The barrel of a Bunsen burner. *(1 mark)*

  (ii) The wing of a supersonic fighter *(1 mark)*

  (iii) The filament in a light bulb. *(1 mark)*

  (iv) A commercial aeroplane. *(1 mark)*

(b) Why is aluminium not used in its pure form to make pans, cans or aeroplanes? *(1 mark)*

(c) Explain in simple terms why alloys are harder and stronger than the pure metal. *(2 marks)*

8 The permitted levels of some metal ions in drinking water are:

copper	1 mg/l
lead	0.05 mg/l
zinc	5 mg/l

(a) From these figures, which metal is most toxic, and which is least toxic? *(1 mark)*

Cattle in the fields around the river shown on the map in Figure 2 became ill and metal poisoning was suspected. Water samples taken from the rivers at A, B, C, D and E were analysed.

Table 3 Results of analysis.

	Concentration mg/l		
	copper	lead	zinc
A	0.05	0.001	0.05
B	5.00	0.1	3.0
C	6.0	5.0	10.0
D	1.00	0.02	0.6
E	1.6	1.0	2.0

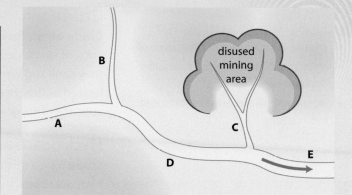

Figure 2 Map of the river system.

(b) **(i)** Which site would give safe drinking water (in terms of metal content)? *(1 mark)*

  (ii) Which site shows the most polluted water? *(1 mark)*

  (iii) Where do you think this pollution has come from? *(1 mark)*

(c) **(i)** The rivers flow from A, B and C to E. Why are the metal levels lower at E than at C? *(1 mark)*

  (ii) From the figures, which river carries more water, the main river at D or the side river from C? *(1 mark)*

(d) **(i)** Site A has a full range of wildlife. How would you expect site E to compare with this? *(1 mark)*

  (ii) You find water snails at site A but they have disappeared from the river at site D. Which metal do you think the snails might be sensitive to? *(1 mark)*

(e) The herd of cattle could only get to the river to drink at E. Which metal is most likely to be responsible for the poisoning? Explain your answer. *(2 marks)*

(f) Scientists have suggested that the pollution problem could be tackled by throwing scrap iron into the rivers at B and C.

  (i) How would this work? *(1 mark)*

  (ii) Which metal would not be affected? *(1 mark)*

  (iii) Which metal ion would increase in the water at E? Would this be a problem? *(2 marks)*

9 Explain why iron and copper can be extracted from their ores by carbon reduction but aluminium cannot be obtained in this way. Describe briefly how aluminium is extracted and explain why this makes aluminium relatively expensive.

In this question you will be assessed on using good English, organising information clearly and using specialist terms where appropriate. *(6 marks)*

Alkanes

Crude oil

Crude oil is a **fossil fuel**. It formed over millions of years from the remains of ancient single-celled plants and animals that lived in the sea. When they died, they sank and were buried in sediments. The remains were compressed and heated as they became more deeply buried. Chemical reactions took place in the absence of air, gradually converting the remains into the oil we use today.

Hydrocarbons

Crude oil is a mixture of a very large number of **compounds**. Most of these are **hydrocarbons**, molecules made from hydrogen and carbon atoms only. The hydrogen atoms are joined to the carbon atoms, and these carbon atoms are joined together in chains. Most of the hydrocarbons in crude oil are **alkanes**.

Alkanes

The atoms in hydrocarbons are joined together by **covalent bonds**, a type of chemical bond. Hydrogen atoms can only make one covalent bond but carbon atoms can each make four. This is why carbon atoms can join up to form chains. All the bonds in alkanes are single covalent bonds. They can be C—C bonds, between two carbon atoms, or C—H bonds, between a carbon atom and a hydrogen atom.

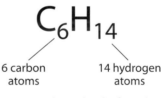

$$C_6H_{14}$$

6 carbon atoms   14 hydrogen atoms

Figure 1 The molecular formula of hexane is written like this.

Crude oil is a liquid containing many different hydrocarbons.

Hexane is an alkane. Its molecules all contain six carbon atoms and 14 hydrogen atoms, so its **molecular formula** is C_6H_{14}. A molecular formula gives information about the number of atoms of each element a molecule contains, but it gives no information about how these atoms are arranged.

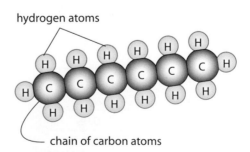

hydrogen atoms

chain of carbon atoms

Figure 2 A simple representation of a hexane molecule.

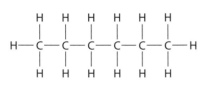

Figure 3 The displayed formula of hexane.

The arrangement of the atoms in a molecule can be shown using a **displayed formula**. In a displayed formula, each atom is shown by its chemical symbol, and each covalent bond by a straight line. It is usual to make all the angles 90° in the displayed formulae of alkanes. It is easiest to write all the carbon atoms first, draw in all the covalent bonds, and then add the hydrogen atoms.

Notice that there are no 'spare' bonds in hexane. The molecule is 'full' of hydrogen and cannot join with any more hydrogen atoms. It is a **saturated** hydrocarbon. All alkanes contain only single bonds, so they are all saturated.

Naming alkanes

The names of alkanes have two parts. The first part shows how many carbon atoms the alkane molecule contains. The second part, 'ane', tells you that the carbon atoms are joined by single bonds. Table 1 shows how this works.

Table 1 Naming alkanes.

Number of carbon atoms	First part of name	Name of alkane	Molecular formula of alkane
1	meth	methane	CH_4
2	eth	ethane	C_2H_6
3	prop	propane	C_3H_8

The number of hydrogen atoms in each alkane is twice the number of carbon atoms plus two. So alkanes have a general formula of C_nH_{2n+2}, where n is the number of carbon atoms.

The alkanes form a **homologous series** or 'family' of compounds. They take part in similar chemical reactions and have a common general formula. The molecular formula of each successive member differs by CH_2.

Questions

1 What are the main compounds in crude oil?

2 Explain what hydrocarbons are.

3 Explain why carbon atoms can form chains of atoms.

4 State the molecular formulae of ethane and propane, and draw their displayed formulae.

5 What is the general formula for alkanes?

6 Icosane contains 20 carbon atoms. Write its molecular formula.

7 The diagram shows the displayed formula of a 'cyclic' alkane.

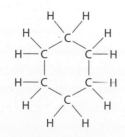

Figure 4 Cyclohexane.

(a) Write the molecular formula of cyclohexane. (b) Describe three similarities between cyclohexane and hexane, and three differences.

8 Explain clearly why the alkanes are saturated and form a homologous series. Give relevant examples in your answer.

Science in action

Alkanes and other compounds that contain carbon are named using a system developed and maintained by IUPAC, the International Union of Pure and Applied Chemistry. Some names might look complex. However, chemists everywhere can work out a compound's structure just from its name.

Examiner feedback

The general formula for alkanes is C_nH_{2n+2}. It helps you recognise alkanes and write their formulae.

Science in action

Instead of a straight chain of carbon atoms, it is possible to have 'branched' alkanes. For example, the six carbon atoms in hexane can be arranged in a chain of five carbon atoms, with one attached as a branch. Branched alkanes and cyclic alkanes (like the one in question 7) are included in petrol to help it burn better in car engines.

Taking it further

Compounds that have the same formula, but whose atoms are arranged differently, as described in 'Science in action', are called isomers. The unbranched and branched alkanes described are examples of chain isomers. The cyclic alkane in question 7 is a functional group isomer. It has single bonds like an alkane, but its formula is the same as the formula of an alkene (see C1 5.1).

Separating crude oil

Science skills

The number of carbon atoms is a categoric variable whose values are restricted to whole numbers. Boiling point is a continuous variable, so the data in Table 1 may be shown as a line graph or as a bar chart.

Table 1 Boiling points of various alkanes.

Formula	Number of C atoms	Boiling point/°C
CH_4	1	−164
C_3H_8	3	−42
C_5H_{12}	5	36
C_7H_{16}	7	98
C_9H_{20}	9	151
$C_{11}H_{24}$	11	196
$C_{20}H_{42}$	20	344

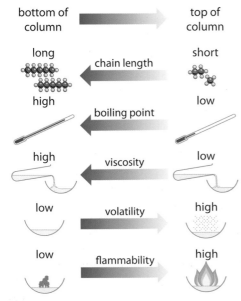

Figure 2 Trends in the physical properties of crude oil fractions.

Separating mixtures

A **mixture** consists of two or more elements or compounds. However, these are not chemically combined. The chemical properties of each substance in a mixture are unchanged, so the substances can be separated by physical methods such as filtration or **distillation**.

Fractional distillation

Fractional distillation is typically used to obtain a liquid from a mixture of liquids that dissolve into one another. The liquid with the lower boiling point **evaporates** first. Its vapours are led away, then cooled and **condensed**, leaving the rest of the mixture behind.

Practical

Ethanol boils at 78 °C and water boils at 100 °C. Fractional distillation is used to separate ethanol from a mixture of the two liquids.

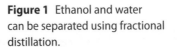

Figure 1 Ethanol and water can be separated using fractional distillation.

Boiling points of alkanes

Short-chain alkanes have low boiling points. They tend to be gases at room temperature. Intermediate-sized alkanes tend to be liquids at room temperature. Long-chain alkanes have the highest boiling points. They tend to be solids at room temperature. This trend in boiling points is the key to separating the hydrocarbons in crude oil using fractional distillation.

Oil fractionation

Crude oil is heated strongly to evaporate it. The hot vapours are led into the bottom of a tower called a **fractionating column**. This is hottest at the bottom and gradually becomes cooler towards the top. The vapours cool down as they rise through it. When the vapours reach a part of the column that is cool enough, they condense. The liquid falls into a tray and is piped out. Fractional distillation is a continuous process. It keeps going as long as vaporised crude oil enters the fractionating column.

Alkanes with the longest chains condense near the bottom of the column. Those with intermediate-sized chains condense at various points further up. Alkanes with the shortest chains remain as gases. They reach the top of the column without becoming cool enough to condense. The separated alkanes are just parts of the original crude oil, so they are called **fractions**. In each fraction, all of the alkane molecules are similar in size.

Useful as fuels?

Alkanes with the longest chains make poor fuels. They are solid at room temperature. They are not very **flammable** and ignite with difficulty, if at all.

Alkanes with the shortest chains, however, are good fuels. They are gases at room temperature and ignite easily. Gases are easily piped to where they are needed but they are bulky to store. So they are often stored under pressure as liquids, like liquefied petroleum gas, LPG.

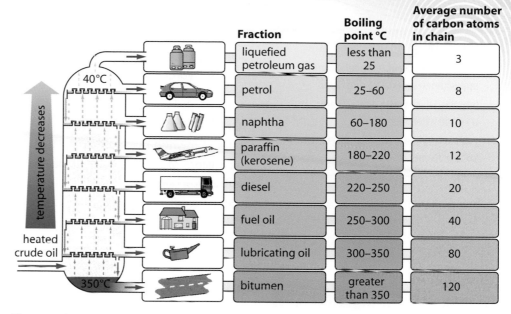

Fraction	Boiling point °C	Average number of carbon atoms in chain
liquefied petroleum gas	less than 25	3
petrol	25–60	8
naphtha	60–180	10
paraffin (kerosene)	180–220	12
diesel	220–250	20
fuel oil	250–300	40
lubricating oil	300–350	80
bitumen	greater than 350	120

Figure 3 The main fractions from crude oil. Naphtha is used in the manufacture of chemicals.

Alkanes with intermediate-length chains also make good fuels. They are liquids at room temperature. They are relatively easy to store and they can be piped to where they are needed. Liquids containing the shorter-chain alkanes are more flammable than those containing the longer-chain alkanes. They are also less **viscous**, which means they flow more easily.

Questions

1. Explain the difference between a compound and a mixture.

2. Why is fractional distillation described as a continuous process?

3. Explain why diesel has a higher boiling point than gasoline (petrol).

4. Use the data in Table 1 on page 132 to plot a line graph of the boiling point against number of carbon atoms. Describe the graph and use it to predict the boiling point of hexadecane, $C_{16}H_{34}$.

5. Describe and explain how crude oil is separated into fractions by fractional distillation.

6. It has been suggested that the sootiness of the flame from a burning alkane depends upon the ratio of carbon atoms to hydrogen atoms. The higher the ratio, the more sooty the flame is. Predict the difference in the flames produced by methane, CH_4, and octane, C_8H_{18}, giving reasons for your answer.

7. **(a)** Explain what is meant by the viscosity of a liquid.

Formula of alkane	Relative viscosity
C_6H_{14}	0.12
C_7H_{16}	0.18
C_8H_{18}	0.24
$C_{10}H_{22}$	0.42
$C_{12}H_{26}$	0.66
$C_{14}H_{30}$	1.00

(b) Plot a suitable line graph of the data in the table.
(c) Describe the relationship between the viscosity of alkanes and the size of their molecules.

8. To what extent does the size of its molecules determine the usefulness of a liquid alkane as a fuel? You should discuss boiling point, viscosity and flammability in your answer.

Burning fuels

Fuel

Fuels are stores of energy that can be released when needed. When they burn, however, hydrocarbons and other chemical fuels also release particles and various gases.

Burning coal

Coal is a brownish black solid fossil fuel that is mostly carbon. It was formed over millions of years from the remains of ancient swamp plants. **Complete combustion** happens when coal burns completely in a plentiful supply of air. Its carbon is **oxidised** to carbon dioxide:

$$\text{carbon} + \text{oxygen} \longrightarrow \text{carbon dioxide}$$
$$C + O_2 \longrightarrow CO_2$$

Incomplete combustion, also called **partial combustion**, happens instead if the supply of air is not plentiful. As the coal burns, its carbon may be oxidised to carbon monoxide instead of carbon dioxide:

$$\text{carbon} + \text{oxygen} \longrightarrow \text{carbon monoxide}$$
$$2C + O_2 \longrightarrow 2CO$$

Carbon monoxide is a colourless and odourless toxic gas. It attaches to the haemoglobin in red blood cells more strongly than oxygen, reducing the amount of oxygen transported in the bloodstream. Carbon monoxide poisoning causes headaches, sickness and fainting, and even death.

Burning natural gas

Natural gas is a fossil fuel that forms in a similar way to crude oil, and is often found with it. Natural gas is mostly methane, with smaller amounts of ethane and other alkanes. The complete combustion of natural gas oxidises its carbon to carbon dioxide, and its hydrogen to water vapour:

$$\text{methane} + \text{oxygen} \longrightarrow \text{carbon dioxide} + \text{water}$$
$$CH_4 + 2O_2 \longrightarrow CO_2 + 2H_2O$$

The partial combustion of natural gas also causes the production of carbon monoxide. This is why gas fires should have good ventilation.

Burning fuels from crude oil

Petrol, paraffin (also called kerosene), diesel and fuel oil are liquid fuels from crude oil. Carbon dioxide and water vapour are produced when they burn completely. For example, petrol contains octane:

$$\text{octane} + \text{oxygen} \longrightarrow \text{carbon dioxide} + \text{water}$$
$$2C_8H_{18} + 25O_2 \longrightarrow 16CO_2 + 18H_2O$$

Carbon monoxide is produced by the partial combustion of these fuels, too.

About 40% of the world's electricity is generated by coal-fired power stations.

Practical

The products of combustion from burning natural gas can be collected and tested. Carbon dioxide turns limewater cloudy white, and water vapour changes anhydrous copper sulfate from white to blue.

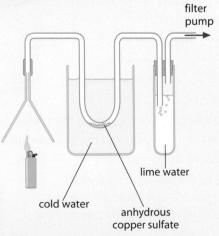

filter pump

lime water

cold water

anhydrous copper sulfate

Figure 1 This apparatus is used to collect and test combustion products.

Examiner feedback

In the balanced symbol equation for the complete combustion of an alkane, the number in front of CO_2 is the number of C atoms in the alkane, and the number in front of H_2O is half the number of H atoms. The number in front of O_2 is the number before CO_2, added to half the number before H_2O.

Particulates

Particulates are solid particles produced when fuels burn. They contain carbon and unburned fuel and are deposited as soot. The carbon is often noticeable as black smoke from diesel-powered vehicles, particularly if the vehicles are old, accelerating or going uphill. High levels of particulates in polluted air can lead to early deaths and extra cases of asthma, bronchitis and other respiratory diseases.

Other oxides from fuels

Fossil fuels often contain small amounts of sulfur. This is naturally occurring and has not been added deliberately. When the fuel burns, the sulfur does too. It forms sulfur dioxide gas, which escapes along with the other waste products:

$$\text{sulfur} + \text{oxygen} \longrightarrow \text{sulfur dioxide}$$
$$S + O_2 \longrightarrow SO_2$$

At the high temperatures in a furnace or engine, nitrogen and oxygen from the air can react together to produce various oxides of nitrogen. Together, these waste gases are called NO_x. The x in the name shows that different molecules with different numbers of nitrogen and oxygen atoms are possible. Sulfur dioxide and NO_x cause acid rain if they escape into the air.

Science in action

Particulates equal to or smaller than 10 µm in diameter, 10 millionths of a metre, are called PM10s. EU legislation limits the concentration of PM10s allowed in air because of their harmful effects on health.

Taking it further

Catalytic converters fitted to car exhaust systems contain the metals platinum and rhodium. These catalysts can convert nitrogen oxides, NO_x (where x can be 1 or 2) into harmless nitrogen and oxygen, helping to reduce the formation of acid rain:

$$2NO_x \longrightarrow N_2 + xO_2$$

Questions

1 State the products formed when hydrocarbons burn completely in air.

2 Write the word equation for the complete combustion of butane.

3 **(a)** Which toxic gas is formed during the incomplete combustion of hydrocarbons? **(b)** Explain why this gas is harmful to health.

4 **(a)** Explain, giving an example, what particulates are. **(b)** Describe the problems caused to human health by particulates.

5 Explain why burning fuels may release sulfur dioxide and NO_x.

6 **(a)** Write a balanced symbol equation for the complete combustion of butane, C_4H_{10}. **(b)** Write a balanced symbol equation for the incomplete combustion of methane, CH_4, where the only products are CO and H_2O.

7 Smog, a mixture of smoke from coal fires and fog, covered London for several days in December 1952. Use information from the graph to answer the questions.

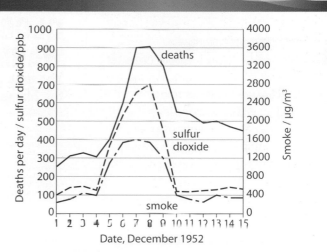

Figure 2 Mortality and atmospheric pollution in London, December 1952.

(a) Explain why sulfur dioxide in the atmosphere is thought to have caused extra deaths. **(b)** Suggest why the death rate remains higher than normal between the 10th and 15th, even though sulfur dioxide levels have returned to normal then. **(c)** To what extent can we be certain that sulfur dioxide was the cause of the extra deaths?

8 To what extent would you agree that it is important to burn fuels in a good supply of air? In your answer, include the products from complete combustion and partial combustion.

Problem fuels

Science in action

Carbon dioxide and water vapour are not the only greenhouse gases. Methane is one, too. It is released as a result of extracting crude oil, natural gas and coal. Large amounts are also produced by cattle and rice paddy fields. Methane has about 25 times the global warming potential of the same amount of carbon dioxide.

Examiner feedback

Take care when answering questions about global warming not to get muddled up with the ozone layer. Carbon dioxide is a greenhouse gas, so it can contribute to global warming but it does not damage the ozone layer.

Examiner feedback

There is no dispute that the Earth and its atmosphere are becoming warmer. It is the causes of this that there is some debate about. You should recognise that people's opinions about global warming may be influenced by economic, ethical, moral, social or cultural considerations. For example, a farmer whose livelihood may be threatened by changing weather patterns might have a different opinion from someone who depends on the oil industry.

Up in smoke?

Around 85 million barrels of oil are used every day in the world, about 2 litres per person on average. Most of it is destined to be burnt as fuel. When it is, waste gases and solid particles are released into the atmosphere. These have significant environmental effects.

Hydrocarbons and global warming

Sunlight reaching the Earth's surface is emitted as **infrared radiation**. **Greenhouse gases** such as carbon dioxide and water vapour absorb this energy, stopping it escaping into space. This is called the **greenhouse effect**: without it our planet would be much colder, and life as we know it would not exist.

Carbon dioxide is released when hydrocarbon fuels burn. Our use of such fuels is releasing carbon dioxide faster than it can be removed by natural processes such as photosynthesis. The concentration of carbon dioxide in the atmosphere increased by around 22% in the second half of the last century. Extra carbon dioxide contributes to an enhanced greenhouse effect, leading to **global warming**.

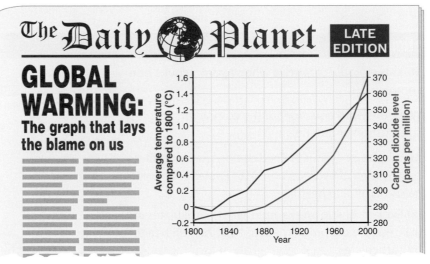

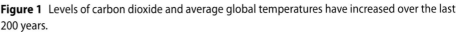

Figure 1 Levels of carbon dioxide and average global temperatures have increased over the last 200 years.

Science in action

Volcanic eruptions and wildfires are natural sources of carbon dioxide. They release much less overall than burning fossil fuels does, but they release it in one go. This makes it difficult for scientists to include the effects of volcanic eruptions and wildfires in computer models of global warming.

Global dimming

Gerry Stanhill, an English scientist working in Israel at the end of the last century, noticed something strange about records of sunlight. Israel received much less sunlight in the 1990s than it did in the 1950s. When he checked, Stanhill found similar results in countries all around the world, with a mean loss of around 2% sunlight per decade. He called this **global dimming**.

Global dimming happens because of the solid particles released from burning fuels. These particles reflect sunlight back into space. In addition, water droplets condense around them, making clouds that are more reflective than normal clouds. Global dimming is likely to have reduced the impact of global warming. It has also reduced the amount of evaporation, interfering with the water cycle and possibly causing droughts.

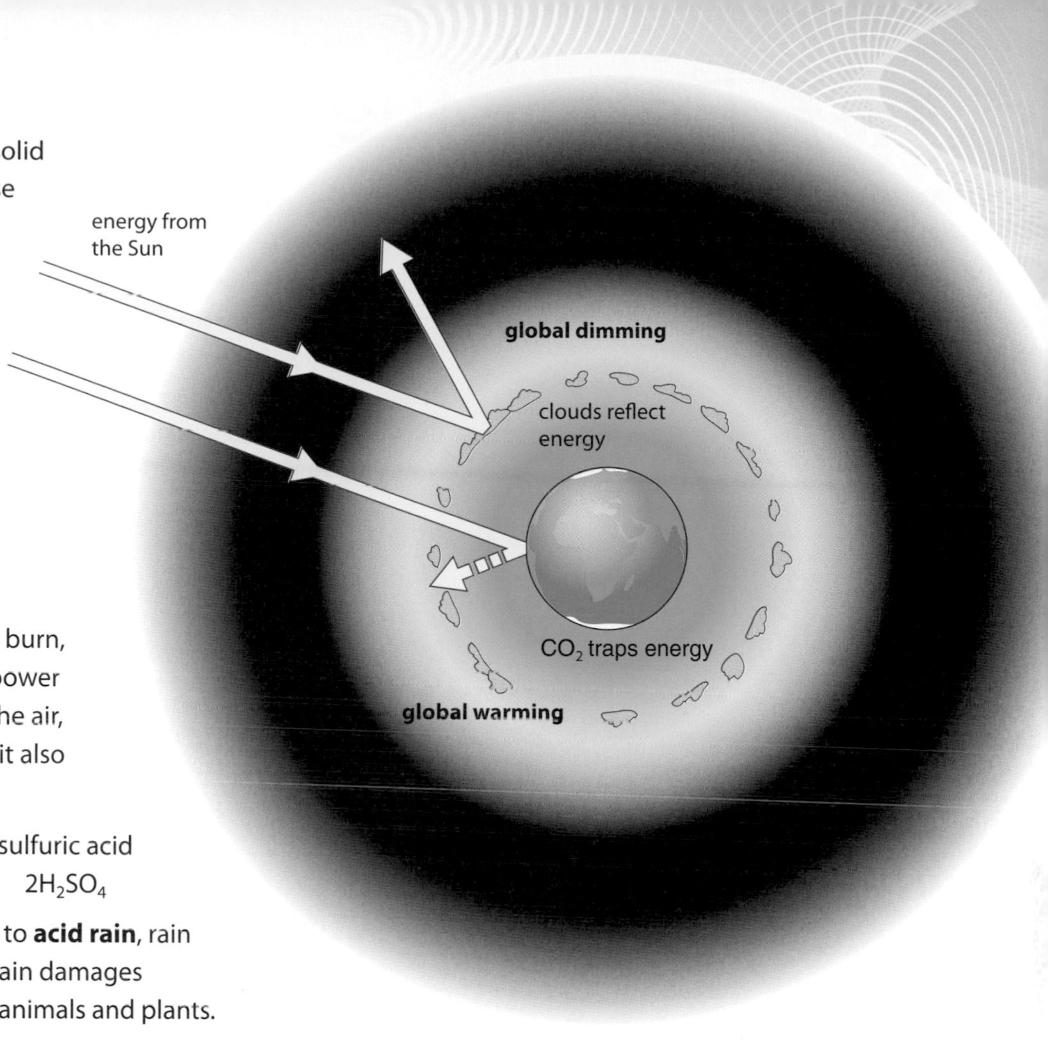

Acid rain

Sulfur dioxide may be released when fuels burn, particularly when coal is used as a fuel in power stations. This gas dissolves in moisture in the air, forming sulfurous acid, and sulfuric acid if it also reacts with oxygen in the air.

$$\text{sulfur dioxide} + \text{water} + \text{oxygen} \longrightarrow \text{sulfuric acid}$$
$$2SO_2 + 2H_2O + O_2 \longrightarrow 2H_2SO_4$$

Sulfurous acid and sulfuric acid contribute to **acid rain**, rain that is more acidic than normal rain. Acid rain damages buildings and rocks and it can harm or kill animals and plants.

Sulfur dioxide can be removed from the waste gases at power stations. It is absorbed by powdered calcium carbonate or damp calcium hydroxide, making a less harmful product, calcium sulfate. Sulfur can also be removed from fuels before they are used. Nowadays low-sulfur diesel and petrol are readily available at filling stations. They release less sulfur dioxide when they burn than fuels did in the past.

energy from the Sun

global dimming

clouds reflect energy

CO_2 traps energy

global warming

Figure 2 Global warming and global dimming (not to scale).

Questions

1. Explain why carbon dioxide is described as a greenhouse gas.
2. Why is global warming believed to be linked to the use of fossil fuels?
3. What is global dimming and how is it caused?
4. Explain why global dimming is likely to have reduced the impact of global warming over the last few decades.
5. How is acid rain formed and what are its effects on the environment?
6. Describe two ways in which the emission of sulfur dioxide may be reduced.
7. **(a)** Write a balanced equation for the reaction between calcium carbonate $CaCO_3$, oxygen and sulfur dioxide, producing calcium sulfate $CaSO_4$ and carbon dioxide. **(b)** Explain why absorbing sulfur dioxide emissions using calcium carbonate may contribute to global warming.
8. Demand for oil continues to increase. Emissions of smoke and soot are decreasing because of clean air laws. Suggest, with reasons, the effect of these two factors on the environment.

Route to A*

Fossil fuels naturally contain sulfur compounds, which produce sulfur dioxide when the fuel is burned. This gas contributes to acid rain. The sulfur can be removed from the fuel and used as a raw material for making sulfuric acid. This acid is an important industrial chemical, and selling it helps to reduce the overall cost of removing sulfur from fuels.

Better fuels

The end of oil

It is difficult to know how much oil is left. The rate at which we use it varies and new sources continue to be discovered. There are also large deposits of unconventional sources of oil, including a mixture of sand, water and very viscous crude oil, called tar sand. Canada, for example, has as much oil in its tar sands as all the world's conventional oil sources put together. Such unconventional sources are more difficult and expensive to exploit than oil wells. However, when oil prices rise it becomes more commercially viable to use them.

What's left?

Crude oil, coal and natural gas are **non-renewable** resources. They cannot be replaced when they have all been used up. Predictions vary as to how long they will last. However, it seems likely that oil and gas will run out in your lifetime. They will last longer if they are used more efficiently. One way to achieve this is to develop and use alternative fuels.

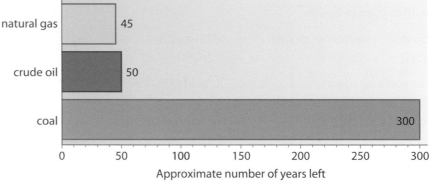

Figure 1 Coal is likely to last much longer than gas and oil at current rates of use.

Ethanol

Ethanol is the alcohol found in alcoholic drinks and antiseptic hand gels. It is also useful as a fuel. Ethanol is made from sugar by a natural process called **fermentation**. The sugar comes from plants, so ethanol made this way is a **renewable** energy resource. Ethanol can be transported, handled and stored in a similar way to petrol and diesel. Modern petrol engines can run on a mixture of 5% ethanol and 95% petrol without any modifications.

The complete combustion of ethanol releases the same products as hydrocarbon fuels do:

$$\text{ethanol} + \text{oxygen} \longrightarrow \text{carbon dioxide} + \text{water vapour}$$
$$CH_3CH_2OH + 3O_2 \longrightarrow 2CO_2 + 3H_2O$$

A proportion of the carbon dioxide released is offset by the carbon dioxide absorbed for photosynthesis by the growing crop plants. So using ethanol as a fuel can help to reduce greenhouse gas emissions. However, different plant species grow at different rates, and produce different amounts of material suitable for fermentation.

There is concern that farmland is increasingly being used to grow crops to fuel vehicles rather than to feed people. Scientists are investigating ways to use waste plant materials to produce ethanol more efficiently.

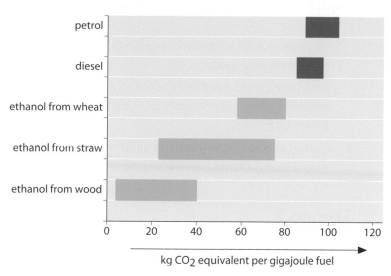

kg CO₂ equivalent per gigajoule fuel

Figure 2 The overall release of carbon dioxide from different fuels used in vehicles.

Hydrogen

Hydrogen gas can be made by passing electricity through water, a process called **electrolysis**. Hydrogen can be a renewable fuel, as long as the electricity to make it is generated using renewable energy resources such as wind or hydroelectric power. When hydrogen burns, the only waste product is water vapour, which does not cause the pollution problems of other fuels. Hydrogen **fuel cells** are the most promising way to use it. These generate electricity by reacting hydrogen with oxygen from the air. The electricity then powers electric motors.

There are problems with the use of hydrogen. It is explosive, so it must be transported and handled with care. Hydrogen is bulky and difficult to store as a gas, so it is usually stored as a liquid at high pressure and low temperature.

This car is fuelled by hydrogen. Its exhaust contains water vapour, but no carbon dioxide.

Questions

1. State which fossil fuel is likely to run out first.
2. Explain why it is difficult to be certain how long fossil fuels will last.
3. State an advantage and a disadvantage of using hydrogen as a fuel.
4. State and compare the combustion products from ethanol, hydrogen and petrol (a hydrocarbon).
5. Write a balanced equation for the production of hydrogen (H₂) and oxygen (O₂) from water by electrolysis.
6. Use the bar chart in Figure 2 to help you answer these questions.
   (a) Explain the benefits of using ethanol from wheat, rather than petrol or diesel, to fuel vehicles. **(b)** Describe two drawbacks of using wheat to produce ethanol. **(c)** Describe two benefits of using straw or wood instead.
7. Hydrogen can be made by reacting steam with coal or natural gas. Suggest why this may not be as sustainable as making it by the electrolysis of water.
8. Describe the benefits and drawbacks of using crops to make fuel such as ethanol. To what extent might this be unethical?

A*

Examiner feedback

You should be able to discuss the benefits and drawbacks of using hydrogen as a fuel in an examination question.

Science in action

Scientists are researching more convenient ways to store hydrogen. For example, metals such as magnesium react with hydrogen to produce metal hydrides. On heating, the reaction reverses and hydrogen is released.

ISA practice: the strength of concrete

Concrete is an important material for all kinds of construction. It is made by mixing water with cement, sand and aggregate – small stones or gravel. Different mixes produce concretes with different strengths. As concrete sets, crystals of calcium carbonate form. These stick the sand and aggregate together. Mortar is made from cement and sand with no aggregate, and is not as strong as concrete.

Your task is to investigate the strength of different concrete mixes. The strength can be measured using a G-clamp to break the concrete. The number of turns needed to crush the concrete is measured and recorded.

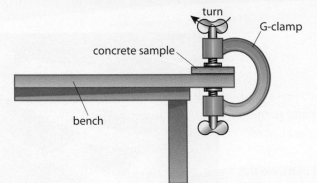

Figure 1 Using a G-clamp to test the strength of concrete.

Hypothesis

It is suggested that there is a link between the mass of aggregate in the concrete and the number of turns of the G-clamp needed to break the concrete.

Section 1

1 In this investigation you will need to control some of the variables.
   (a) Name one variable you will need to control in this investigation. *(1 mark)*
   (b) Describe briefly how you would carry out a preliminary investigation to find a suitable value to use for this variable. Explain how the results will help you decide on the best value for this variable. *(2 marks)*

2 Describe how you would carry out the investigation. You should include:
   • the equipment that you would use
   • how you would use the equipment
   • the measurements that you would make
   • a risk assessment
   • how you would make it a fair test.

You may include a labelled diagram to help you to explain your method.

In this question you will be assessed on using good English, organising information clearly and using specialist terms where appropriate. *(9 marks)*

3 Design a table that will contain all the data that you would record during the investigation. *(2 marks)*

Total for Section 1: 14 marks

Section 2

Two students, Study Group 1, carried out an investigation into the hypothesis. Figure 2 gives the results of their investigation. They used the same volume of water to make each concrete mix, and the clamp method to break the beams they made.

> Mixtures used
>
> A: 100 g cement, 300 g sand, 600 g aggregate
> B: 100 g cement, 400 g sand, 500 g aggregate
> C: 100 g cement, 500 g sand, 400 g aggregate
>
> Turns needed to break concrete
>
> A: 7.8 turns
> B: 6.1 turns
> C: 4.2 turns

Figure 2 Results from Study Group 1's investigation.

4 **(a) (i)** What is the independent variable in this investigation?
       (ii) What is the dependent variable in this investigation?
       (iii) Name one control variable in this investigation. *(3 marks)*
   (b) Put the results from Study Group 1 into the table you designed in answer to question 4. Plot a graph to show the link between the mass of aggregate used and the number of turns needed to break the beam. *(4 marks)*
   (c) Do the results support the hypothesis? Explain your answer. *(3 marks)*

Below are the results of three other study groups. Study Group 2 is another pair of students. Their results are given in Figure 3.

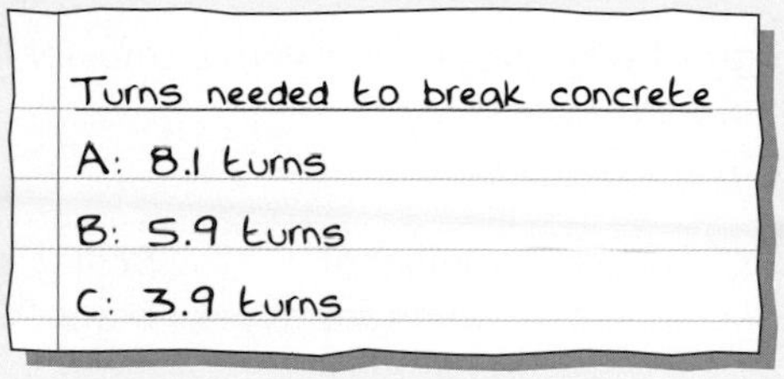

> Turns needed to break concrete
>
> A: 8.1 turns
>
> B: 5.9 turns
>
> C: 3.9 turns

Figure 3 Study Group 2's results.

Table 1 shows the results from a third student group, Study Group 3.

Table 1 Results from Study Group 3.

Mix	Cement/ g	Sand/ g	Aggregate/ g	Turns to break beam			
				Test 1	Test 2	Test 3	Mean of Tests 1–3
A	100	100	800	6.1	5.8	5.8	5.9
B	100	300	600	6.3	8.2	6.4	6.9
C	100	500	400	4.8	5.0	4.6	4.8

Table 2 shows the results from Study Group 4, a group of scientists in a building research laboratory. The scientists are testing the force need to break the concrete and the time it takes.

Table 2 Results from Study Group 4.

Mix	Aggregate in the mixture (%)	Force needed to break the beam/N				Time to break/s
		Test 1	Test 2	Test 3	Mean force/ N	
A	75	255	247	245	249	59
B	60	191	201	202	198	42
C	45	138	144	150	144	36

5 Describe one way in which the results of Study Group 2 are similar or different to those of Study Group 1, and give one reason for this similarity or difference.

(3 marks)

6 **(a)** Draw a sketch graph of the results from Study Group 3. *(3 marks)*

(b) Does the data support the hypothesis being investigated? To gain full marks you should use all the relevant data from the first set of results and Studies 2 and 3 to explain whether or not the data supports the hypothesis. *(3 marks)*

(c) The data from the other groups only gives a limited amount of information. What other information or data would you need in order to be more certain as to whether or not the hypothesis is correct? Explain the reason for your answer. *(3 marks)*

(d) Use the results from Study Groups 2, 3 and 4 to answer this question. What is the relationship between the mass of aggregate in concrete and its strength? How well does the data support your answer? *(3 marks)*

7 Look back at the investigation method of Study Group 1. If you could repeat the investigation, suggest one change that you would make to the method, and explain the reason for the change. *(3 marks)*

8 A company provides concrete for builders to use for driveways. They need to know the best concrete to use. How could the results of this investigation help the company to decide on the best mixture of concrete for a driveway? *(3 marks)*

Total for Section 2: 31 marks
Total for the ISA: 45 marks

Assess yourself questions

1 (a) State *one* difference between a mixture and a compound (1 mark)
  (b) Name the main type of compound present in crude oil. (1 mark)
  (c) Name the method used to separate the compounds present in crude oil. (1 mark)

2 Alkanes are saturated hydrocarbons.
  (a) What is a hydrocarbon? (2 marks)
  (b) Ethane and propane are two alkanes. What feature of their molecules makes them saturated? (1 mark)
  (c) Table 1 shows the molecular formulae of four alkanes.

Table 1 Molecular formulae for the first four alkanes.

Alkane	Molecular formula
methane	CH_4
ethane	C_2H_6
propane	C_3H_8
butane	C_4H_{10}

  (i) Which alkane has four carbon atoms in its molecules? (1 mark)
  (ii) Dodecane molecules each contain 12 carbon atoms. Write the molecular formula of dodecane. (1 mark)
  d Figure 1 shows the displayed structural formula of ethane.

Figure 1 The displayed structural formula of ethane.
  (i) What does each line in the formula represent? (1 mark)
  (ii) Draw the displayed structural formula of propane. (1 mark)
  (iii) What extra information does a displayed structural formula give compared with a molecular formula? (1 mark)

3 Crude oil is separated into fractions using fractional distillation. Table 2 shows six fractions from a fractionating column.
  (a) Describe the main processes that happen in a fractionating column. (4 marks)
  (b) What is the relationship between the boiling point range and the number of carbon atoms per molecule? (1 mark)
  (c) Why does the table show a range of boiling points for each fraction? (2 marks)
  (d) In which fraction would you expect to find propane? Give a reason for your answer. (2 marks)

(e) Which fraction would you expect to be:
  (i) The least volatile? (1 mark)
  (ii) The least viscous liquid? (1 mark)
  (iii) The easiest to set alight? (1 mark)
(f) Suggest why diesel burns with a smokier flamer than petrol does. (2 marks)

Table 2 Crude oil fractions.

Name of fraction	Number of carbon atoms per molecule	Boiling point range/°C
refinery gas	1–4	less than 25
petrol	5–10	25–100
paraffin	11–15	100–250
diesel	16–20	250–350
lubricating oils	21–35	350–500
bitumen	more than 35	more than 500

4 Candles contain waxes, which are hydrocarbons. Figure 2 shows an experiment to collect and test the products formed when a candle burns.

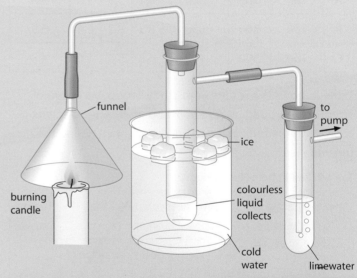

Figure 2 An experiment to collect and test the products formed when a candle burns.
  (a) (i) Name the liquid that collects in the chilled boiling tube. (1 mark)
    (ii) Name the gas detected by the limewater. (1 mark)
    (iii) Which gas is produced when burning happens in a limited supply of air? (1 mark)
    (iv) How could you tell, safely, that heat is released in the reaction? (1 mark)
  (b) Write a word equation for the complete combustion of methane. (2 marks)
  (c) Balance this equation for the combustion of propane:
$$C_3H_8 + O_2 \longrightarrow H_2O + CO_2$$
(1 mark)

(d) Nitrogen oxides are produced when petrol burns in a car engine. State one condition needed for these compounds to form. *(1 mark)*

5 Figure 3 shows information about the production of sulfur dioxide from burning different fuels.

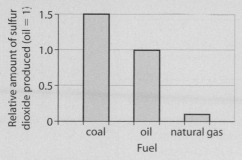

Figure 3 Bar chart of sulfur dioxide released from burning different fuels.

(a) Compare the amount of sulfur dioxide produced by each fuel. Include information from the bar chart in your answer. *(2 marks)*

(b) What environmental problem does sulfur dioxide cause? *(1 mark)*

(c) Describe two ways in which the amount of sulfur dioxide released into the atmosphere may be reduced. *(2 marks)*

(d) What is global dimming and how is it caused? *(3 marks)*

6 Table 3 shows information about the complete combustion of some alkane fuels.

Table 3 Combustion data for alkanes.

Number of carbon atoms per alkane molecule	kJ of energy released per g of fuel burned	mg of CO_2 produced per kJ of energy released
1	55.6	49.4
2	52.0	56.4
3	50.4	59.5
4	49.6	61.2
5	48.7	62.7
6	48.4	63.4

(a) Plot a suitable graph to show how the data in the two right-hand columns depend upon the number of carbon atoms per alkane molecule. Choose a suitable vertical scale so that both sets of data are shown on the same graph, and that the plotted points occupy at least half of the area of the graph. *(5 marks)*

(b) Describe in detail the relationship between the following:

(i) The energy released per gram of fuel burned and the number of carbon atoms per molecule of fuel. *(2 marks)*

(ii) The mass of carbon dioxide produced per kJ of energy released and the number of carbon atoms per molecule of fuel. *(2 marks)*

(c) Use your answers to part (b) to explain why methane, with one carbon atom per molecule, could be considered a better fuel than hexane, with six carbon atoms per molecule. *(2 marks)*

(d) Methane is a gas at room temperature but hexane is a liquid. To what extent might this information alter your answer to part (c)? *(2 marks)*

(e) Coal is a solid fuel that is almost pure carbon. When one gram of coal burns completely, it releases 32.8 kJ. Burning carbon produces 112 mg of carbon dioxide per kJ of energy released.

Suggest why hydrocarbons could be considered better fuels than coal. *(2 marks)*

(f) Table 4 shows information about the combustion of ethanol and hydrogen.

Table 4 Combustion of ethanol and hydrogen.

Fuel	kJ of energy released per g of fuel burned	mg of CO_2 produced per kJ of energy released
ethanol	29.7	28.4
hydrogen	143	0

(i) Use the table to help you describe at least three advantages of using ethanol and hydrogen in cars rather than hydrocarbon fossil fuels, such as petrol. *(3 marks)*

(ii) Describe at least two disadvantages of using ethanol or hydrogen in cars instead of hydrocarbon fossil fuels such as petrol. *(2 marks)*

(iii) Ethanol and hydrogen may be described as renewable resources. What does this mean? To what extent may these two fuels be described in this way? *(4 marks)*

7 Read the information about benzene, and then answer the questions.

Benzene is a toxic, colourless liquid with a sweet smell. Its chemical formula is C_6H_6. It was first isolated in 1825 by Michael Faraday from whale oil, which was used as a fuel for lamps. Nowadays most benzene is made from crude oil. It is added to petrol to increase the fuel's 'octane rating'. This helps car engines run more smoothly.

(a) Explain how you know that benzene is a hydrocarbon but not an alkane. *(3 marks)*

(b) Suggest reasons for the following:

(i) Benzene is made from crude oil rather than whale oil. *(2 marks)*

(ii) Benzene is becoming increasingly expensive. *(2 marks)*

(c) Write a word equation and a balanced equation for the complete combustion of benzene. *(3 marks)*

(d) Explain why the amount of benzene allowed in petrol is limited by law. *(1 mark)*

Here are three students' answers to the following question:

Titanium is as strong as steel but much lighter. It is produced from titanium dioxide using a batch process that may take several days to complete. The flow chart summarises the main stages involved.

About 1 tonne of titanium is produced per day by a titanium reactor. About 13 000 tonnes of iron is produced per day by a blast furnace, most of which is converted into steel.

Explain why titanium costs more than steel to produce, and why it is better to recycle these metals.

In this question you will be assessed on using good English, organising information clearly and using specialised terms where appropriate. (6 marks)

Titanium dioxide and chlorine gas react together to produce titanium chloride.

↓

Titanium chloride reacts with magnesium at 900 °C in a closed reactor for several days, producing titanium.

↓

The reactor is opened after it cools down, then the titanium is separated by hand from the magnesium chloride.

Figure 1 Flow chart of titanium production.

Read the answers together with the examiner comments. Then check what you have learnt and try putting it into practice in any further questions you answer.

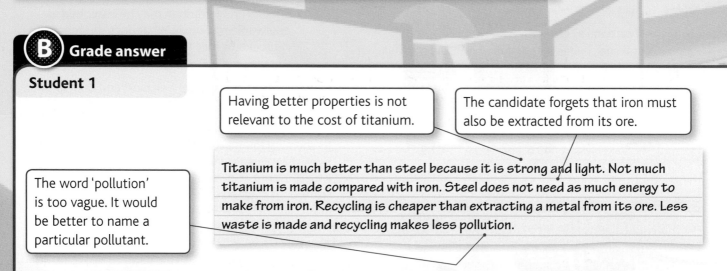

B Grade answer

Student 1

Having better properties is not relevant to the cost of titanium.

The candidate forgets that iron must also be extracted from its ore.

The word 'pollution' is too vague. It would be better to name a particular pollutant.

Titanium is much better than steel because it is strong and light. Not much titanium is made compared with iron. Steel does not need as much energy to make from iron. Recycling is cheaper than extracting a metal from its ore. Less waste is made and recycling makes less pollution.

Examiner comment

The candidate has not answered the question carefully enough. Their answer lacks detail and they make little use of relevant specialist terms. They mistakenly think that a metal will be expensive just because it has desirable properties. The candidate makes a valid point about the amount of titanium produced compared with the amount of iron and steel produced. However, steel is made from iron, and this also has to be extracted from its ore.

The candidate mentions the cost of recycling but they could have explained why it is cheaper than extracting metals from their ores. They make an attempt to explain why recycling is better for sustainable development. Fewer waste materials will be produced because of recycling, but the candidate should have taken care to name a relevant pollutant, such as carbon dioxide.

A Grade answer

Student 2

'It' (titanium) is not extracted using electrolysis.

The candidate mentions limited resources and gives a correct example.

Titanium takes a long time to produce. Even then it is produced in much smaller amounts than iron and steel. It is expensive to produce because it is extracted by electrolysis, which uses a lot of electricity. Recycling saves limited resources such as metal ores. It is cheaper than extracting the metal from its ore because it saves energy needed for extraction and processing.

The candidate gives a correct reason why recycling is cheaper.

Examiner comment

The candidate has answered in detail, giving reasons for several of their statements. Their answer has a clear structure and they use some relevant specialist terms. They give a correct reason why electrolysis is expensive. Unfortunately, they incorrectly state that titanium, rather than the magnesium needed, is extracted using electrolysis. The candidate gives a correct example of a limited resource that will be conserved by recycling. They mention cost and give a correct reason why recycling is cheaper. However, they do not link saving energy to sustainable development, when they could have easily done so.

A* Grade answer

Student 3

A correct use of information given in the stem of the question.

A more precise word than 'saves'.

Titanium is produced in a batch process while iron is produced in a continuous process. The production of titanium is more labour intensive and requires several stages. Magnesium and chlorine are needed to produce titanium, and these are expensive.
Recycling conserves limited resources such as metal ores. Less energy is needed for recycling than for extracting the metal from its ore. Fewer waste materials are produced by recycling.

Specific expensive materials are mentioned.

Examiner comment

The candidate has answered in great detail, taking care to give correct reasons for their statements. Their answer is structured well, with a good range of relevant specialist terms used accurately. They correctly compare the different types of processes used to extract titanium and iron. They give two different statements in their second sentence, each of which is valid. They then go on to make a correct statement about the cost of magnesium and chlorine. However, this would be unnecessary and the candidate could run out of time if they wrote too much.

The candidate gives a clearly structured answer about recycling, too. They focus on sustainable development rather than cost, but do make three valid points about this. They give a correct example of a conserved resource, they correctly explain why less energy is used without going into too much detail, and they mention the reduction in waste.

MOVING UP THE GRADES

- Read the whole question carefully.
- Put your answers into a logical sequence and make sure you include relevant specialist terms.
- Use both your knowledge *and* the information given to you in the question.
- Make sure you explain all the ideas asked of you in the question.
- Take care not to give an incorrect explanation for a correct statement.

Oils, Earth and atmosphere

The Earth's crust, the oceans and the atmosphere are our only source of raw materials to make everything that we have. In this unit you will learn about the structure of the Earth, including looking at how scientists came to realise that the continents are moving. You will also study the gases in the atmosphere today and how the atmosphere has changed since the Earth was young.

Crude oil is extracted from the Earth's crust. As well as being used as a fuel, crude oil is a very valuable resource from which many important substances are made. In this unit you will look at how chemicals called alkenes are made from crude oil fractions and how these alkenes are used to make polymers. You will look at the properties and uses of different polymers, and how polymers are disposed of. You will also consider the social, economic and environmental advantages and disadvantages of the production and disposal of polymers. The alcohol ethanol, used as a fuel, can also be made from alkenes and you will compare this with its production from crops such as sugar cane.

Crops are also a source of vegetable oils. You will study how these oils are extracted and their uses in food, cooking and to make biodiesel. You will compare the structure and properties of different vegetable oils and their effects on health. You will see how emulsifiers, as an example of food additives, allow vegetable oils and water to mix.

Test yourself

1   What is different about the fractions separated from crude oil?
2   Why is the burning of many fuels thought to be leading to global warming?
3   What gases are in the air?
4   Why do volcanoes and earthquakes only occur in certain parts of the world?
5   Why are additives put in some foods?

Objectives

By the end of this unit you should be able to:

- describe what alkenes are and how they are made from fractions of crude oil by cracking
- describe what polymers are and some of their uses
- evaluate social, economic and environmental advantages and disadvantages of the production and disposal of polymers
- compare the production of ethanol from crops such as sugar cane with its production from an alkene
- describe how vegetable oils are extracted from crops and some of their uses in food and to make fuels
- explain the difference between saturated and unsaturated vegetable oils and describe their relative effects on human health
- describe the structure of the Earth and explain how scientists came to accept the theory that the continents are moving
- explain why scientists are uncertain about changes that have happened to the atmosphere since the Earth was young.

Cracking

How useful is crude oil?

Crude oil is a mixture (see lesson C1 4.1) and is separated into useful **fractions** at oil refineries (see lesson C1 4.2). Some fractions, such as petrol, are in greater demand than the supply. Other fractions, such as fuel oil, are in less demand than the supply. This problem is solved by **cracking** the large, less sought-after molecules into smaller, more useful ones. The process of cracking also produces chemicals called **alkenes**, from which many plastics are made.

Supply and demand issues

Crude oil is a mixture of mainly **hydrocarbons**, most of which are **alkanes**. Hydrocarbons are compounds containing carbon and hydrogen only. The different fractions separated from crude oil by **fractional distillation** have different properties. Fractions containing shorter alkanes are less viscous, are easier to ignite and burn with a cleaner flame. These properties make them more useful as fuels and so more valuable than fractions containing longer alkane molecules.

The more useful fractions containing shorter alkanes, for example petrol, are in greater demand than the supply. In other words, more petrol is needed than we produce by the fractional distillation of crude oil. Fractions containing longer alkanes, for example fuel oil, are less useful as fuels, and the supply is greater than the demand.

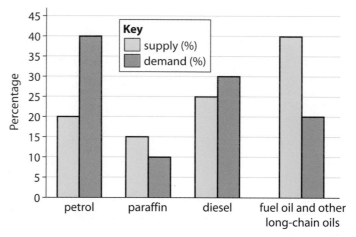

Figure 1 Supply and demand for some fractions of crude oil.

Cracking

To solve the supply and demand problem, longer hydrocarbons can be broken down into smaller ones by a process called cracking. Cracking involves breaking C—C covalent bonds.

The hydrocarbons are heated to vaporise them, turning them into gases. They are then passed over a hot **catalyst**. A catalyst is a chemical that speeds up a reaction but does not get used up. Cracking is a **thermal decomposition** reaction in which molecules are broken down by heating them. Much energy is needed to break the covalent bonds between the atoms.

Cracking produces shorter alkanes and alkenes. The shorter alkanes are used to help meet the demand for fuels such as petrol. The alkenes are used to make plastics.

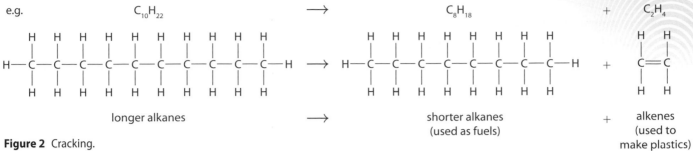

e.g. $C_{10}H_{22}$ $\rightarrow$ C_8H_{18} + C_2H_4

longer alkanes $\rightarrow$ shorter alkanes (used as fuels) + alkenes (used to make plastics)

Figure 2 Cracking.

Alkenes

Alkanes are **saturated** hydrocarbons. These contain no C=C double bonds, and have the general formula C_nH_{2n+2}. Alkenes are **unsaturated** hydrocarbons that contain one or more C=C double bonds. Alkenes have the general formula C_nH_{2n}.

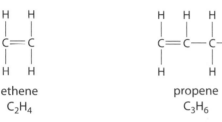

ethene
C_2H_4

propene
C_3H_6

Figure 4 Alkenes.

A simple test for unsaturated hydrocarbons uses bromine water. Bromine water is an orange colour. It turns colourless when it reacts with unsaturated hydrocarbons. However, if it is added to a saturated hydrocarbon, there is no reaction so it stays an orange colour.

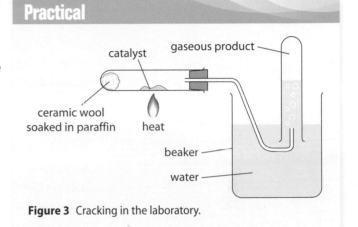

catalyst — gaseous product

ceramic wool soaked in paraffin — heat

beaker

water

Figure 3 Cracking in the laboratory.

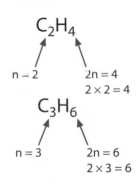

C_2H_4

n − 2   2n = 4
$2 \times 2 = 4$

C_3H_6

n = 3   2n = 6
$2 \times 3 = 6$

Figure 5 The general formula of alkenes.

Questions

1. Why are shorter alkanes in more demand as fuels than longer alkanes?

2. **(a)** Name one fraction of crude oil that is in greater demand than the supply. **(b)** Name one fraction of crude oil that is in greater supply than the demand.

3. **(a)** Why is cracking carried out? **(b)** How is cracking carried out?

4. What are the main two types of products of cracking and what are they used for?

5. An alkane has the formula $C_{12}H_{26}$. Write a balanced equation to show this alkane being cracked to form propene and an alkane.

6. **(a)** Alkenes are unsaturated molecules. What is an unsaturated molecule? **(b)** Describe a test to show that a molecule is unsaturated. Give the result of this test.

7. **(a)** Give the formula of an alkene that contains four C atoms. **(b)** Draw the structure of an alkene that contains four C atoms.

8. Explain as fully as you can the economic reasons for cracking crude oil.

The bromine reacts with the double bond in the alkenes.

Polymers

Versatile polymers

Polymers (plastics) are used to make many things, including bags, bottles, CDs, DVDs and the casings of electrical items such as computers and mobile phones. Many polymers are made from alkenes. This lesson looks at what polymers are.

Polymerisation

Polymers are large molecules made from lots of small molecules joined together. The small molecules that are joined together are called **monomers**. The process in which monomers join together is called **polymerisation**. For a simple model of polymerisation, imagine joining paper clips together to make a long chain.

Figure 1 A simple model for polymerisation.

Taking it further

Alkenes polymerise by a process called addition polymerisation in which the polymer is the only product. There is another type of polymerisation called condensation polymerisation where small molecules, such as water, are produced, as well as the polymer. Many artificial polymers, such as nylon and polyesters, are condensation polymers, as well as many natural polymers, such as starch and cellulose.

Making poly(ethene)

Many common polymers are made from alkenes. This is because the double bond in alkenes can be broken and used to join the molecule to another alkene molecule. For example, **poly(ethene)** is made by joining several thousand ethene molecules together into long-chain molecules. This polymer is better known as polythene and is used to make things like plastic bags and plastic bottles.

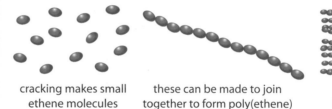

cracking makes small ethene molecules

these can be made to join together to form poly(ethene)

the chains stack up like molecular spaghetti

Figure 2 Making a polymer.

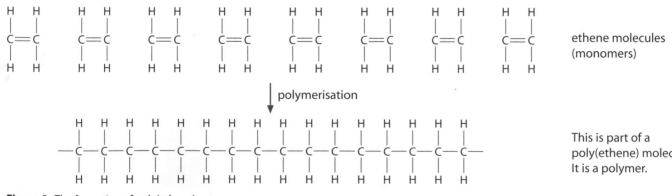

ethene molecules (monomers)

polymerisation

This is part of a poly(ethene) molecule. It is a polymer.

Figure 3 The formation of poly(ethene).

Figure 4 Equation for the formation of poly(ethene).

A balanced equation can be written for the polymerisation of ethene, as shown in Figure 4. The number of molecules that join together is very large. It is often several thousand but the exact number varies. We can use n to mean a large number.

Making other polymers

Poly(propene) is formed from the polymerisation of the alkene propene. Poly(propene) is used to make plastic crates, bins and rope. Polymers are often named by putting *poly* before the name of the monomer in brackets. Poly(propene) is a good example of this.

Many other polymers can be made from monomers containing C=C double bonds. Some examples are shown in the table below. Different polymers have different properties and uses. You will not have to learn these for the exam.

Figure 5 Equation for the formation of poly(propene).

Table 1 Useful polymers.

Polymer	Common name	Properties	Uses
poly(ethene)	polythene	flexible	bags, cling film
poly(propene)	polypropylene	flexible	crisp packets, crates, ropes, carpets
poly(chloroethene)	PVC	tough, hard	window frames, gutters, pipes
poly(tetrafluoroethene)	Teflon/PTFE	tough, slippery	frying pan coatings, stain-proof carpets
poly(methyl 2-methylpropenoate)	Perspex	tough, hard, clear	shatter-proof windows
poly(ethenol)	PVA	dissolves in water	hospital laundry bags

Questions

1. **(a)** What is a polymer? **(b)** What is a monomer? **(c)** Name five different polymers mentioned on these pages.

2. Polymers can be made from alkenes. What do alkenes contain that allow them to react to form polymers?

3. **(a)** Give two uses for poly(ethene). **(b)** Give two uses for poly(propene).

4. Describe what happens to molecules in a polymerisation reaction.

5. **(a)** Name the polymer made from phenylethene. **(b)** Name the polymer made from methylpropene.

6. Draw a diagram to show four molecules of propene reacting together to form poly(propene).

7. Write an equation for the formation of poly(chloroethene), also called PVC, from chloroethene. The structure of chloroethene is shown in Figure 6.

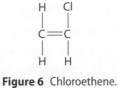

Figure 6 Chloroethene.

8. Explain what polymers are and how they are formed from alkenes. Use poly(ethene) as an example in your answer.

Examiner feedback

Examination questions often provide the structure of an alkene and ask for the polymer formed from this alkene to be drawn.

When drawing the structures of polymers, make sure that the brackets go through the bonds at the ends, that there is a single (not double) bond between the two carbon atoms and that you put the n after the bracket.

New uses for polymers

Learning objectives

- describe how new polymers are being developed with new uses
- evaluate the advantages of some new uses of polymers.

Designing materials

Many new uses are being found for polymers, and new types of polymers are being developed all the time with new and exciting properties and applications. Scientists are even designing polymers to have very specific properties. This lesson looks at some new uses for polymers.

Smart polymers

Smart materials are materials that have one or more properties, for example shape, colour or size, that change with a change in conditions, for example temperature or pH.

Shape-memory polymers are one example of a smart material. They can change shape as the temperature changes. However, they can go back to their original shape when the temperature gets high enough. Two uses of shape-memory polymers are in heat-shrink wrapping used for packaging and in heat-shrink tubing used to cover bundles of electrical wires. The wrapping or tube is placed over what it is being used to protect and then heated. As it is heated, it shrinks to a tight fit.

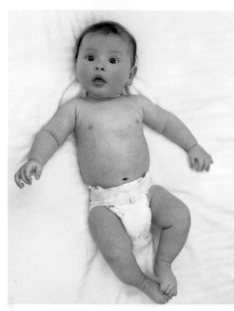
Hydrogels keep urine from the baby's skin.

Polymers that change colour with changes in temperature or light are also smart materials. For example, you can buy plastic bowls and spoons for baby food that change colour if the food is too hot. There are also light-sensitive lenses for spectacles that darken in bright light.

The colour of the spoon changes from purple to pink to show that the food is too hot.

Hydrogels

Hydrogels are polymers that can absorb a lot of water and turn into a gel. One use is in disposable nappies; the hydrogel is built into the nappy to absorb urine to prevent nappy rash. Hydrogels are also used to make soft contact lenses. They are also used in garden plant containers and hanging baskets. The hydrogel is mixed into the compost and absorbs a lot of water. This means that the plants can survive longer without being watered.

Waterproof coatings for fabrics

For many years, waterproof coatings for clothing fabrics kept water out but would not allow water vapour released by the body to escape. This meant that the fabric still got wet on the inside. However, light, waterproof coatings for fabrics have been developed that will keep water out but allow water vapour from sweat to escape, keeping the wearer dry. The pores in the coating are too small to allow droplets of liquid water through but large enough to allow individual molecules in water vapour out.

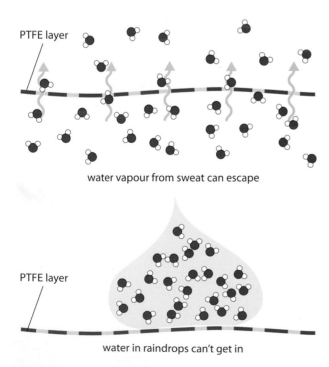

Figure 1 How a breathable PTFE coating works.

Dental polymers

For many years, dentists have used metal alloys to fill cavities in teeth. Over the last few years new tooth-coloured materials have been developed using polymers. These are mixtures of chemicals that include a monomer plus chemicals to start the polymerisation reaction. The mixture is placed in the cavity in the tooth and a very bright blue light is shone onto the mixture. This starts the polymerisation reaction making a polymer in the tooth.

Packaging materials

Plastics have been used in many ways in recent years for packaging. However, there have been several new developments. For example, there is now food packaging that changes colour if the food starts to go off, making consumers more aware and helping to prevent food poisoning. The addition of antimicrobial chemicals to some packaging prevents bacterial growth and has extended the shelf-life of some foods. There are also some biodegradable plastics used now that decay naturally over time, reducing long-term waste problems.

Wound dressings

Wound dressings are needed to protect wounds while they heal. New polymers have allowed the development of improved dressings that can now include antibacterial barriers, hydrogels and waterproof but breathable films. These help to protect the wound as it heals.

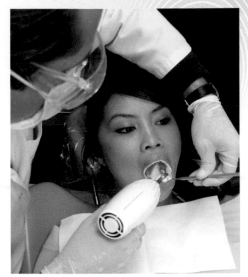

A dentist starts the polymerisation reaction in a new filling.

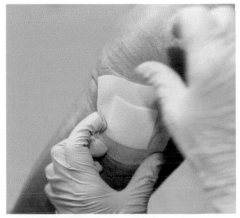

A wound dressing.

Questions

1   **(a)** What is a smart material? **(b)** What is a shape-memory polymer? **(c)** Give two different uses for smart materials.

2   **(a)** What is a hydrogel? **(b)** Why are hydrogels used in nappies?

3   Why do people prefer to have polymer fillings in front teeth instead of metal fillings?

4   Explain how a cavity in a tooth can be filled with a polymer.

5   Give two recent developments in packaging materials and explain why these are useful developments.

6   **(a)** What advantage does a waterproof coating on a wound dressing have? **(b)** What advantage does an antibacterial coating on a wound dressing have?

7   Modern waterproof coatings for fabrics are now 'breathable'. **(a)** Explain what this means and why this is an improvement on waterproof coatings that were not breathable. **(b)** Explain how breathable coatings work.

8   There have been many developments in the production and uses of polymers in recent years. Choose three separate examples and explain why these developments are helpful to our everyday lives.

Examiner feedback

You will not need to remember all the details about these uses of polymers. However, you may be provided with information about them in an exam question and asked to comment on their uses.

Practical

Nappies contain hydrogels. Plan and carry out an experiment to compare the absorbency of different brands of nappies.

Disposing of polymers

Where does all the rubbish go?

We produce about 5 million tonnes of polymers each year in the UK. What happens to them when we have finished with them?

Waste plastic made from polymers. We produce 20 times more plastic today than we did 50 years ago.

Burial in landfill

Most discarded polymers are buried with domestic and industrial waste in **landfill sites**. This is easy and cheap but can cause many problems. It is becoming difficult to find places to put more landfill sites as we bury more and more rubbish. Also, most polymers are not **biodegradable**. This means that microorganisms cannot break them down, so they will not **decompose**, or rot away, if they are buried. It also means that they will not rot away if dropped as litter.

New polymers are being developed that are broken down by microorganisms, that is, they are biodegradable. Some biodegradable polymers, such as PLA, are being made from cornstarch. Cornstarch is extracted from corn (maize). Some food packaging and shopping bags are made from polymers made from cornstarch. At the moment, these polymers are more expensive to produce than polymers made from chemicals derived from crude oil. However, some companies are using them as they are more environmentally friendly. Their use will increase if their cost becomes more competitive compared with other polymers, if there is more public pressure for their use or if there is government legislation.

Landfill sites in the UK are regulated to minimise their impact on the environment.

Incineration

Polymers burn well and a lot of polymer waste is burned in **incinerators**. However, carbon dioxide (CO_2) is formed when polymers are burned. This is a greenhouse gas and contributes to global warming. In addition, unless the conditions in the incinerator are carefully controlled to ensure there is enough oxygen from the air, toxic gases including carbon monoxide (CO), hydrogen chloride (HCl) and hydrogen cyanide (HCN) can also be produced.

Modern incinerators are designed to burn waste very efficiently with few harmful emissions except the greenhouse gas carbon dioxide. The heat released can also be used to generate electricity.

Waste incinerators are more common in countries, such as Japan, where land is a scarce resource.

Recycling

Most polymers can be recycled. Less crude oil and energy are used in **recycling** polymers than in making them from the raw materials. Also, recycling avoids the problems caused by burying or burning waste polymers.

Polymers are often recycled by chopping up the polymer into pellets that can be melted and moulded into new products. Polymers that can be melted and remoulded are called thermosoftening polymers. New developments allow some polymers to be broken back down into monomers, from which new polymers can then be made.

Polymers must be separated out into their different types to be recycled. Most polymer products have a symbol to show which polymer they are made from. However, the sorting has to be done by hand, which is time-consuming and expensive. New technologies are being developed to sort polymers by machine.

Using products made from crude oil

Polymers are one of several groups of products made from crude oil. Other examples include medicines, solvents and detergents. We also make much use of crude oil for fuels to keep us warm and for transport. Fuels, polymers and other substances made from crude oil have greatly improved our everyday lives. However, the problems of the disposal of some of these substances made from crude oil and the pollutants produced from their use as fuels are problems we have to deal with as a society.

PETE
polyethylene terephthalate

HDPE
high-density poly(ethene)

PVC
poly(chloroethene) (also called polyvinyl chloride, PVC)

LDPE
low-density poly(ethene)

PP
poly(propane)

PS
polystyrene

OTHER
other polymers

Figure 1 Polymers and their recycling symbols.

Questions

1. What are the three ways of disposing of used polymers?

2. How are most polymers disposed of?

3. **(a)** Most polymers are not biodegradable. What does this mean?
   (b) Why is this a problem for burying polymers in landfill?

4. **(a)** Give two disadvantages of burning polymers. **(b)** Give two advantages of burning polymers.

5. **(a)** What must be done to polymers before they can be recycled?
   (b) How is this usually done? **(c)** How are advances in technology helping to solve this problem?

6. **(a)** There are about 60 million people in the UK. At its peak a few years ago, 17.5 thousand million supermarket bags were used per year in the UK. Calculate the average number of bags that were used per person in the UK that year. **(b)** Why has the number of bags used per person started to fall?

7. Draw a table to list the three ways of disposing of polymers. Include columns to show: **(a)** what is done in each method; **(b)** the advantages and disadvantages of each method; **(c)** whether each method has any social, economic or environmental issues.

8. Chemicals made from crude oil have improved modern life, but also bring problems. Discuss some of the advantages and disadvantages of using chemicals made from crude oil.

Taking it further

Polymers made from alkenes are not biodegradable because the electrons in the covalent bonds between the carbon atoms in the polymer chain are shared equally between the atoms, making the bonds very strong. In biodegradable polymers, the electrons in the bonds between different atoms in the chain are not shared equally between atoms, making it easier for other molecules to react with the polymer and break down the chain.

Ethanol production

Ethanol is used to fuel some cars in Brazil.

Examiner feedback

When comparing the two methods of making ethanol, make sure that you write about both methods, e.g. fermentation is slow, hydration is fast; or make statements that are comparative, e.g. hydration is faster than fermentation.

Practical

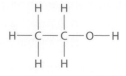

conical flask — delivery tube

50 cm³ of water + 2 teaspoons of sugar + yeast

test tube of limewater

Figure 2 Fermentation in the laboratory.

What is ethanol?

There are many different types of alcohol. The commonest alcohol is ethanol, C_2H_5OH, the alcohol found in alcoholic drinks.

Figure 1 The structure of ethanol.

Uses of ethanol

Ethanol is used as a solvent to make many common substances, such as detergents and medicines, which are also called pharmaceuticals.

Ethanol can also be used as a fuel, as explained in lesson C1 4.5. In Brazil, for example, it is used as fuel for some cars. In the UK, 5% of the mixture sold as petrol at petrol stations is now ethanol. The ethanol used in fuels is made from fermentation of crops and is a biofuel.

Over 330 000 tonnes of ethanol are made in the UK every year. There are two main ways to make ethanol: by fermentation of carbohydrates and by hydration of ethene. This lesson describes and compares these two methods.

Making ethanol by fermentation

Over 90% of the world's ethanol is produced by fermentation from crops that contain a lot of carbohydrates, such as sugar cane, sugar beet, rice and maize. These are **renewable** raw materials because we can grow more to replace those that have been used. The ethanol made by fermentation is not pure; it also contains water. Fractional distillation is used to make pure ethanol from the mixture.

Alcoholic drinks are made by fermentation. Different drinks are made by fermenting different crops. For example, wine is made from grapes and beer from malted barley. Yeast and water are added to the grapes or barley and the mixture left, in the absence of air, for fermentation to produce the drink.

Making ethanol by hydration of ethene

The other main method of producing ethanol is by reaction of ethene with steam. This reaction is called **hydration**. Ethene is made from crude oil. Crude oil is a non-renewable resource (see lesson C1 4.5), which means that it cannot be replaced when we use it. However, the ethanol produced this way is pure.

Table 1 compares the two ways of making ethanol.

Table 1 Comparison of two ways of making ethanol.

	Fermentation of carbohydrates	Reaction of ethene with steam
raw materials	source of carbohydrates (e.g. sugar cane, sugar beet, rice and maize) 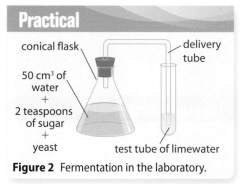	crude oil
type of raw materials	renewable	non-renewable

	Fermentation of carbohydrates	Reaction of ethene with steam
reaction	sugar $\longrightarrow$ ethanol + carbon dioxide $C_6H_{12}O_6 \longrightarrow 2C_2H_5OH + \quad 2CO_2$	ethene + steam $\longrightarrow$ ethanol $CH_2=CH_2 + H_2O \longrightarrow C_2H_5OH$
temperature	30–40 °C	300 °C
pressure	normal pressure (1 atm)	high pressure (60–70 atm)
catalyst	enzymes in yeast	concentrated phosphoric acid
other essential conditions	aqueous (in water) anaerobic (no oxygen present)	
type of process	batch process (stop–start process, which is labour intensive, i.e. requires a lot of workers)	continuous process (process kept running 24 hours a day, seven days a week, less labour intensive)
comparison of reaction rates (speed)	slow reaction	fast reaction
comparison of ethanol made	impure ethanol (purified by fractional distillation)	pure ethanol
energy use	lower energy use	high energy use for high temperature and pressure
sustainability	raw materials are renewable so sustainable process	raw materials are not renewable so this is not a sustainable process
comparison of economic factors	cheaper process (but the main cost is the energy needed to separate the ethanol by fractional distillation)	more expensive process due to high temperatures and pressures
comparison of environmental factors	uses renewable raw materials and less energy	uses non-renewable raw materials and more energy

Questions

1 Give the formula of ethanol.

2 Give three uses for ethanol.

3 (a) What are the raw materials used to make ethanol by fermentation? (b) What conditions are used to make ethanol by fermentation? (c) How is the impure ethanol that is made purified?

4 (a) What is the raw material used to make ethanol by hydration? (b) What conditions are used to make ethanol by hydration?

5 Explain why fermentation has lower energy costs than hydration.

6 (a) What is the difference between batch and continuous processes? (b) Which process has the highest labour costs and why?

7 Compare the two methods of producing ethanol in terms of rate, purity of ethanol, use of raw materials, type of process and energy costs.

8 Compare the two methods of making ethanol in terms of sustainability and economic and environmental factors.

Vegetable oils and biodiesel

Rapeseed oil is used for cooking and for making biodiesel.

Stored chemical energy from light

Plants make glucose from carbon dioxide and water by photosynthesis. Plants need glucose for respiration, and they may convert it into starch or vegetable oils for storage. These oils provide energy to the plants but we can use them for food and fuel.

Extracting oils

Some plant materials contain enough oil to make extraction worthwhile. These include nuts from hazel bushes, fruit from oil palms and olive trees, and seeds from rapeseed plants and sunflowers. After harvesting, the plant material is sieved to remove stones and other objects that might damage machinery. It is then crushed to break open the oil-containing cells, forming a mixture of oil and broken plant material.

The oil can then be separated from the broken plant material by squashing or **pressing** the mixture. Oil and water are released. They separate into two layers. The water is drained away leaving the oil behind. It may still contain water and other impurities such as plant resins and bacteria, so it is filtered, then heated to drive off the remaining water and kill bacteria.

Vegetable oils may also be separated using a solvent. They dissolve in the solvent to form a solution that is more easily removed from the broken plant material. The solvent is distilled from the mixture, leaving the oil behind. Hexane is a solvent commonly used to dissolve vegetable oils for extraction on industrial scales. It is a hydrocarbon produced from crude oil but it is recycled so that it can be used again.

Practical

Oils that degrade when heated strongly, such as limonene from orange peel, may be separated using steam distillation. Steam is passed through the broken plant material, carrying the oil with it. When the mixture is cooled, a layer of water forms with the oil floating on top.

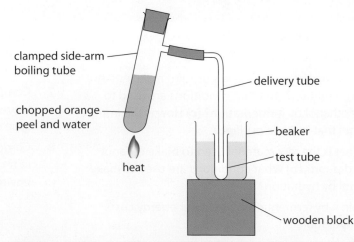

Figure 1 Volatile oils can be separated by steam distillation.

Biodiesel

Fuels for vehicles can be produced from vegetable oils. **Biodiesel** is a biofuel made from vegetable oils such as rapeseed oil. The oil is converted into biodiesel by reacting it with methanol and sodium hydroxide. Biodiesel can be used alone or blended with ordinary diesel. Diesel-engined vehicles can use it without modification.

Unlike diesel from crude oil, biodiesel molecules contain oxygen atoms. A vehicle running on biodiesel produces fewer particulates and less carbon monoxide than it does using ordinary diesel. Particulates increase the risk of respiratory disease and carbon monoxide is toxic. Biodiesel naturally contains very little sulfur, so it releases almost no sulfur dioxide when in use. This helps to reduce production of acid rain.

A **carbon-neutral** fuel has no life-cycle emissions of carbon dioxide. The amount of carbon dioxide released in its production and use is the same as the amount absorbed by the growing plants during photosynthesis. Biodiesel should be a carbon-neutral fuel. However, this is not the case, because fossil fuels are used in its production.

For example, fertilisers and methanol are manufactured from coal or natural gas. Sodium hydroxide is manufactured by passing electricity through sodium chloride solution, and at the moment two-thirds of the world's electricity is generated using fossil fuels. These fuels may also be used by the machinery needed to harvest the plants, process the oil and deliver the biodiesel. Even so, the life-cycle emissions of carbon dioxide from biodiesel are much less than those from ordinary diesel.

Biodiesel is a low-sulfur fuel with lower life-cycle emissions of carbon dioxide.

Science skills

Malaysia is the world's largest exporter of palm oil. Large areas of forest are cleared for palm oil production. This displaces local people and damages the habitats of wildlife. The table below shows how long it takes for the savings in carbon dioxide emissions to equal the amount of carbon dioxide released when land is cleared for biodiesel production.

Payback time/years	
Grassland	Forest
0–11	18–38

a Use the data to explain why it may be preferable to use cleared grassland rather than cleared forest to grow biofuel crops.

b How do the data affect the description of biodiesel as a carbon-neutral fuel?

c Use the information above to discuss the benefits and drawbacks of using existing farmland, which may have been cleared many years ago.

Questions

1 State three sources of vegetable oils.
2 Suggest why some plant materials may not be useful sources of vegetable oils.
3 Describe how vegetable oils may be extracted using mechanical methods.
4 Suggest a hazard associated with using a solvent such as hexane in the extraction of vegetable oils.
5 Describe two benefits of using biodiesel instead of ordinary diesel.
6 This table shows information about the production of bioethanol. The reduction in greenhouse gas emissions does not include emissions as a result of clearing land, and the **payback time** refers to clearing grassland.

Country	Crop	Greenhouse gas reduction (%)	Payback time/years
UK	wheat	28	20–34
Brazil	sugar cane	71	3–10

(a) Explain which country's bioethanol is likely to produce the greatest reduction overall in greenhouse gas emissions, assuming that both are made from crops grown on cleared grassland. (b) Suggest why wheat for UK bioethanol is grown on existing farmland or land previously taken out of farming.

7 What is a carbon-neutral fuel? Apart from issues surrounding land use or re-use, explain why biodiesel is not a carbon-neutral fuel.

A*

Examiner feedback

You do not need to memorise how biofuels are manufactured but you may be asked to use information given to you on the topic.

Emulsions

Learning objectives

- describe how emulsions form
- describe the properties of emulsifiers
- explain the uses of emulsions in food, cosmetics and paints.

Paints plus

You may have used **emulsion** paint to decorate your bedroom walls. It is a complex mixture of water, coloured pigments and various other chemicals. These stick the pigment to the wall and make the paint easier to use and store. However, there is more to emulsions than paint. Emulsions are mixtures with special properties.

Making emulsions

Oil and water are **immiscible**. They do not dissolve into one another when they are mixed. You have probably seen how a layer of oil can spread over the top of a puddle of water. When oil and water are shaken vigorously together, tiny oil droplets spread evenly through the water to produce a mixture called an emulsion. Traditional oil-and-vinegar salad dressings are made this way. Milk is an oil-in-water emulsion. It comprises a watery liquid mixed with tiny droplets of butterfat. In full fat milk, the water and butterfat gradually separate, and the butterfat rises to the surface to form a layer of cream. The cream can be processed to make butter, which is a water-in-oil emulsion.

Emulsifiers

Emulsions are not stable mixtures. An oil and water emulsion eventually separates out if it is left to stand. Oil is less dense than water so it floats on top. This happens with oil-and-vinegar salad dressing. Chemicals called **emulsifiers** can be added to make emulsions more stable, extending their shelf-life.

Milk contains proteins that act as emulsifiers. Mayonnaise is made from oil and vinegar, just like a traditional salad dressing, but it also contains a little egg yolk to stop it separating out. Egg yolk contains an emulsifier called lecithin.

Emulsifier molecules have two different properties. One end of the molecule is **hydrophobic**, which means 'water-hating'. It dissolves in the oil in the emulsion. The other end of the molecule is **hydrophilic**, which means 'water-loving'. It dissolves in the water in the emulsion. In an oil-in-water emulsion, for example, the emulsifier molecules surround the tiny droplets of oil. The hydrophobic 'tails' dissolve in the oil, and the hydrophilic 'heads' dissolve in the water. This stops the oil droplets joining together again to form a layer of oil, and so stabilises the emulsion.

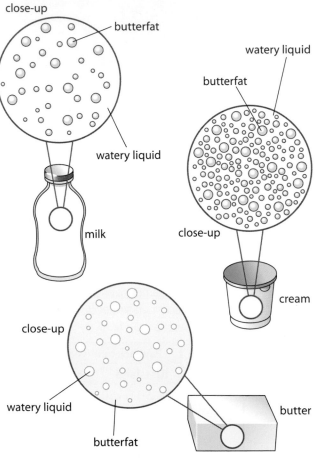

Figure 1 Milk, cream and butter are three different emulsions of the same ingredients.

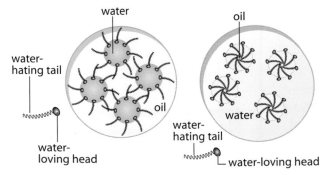

Figure 2 Emulsifier molecules have a hydrophilic 'head' and a hydrophobic 'tail'.

Examiner feedback

The link between the structure of emulsifiers and their properties is the sort of difficult chemistry tested in questions aimed at more able candidates.

Emulsions are very versatile but even with emulsifiers they do eventually separate out.

Mayonnaise without emulsifier quickly seperates into two layers.

Mayonnaise with emulsifier forms a stable emulsion.

Taking it further

Lecithin is a mixture of substances commonly used as an emulsifier in food. Originally, the name was used for a pure substance, now called phosphatidylcholine. This compound is an example of a phospholipid: phospholipids are one of the main components of all cell membranes, which rely on their head-and-tail structure.

Uses of emulsions

Emulsions are thicker, or more viscous, than the oil and water they contain. This special property is important for non-drip emulsion paints, which stay on the brush but spread easily on the walls without running. This thick texture is also important in cosmetics such as skin creams, to make them look and feel better, and in foods for the same reasons. Mayonnaise is thicker than traditional salad dressings. It does not pour and it coats food well. Similarly, cream is thicker than milk. It coats strawberries and other food in an attractive way and has a pleasant texture on the tongue.

Ice cream is a frozen emulsion that contains ice crystals and tiny bubbles of air. Like butter, soft margarines are water-in-oil emulsions. The oils alone would be too runny to make a pleasant spread for your toast. As part of an emulsion, they form a product that does not create an oil slick on your toast but is soft enough to spread straight from the fridge.

Science in action

Cosmetics contain different types of emulsion. Hand creams and shaving creams are usually oil-in-water emulsions because they need relatively little oily material to work. Sun blocks and moisturising creams are usually water-in-oil emulsions. They contain more oily material, so they have a greasy feel and stay on the skin for longer.

Questions

1. When oil and water are mixed and left to stand, why does the oil form a layer on top of the water?

2. What is an emulsion?

3. Compare the viscosity of an emulsion with the viscosity of the oil and water it contains.

4. Describe three different uses for emulsions.

5. In terms of the type of emulsion, what is the difference between milk and margarine?

6. Explain how emulsifiers work.

7. Detergent molecules have a similar structure to emulsifier molecules. Suggest how this helps washing machine detergents remove greasy stains from clothes.

8. Mayonnaise is an emulsion. Explain how it is made and how it is stabilised. To what extent would you agree that an emulsifier is used to extend the shelf-life of the mayonnaise?

Hardening vegetable oils

Animal fats tend to be more saturated than vegetable oils.

Vegetable oils in food

Vegetable oils may be altered chemically to change their properties, making them more suitable for some food uses.

Fats and oils

Many animal fats and some vegetable oils are saturated. The carbon atoms in their chains are joined to each other by single chemical bonds. This makes the chains flexible, so they can line up closely next to each other. As a result, the attractive forces between individual molecules are relatively strong and a lot of energy is needed to overcome them. This means that the melting point of saturated fats tends to be high, making them solid at room temperature.

In contrast, most vegetable oils are unsaturated. Some of their carbon atoms are joined by double bonds instead of single bonds. These C=C bonds can put a rigid bend in the carbon chain. Bent chains cannot line up as closely as unbent chains. As a result, the attractive forces between individual molecules are relatively weak and the molecules are easily moved apart. This means that the melting point of unsaturated oils tends to be low, making them liquid at room temperature.

How unsaturated?

Unsaturated vegetable oils contain C=C bonds, just as alkenes do. The presence of these double bonds can be detected using bromine. Bromine water, an orange solution of bromine, turns colourless when it is mixed with alkenes or with unsaturated vegetable oils.

Like alkanes, completely saturated oils or fats have no double bonds. They cannot decolourise bromine water. The more double bonds present, the greater the volume of bromine water that can be decolourised. Sunflower oil is more unsaturated than olive oil. It can **decolourise** about two-thirds more bromine water than olive oil can.

A **burette** is used to add precise volumes of bromine water to an oil.

Adding hydrogen

Chemists working in the food industry have developed a way to convert cheap unsaturated vegetable oils into solid fats. Hydrogen is used to change carbon–carbon double bonds into single bonds, turning an unsaturated oil into a saturated fat. This process is called **hardening**.

Vegetable oils are hardened by warming them to about 60 °C and bubbling hydrogen gas through them. One of the bonds in a C=C bond breaks, and so does the H—H bond in a hydrogen molecule. Then each of the two carbon atoms makes a new bond with a hydrogen atom. This sort of reaction is called **hydrogenation**. Nickel is used as a catalyst to speed up the reaction. It is not used up in the reaction and is easily separated from the product.

Hydrogenated vegetable oils can be used for cooking in the same way as animal fats, but they are suitable for vegetarians. For example, they can be used in baking to make cakes and pastries. Solid vegetable fats and liquid vegetable oils can be blended together for use in the manufacture of margarine and chocolate.

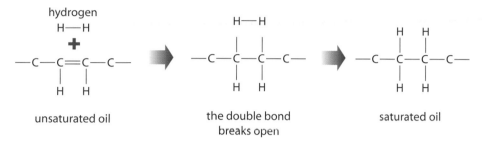

unsaturated oil   the double bond breaks open   saturated oil

Figure 1 Hydrogenation converts unsaturated oil into saturated fat at 60 °C in the presence of a nickel catalyst.

Hydrogenated vegetable oils are used to make spreads, cakes and pastries.

Questions

1   In general, what is the physical state of saturated fats at room temperature, and what is the state of unsaturated oils?

2   Which chemical bonds are present in both unsaturated oils and saturated oils?

3   Explain how vegetable oils may be hardened by hydrogenation.

4   State three uses of hydrogenated vegetable oils.

5   What would you see if bromine water were added to sunflower oil? What difference would you see if it were added to olive oil?

6   Describe the main difference in chemical structure between saturated fats and unsaturated oils.

7   **Table 1** Results of experiments to determine the melting points of some vegetable oils, and their amount of unsaturation.

Oil	Melting point/°C	Volume of bromine water reacted/cm³		
		Test 1	Test 2	Test 3
olive	−5	22.6	22.4	22.2
palm	34	14.7	15.3	15.0
sunflower	−16	34.6	34.5	34.7
rapeseed	−11	27.1	27.5	26.7

(a) Calculate the mean volume of bromine water reacted for each vegetable oil. **(b)** Explain which oil is the most unsaturated. **(c)** Describe the relationship between the amount of unsaturation of a vegetable oil and its melting point. **(d)** Coconut oil reacted with 3.2 cm³ of bromine water in a similar experiment. Predict, giving your reasons, the state of this oil at room temperature.

8   Explain the link between the degree of unsaturation of a vegetable oil and its melting point and chemical properties.

Oils and fats in our diet

Food cooks more quickly in oil than it does in water.

Nutritious oils and fats

Oils and fats are an important part of a healthy diet. They are a good source of energy and provide us with **nutrients**. For example, vitamins A, D and E are fat-soluble. They are found in foods such as dairy products, eggs, oily fish and nuts. However, too much oil and fat in the diet is harmful to health.

Cooking with oil

Water boils at 100 °C. This is hot enough to scald you but it limits how quickly food can be cooked by boiling. Food cooks much faster when it is fried. Vegetable oils have higher boiling points than water, so food can be fried at higher temperatures than it can by boiling in water. For example, olive oil can reach around 240 °C before it begins to smoke and decompose. Food cooks more quickly in vegetable oils, and different flavours are produced. Fried or fatty food may have a pleasant taste or texture, which can encourage people to over-eat.

Oils for energy

Fats and oils have a high energy density compared with proteins and carbohydrates. Table 1 shows the typical amounts of energy available to the body from these nutrients. This can be less than the energy released by burning because, for example, not all the products of digestion may be absorbed.

Table 1 Oils for energy.

Nutrient	Energy released in the body / kJ/g	Energy released when burnt / kJ/g
fat	37	39
protein	17	24
carbohydrate	17	17

As oils and fats contain a lot of energy, eating too much fried or fatty food may make you overweight, or even obese.

Taking it further

Cholesterol is an important part of cell membranes. It is made by the liver and carried in the bloodstream by substances called lipoproteins. High levels of cholesterol in the blood increase the risk of blocked arteries and heart disease.

Saturated fats can raise the level of cholesterol in the blood. They are found mainly in meat and dairy products. Too much saturated fat in the diet can also make us gain weight.

Unsaturated fats and oils tend to be more healthy options than saturated fats. They are found mainly in vegetables, nuts and fruits, and are less likely to raise cholesterol levels. One type, called omega-3, comes mainly from oily fish such as sardines and mackerel. Omega-3 oils are thought to help prevent heart disease and may even help to improve brain function.

Too much fatty food can make you obese and cause health problems.

Figure 1 Typical dietary advice concerning fats and oils.

Examiner feedback

You may be given information about fats and oils to evaluate and compare in the examination.

Science skills

In 2009, American scientists reported the results of a study involving over half a million adults. The volunteers completed questionnaires about their diet, and their medical records were checked over the next six years to see if they later developed cancer of the pancreas. The intake of polyunsaturated fats made no significant difference to cancer rates.

Table 2 A summary of some more results from the study.

	Cases of pancreatic cancer per 100 000 people per year	
	Most fat in the diet	Least fat in the diet
Total fat	46.8	33.2
Saturated fat	51.5	33.1
Monounsaturated fat	46.2	32.9
Saturated fat from animals	52.0	32.2

a What type of graph is most suitable to display these results, and why?

b The scientists concluded that saturated fat in the diet, particularly from meat and dairy products, can increase the risk of getting pancreatic cancer. Explain how their results support this conclusion.

Science in action

Bromine water is used to detect the presence of carbon–carbon double bonds. Bromine and iodine are in the same group in the periodic table, so their chemical reactions are similar.

Iodine can also be used to detect the presence of C=C bonds. As iodine is less reactive than bromine, iodine monochloride may be used instead of iodine.

An 'iodine value' is the number of grams of iodine that would react with 100 g of an oil or fat. Sunflower oil has a typical iodine value of about 135 and olive oil about 82. Iodine values are used in the food industry to measure the degree of unsaturation of a fat or oil.

Questions

1 State two sources each of saturated fats and unsaturated oils.

2 Explain why you should include fats and oils in your diet.

3 Outline why people might fry food in vegetable oil rather than boil it in water.

4 Explain why you might have to reduce your intake of saturated fat.

5 Study the data in Table 1. **(a)** Which nutrient has the highest energy density? **(b)** Suggest why fats and carbohydrates are more efficient sources of energy than proteins.

6 Compare the health benefits and problems of different types of fats and oils.

Assess yourself questions

1   Fuel oil is a fraction of crude oil. More fuel oil is produced at oil refineries than can be sold. To solve this problem, some fuel oil is cracked by passing vaporised fuel oil over a hot catalyst. This cracking produces more-useful shorter chain alkanes and alkenes.

(a) Explain in terms of supply and demand, and the production of fuels, such as petrol, why some of the fuel oil produced at an oil refinery is cracked.
(2 marks)

(b) Cracking is a thermal decomposition reaction. What happens in decomposition reactions? *(1 mark)*

(c) What are the alkenes made in cracking used to make? *(1 mark)*

(d) Alkenes are unsaturated molecules. What is meant by the term *unsaturated* in this context? *(1 mark)*

(e) Which of the following molecules are alkenes?

C_3H_6, C_3H_8, C_6H_{14}, C_8H_{18}, $C_{10}H_{20}$ *(1 mark)*

2   Figure 1 shows the cracking of some paraffin that contains the alkane $C_{12}H_{26}$.

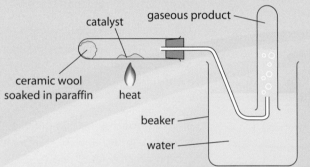

Figure 1 Cracking.

(a) Why was the paraffin soaked in ceramic wool?
(1 mark)

(b) Why are the first few bubbles of gas not collected?
(1 mark)

(c) Describe a test that you could do to prove that alkenes have been produced. Give the results of the test. *(2 marks)*

(d) The cracking of $C_{12}H_{26}$ in the paraffin produced ethene, C_2H_4, and another alkane. Write a balanced equation for this reaction. *(1 mark)*

3   (a) The structure of propene is shown in Figure 2.

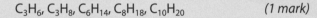

Figure 2 Propene.

Which of the following structures is the correct structure of the polymer made from propene?
(1 mark)

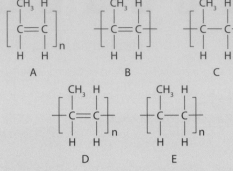

Figure 3

(b) What is the name of the polymer that is made from propene? *(1 mark)*

(c) Draw the structure of the polymer formed from each of the monomers in Figure 4. *(3 marks)*

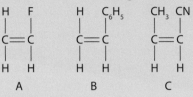

Figure 4 Monomers.

4   Table 1 shows data about waste in the UK over recent years.

Table 1 Data from DEFRA.

	1994	1999	2004	2009
Annual waste per person/kg	445	482	520	473
Annual waste that was recycled or composted per person/kg	15	44	75	178

(a) Describe how the total amount of waste in the UK per person has changed over the years since 1994.
(1 mark)

(b) Draw a bar chart showing how the amount of waste that has been recycled or composted has varied since 1994. *(2 marks)*

(c) (i) Describe how the total amount of waste in the UK that has been recycled or composted per person has changed over the years since 1994. *(1 mark)*

(ii) Suggest a reason why this might have changed since 1994. *(1 mark)*

(d) Many polymers are non-biodegradable. Explain why this is a problem. *(2 marks)*

5   Ethanol is a good fuel that can be used in cars. Petrol sold in the UK contains 5% ethanol made from carbohydrates such as sugar and maize. Ethanol can also be made from ethene. Ethene is made from crude oil.

(a) Ethanol made from carbohydrates is a carbon-neutral fuel. Explain what *carbon neutral* means here. *(2 marks)*

(b) Give an advantage of making ethanol from carbohydrates compared with ethene in terms of the use of raw materials. *(2 marks)*

6 Crude oil is a very important raw material with many uses. Give some social and economic advantages and disadvantages of using products from crude oil as fuels or as raw materials for plastic and other chemicals.

In this question you will be assessed on using good English, organising information clearly and using specialist terms where appropriate. *(6 marks)*

7 Vegetable oils can be extracted from plant materials. Table 2 shows the stages in extracting sunflower oil. Copy the table and write numbers in the boxes to put the stages into the correct order. Two have been done for you. *(4 marks)*

Table 2 The stages in the extraction of sunflower oil.

seeds are harvested	1
seeds are crushed	
the mixture is pressed	
water and other impurities are removed	
water and sunflower oil are obtained	
water and sunflower oil separate into two layers	
water is added to the crushed seeds	
sunflower oil is collected	8

8 **Table 3** The properties of some vegetable oils.

Oil	Melting point/°C	Smoke point/°C	Iodine number	Energy provided / kJ/g
corn oil	−15	240	120	37.1
olive oil	−12	220	60	33.8
rapeseed oil	5	235	100	36.9
sunflower oil	−18	245	130	37.0

The smoke point is the temperature at which the oil begins to give off smoke, risking it setting alight. The higher the iodine number, the more double bonds the oil contains.

(a) Which oil is most likely to become solid in a refrigerator? *(1 mark)*

(b) Which oil is the most likely to ignite? *(1 mark)*

(c) Which oil is the most unsaturated? *(1 mark)*

(d) Which oil provides the least energy? *(1 mark)*

9 Vegetable oils that are unsaturated can be hardened using hydrogen in the presence of nickel at about 60 °C.

(a) What is the purpose of the nickel? *(1 mark)*

(b) Explain what happens to the carbon–carbon double bonds during hardening. *(1 mark)*

Table 4 shows the melting points of four fatty acids, which are components of vegetable oils. Linolenic acid can be converted into stearic acid by hydrogenation.

Table 4 Melting points.

Fatty acid	Number of C=C bonds	Melting point/°C
stearic acid	0	70
oleic acid	1	16
linoleic acid	2	−5
linolenic acid	3	−11

(c) **(i)** Plot a suitable graph to show the data in the table. *(4 marks)*

(ii) Describe the trend in melting point. *(1 mark)*

(iii) What is the effect of hydrogenation on the melting point of vegetable oils? Use data from the table in your explanation. *(2 marks)*

(d) Describe two food uses of hydrogenated vegetable oils. *(2 marks)*

10 Figure 5 shows information about four different vegetable oils. Use information from it to answer the questions.

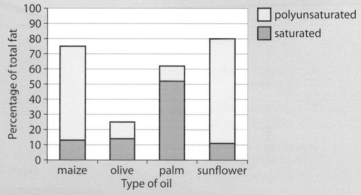

Figure 5 Composition of various vegetable oils.

(a) Which oil contains the most saturated fat? *(1 mark)*

(b) Explain which oil appears to be best for the heart. *(3 marks)*

(c) Studies have shown beneficial health effects from consuming monounsaturated fats. Assuming that the other type of fat in each oil is monounsaturated fat, which is most likely to provide these beneficial health effects and why? *(2 marks)*

(d) Describe two benefits and a drawback of cooking food with vegetable oils rather than with water. *(3 marks)*

11 Describe what emulsifiers are and how they work.

In this question you will be assessed on using good English, organising information clearly and using specialist terms where appropriate. *(6 marks)*

The structure of the Earth

Learning objectives

- describe the structure of the Earth
- describe what tectonic plates are
- explain how convection currents inside the Earth move tectonic plates
- explain where earthquakes and volcanoes are most likely to occur
- explain why scientists cannot accurately predict when earthquakes and volcanic eruptions will occur.

The structure of the Earth

At the centre of the Earth is the **core**, which is made mostly of molten iron. Next is the **mantle**, which is mainly solid rock but can flow due to some parts that have partially melted. The outer layer of the Earth is called the **crust** and is very thin compared with the mantle and core. Above the crust is the **atmosphere**, which is a layer of gases.

The Earth's crust, the atmosphere and the oceans are our only source of all the minerals and other resources that we use to make the substances that we need for everyday life.

Tectonic plates

The outer part of the Earth is called the **lithosphere** and includes the crust and the upper part of the mantle. It is cracked into a number of huge pieces called **tectonic plates**. These plates are moving at a speed of a few centimetres each year.

The slow movement of tectonic plates is caused by very powerful **convection currents** in the mantle.

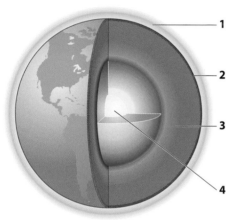

1 atmosphere – a layer of gases (100 km thick)

2 crust – a relatively thin layer of rock (5–70 km thick)

3 mantle – a thick layer of rock (2900 km thick) at over 1000 °C and able to move very slowly

4 core – a central ball of iron (3400 km radius) at about 4000 °C

Figure 1 The structure of the Earth.

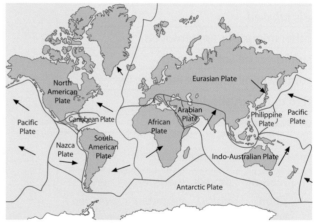

Figure 2 Tectonic plates.

Examiner feedback

Tectonic plates are made up of the crust and the top part of the mantle.

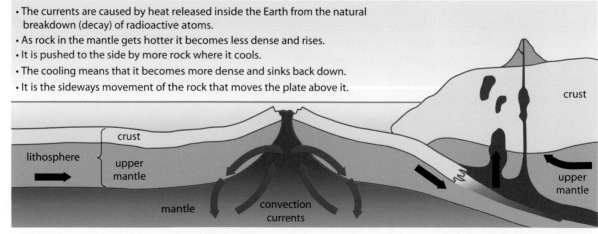

- The currents are caused by heat released inside the Earth from the natural breakdown (decay) of radioactive atoms.
- As rock in the mantle gets hotter it becomes less dense and rises.
- It is pushed to the side by more rock where it cools.
- The cooling means that it becomes more dense and sinks back down.
- It is the sideways movement of the rock that moves the plate above it.

Figure 3 Convection currents in the mantle.

Earthquakes and volcanoes

Earthquakes and **volcanoes** occur at the boundaries between tectonic plates, as shown in Figure 4. As plates move past, over, under or apart from each other, hot magma from the mantle can escape, resulting in a volcano. Friction between the moving plates can make them move in sudden jerks, producing earthquakes.

Scientists know where earthquakes are likely and many buildings in danger zones are built with special foundations to help them withstand earthquakes.

There are some warning signs before an earthquake:

- increased seismic activity: small shocks that may not be detected by people, but are detected by scientific instruments
- water levels in wells fall
- some animals act strangely.

Volcanic eruptions are much easier to predict than earthquakes because we know where volcanoes are and the signs are more definite. Warning signs before an eruption include:

- increasing temperature of the volcano due to magma moving underground
- rising ground level due to the build-up of magma
- more sulfur dioxide (SO_2) gas given out.

When these warning signs appear, people can be moved to safety. However, scientists cannot reliably predict major earthquakes or volcanic eruptions exactly. It is difficult to predict when there will be enough pressure for plates to slide or for magma to burst through the crust. If scientists issue false warnings, people could be moved when not necessary, causing disruption and economic loss, and reducing trust in future warnings.

Examiner feedback

When two objects rub together there is friction, which makes it difficult for the objects to move against each other. Only when the pressure reaches a certain point is the force of friction overcome very suddenly. This is why we cannot predict when an earthquake will occur, where it will happen or how violent it will be.

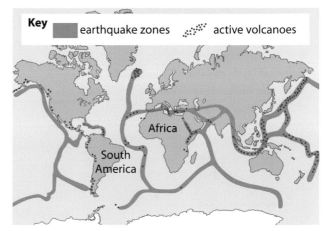

Figure 4 Earthquakes and volcanoes.

Questions

1. What are our three sources of minerals and other resources to make all the substances that we use in modern life?

2. Sketch a diagram of the Earth and label the main layers.

3. A hard-boiled egg is a simple model of the structure of the Earth. Compare the structure of an egg with the structure of the Earth. Give some weaknesses with this model.

4. **(a)** What are tectonic plates? **(b)** How fast do tectonic plates move? **(c)** What causes tectonic plates to move?

5. Why do volcanoes and earthquakes develop at plate boundaries?

6. Why is it important that scientists do not make predictions about earthquakes and volcanic eruptions that do not happen?

7. **Table 1** Data on how much lava has erupted from Mount Etna, Sicily, in recent eruptions.

Year	Number of years since previous eruption	Amount of lava erupted /millions of m³
1794	34	27
1858	64	120
1872	14	20
1906	34	80
1929	23	12
1944	15	25

(a) Use the data in Table 1 to plot a graph with a line of best fit to show how the amount of lava changes with the number of years since the last eruption.
(b) What is the relationship shown by the graph?
(c) Estimate how much lava would erupt if Mount Vesuvius erupted tomorrow.

8. Give some warning signs that earthquakes and volcanic eruptions may be about to happen. Explain why it is important to predict earthquakes and volcanic eruptions but why it is hard to predict exactly when they may happen.

Continental drift

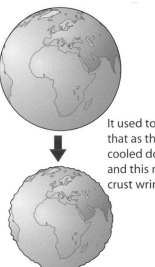

It used to be thought that as the Earth cooled down it shrank and this made the crust wrinkle.

Figure 1 The shrinking Earth theory.

Shrinking apples

Scientists have wondered for centuries how features on the Earth such as mountains formed. A popular theory for many years was that as the hot, young Earth cooled down, the crust shrank and wrinkled, forming mountains. This wrinkling is similar to the way that apples shrink and the peel wrinkles up as they get older.

We now explain the formation of mountains by rock being forced upwards where tectonic plates collide. However, this theory of plate tectonics is a relatively new theory and was not accepted for some time.

Alfred Wegener's ideas

In 1911, Alfred Wegener read that the fossils of identical creatures had been found both in South America and Africa, and that these were creatures that could not swim. People at the time said that there must have been a piece of land between the continents that was now covered by the Atlantic Ocean.

Wegener came up with the idea of **continental drift**. He believed that the continents were moving around and were once joined together in a big landmass. In 1915, Wegener published a book about his theory.

The main evidence in Wegener's book was as follows:

- The continents appear to fit together like a jigsaw.
- The west coast of Africa and the east coast of South America have the same patterns of rock layers.
- These two coasts have the same types of plant and animal fossils.
- Some of these animals are found only in these parts of the world and their fossils show they could not swim or fly.
- There could not have been a piece of land connecting South America and Africa, which means they must have been joined together.

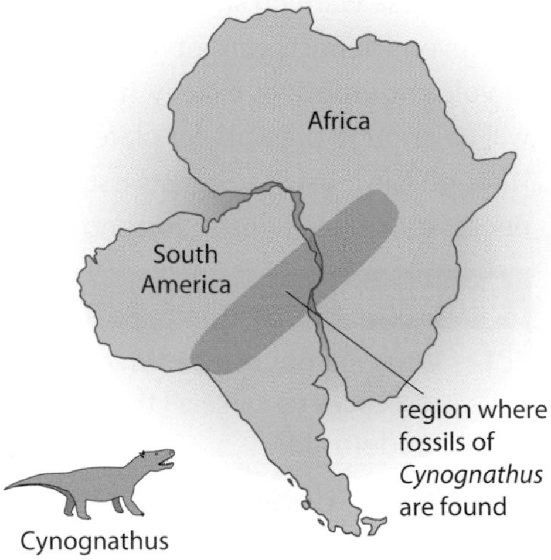

Figure 2 Positions of the continents millions of years ago and the region in which fossils of *Cynognathus* have been found.

Cynognathus

region where fossils of *Cynognathus* are found

When Wegener died in 1930 his ideas had still not been accepted. The main reason for this was that there was no explanation for how the continents moved. Another reason was that, although he was a scientist, Wegener was known as a meteorologist rather than a geologist. This meant that his ideas were given less credit among geologists.

The acceptance of Wegener's ideas

Wegener's ideas were not accepted until the 1960s, when the Atlantic Ocean floor was surveyed in detail and the Mid-Atlantic Ridge was found. This is a range of underwater mountains and volcanoes in the middle of the Atlantic Ocean.

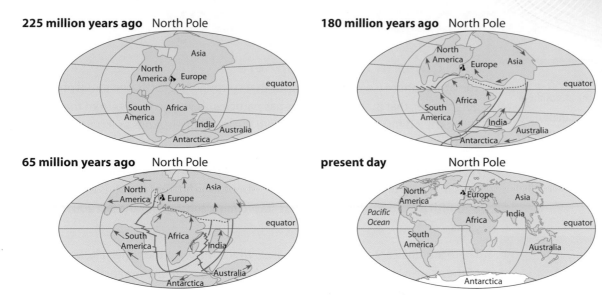

225 million years ago   North Pole

180 million years ago   North Pole

65 million years ago   North Pole

present day   North Pole

Figure 3 The moving continents.

Soon afterwards, it was discovered that the rock In the ocean floor is younger than the rock in the continents. The rock closest to the ridge is youngest as new rock is formed there from magma coming up from inside the Earth. The older rock is pushed away. This new evidence fitted in with the theory of continental drift.

Scientists also discovered that this movement could be caused by very powerful convection currents in the mantle. This prompted the new theory of plate tectonics, which is based on Wegener's theory of continental drift.

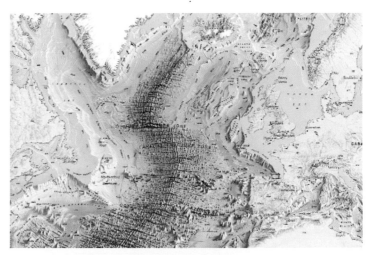

Figure 4 Map showing the Mid-Atlantic Ridge.

Questions

1   **(a)** How did people believe mountains were formed before the theory of plate tectonics? **(b)** How is the formation of mountains explained now with the theory of plate tectonics?

2   What is meant by continental drift?

3   **(a)** How do rock patterns support Wegener's ideas? **(b)** How do fossils of *Cynognathus* support Wegener's ideas? **(c)** What other evidence did Wegener use to support his ideas?

4   A common theory for fossils of creatures that cannot swim or fly, such as *Cynognathus*, on different continents was a piece of land joining continents that is now underwater. What piece of Wegener's ideas did not fit in with this idea?

5   What evidence was discovered that led to Wegener's ideas being accepted?

6   A key reason why Wegener's ideas were not accepted

at the time was that no one could explain what moved the continents. What do scientists now believe is moving the continents?

7   The following data show the age of the rocks on the floor of the Atlantic Ocean between the Mid-Atlantic Ridge and Brazil.

Distance from the Mid-Atlantic Ridge/km	500	1000	1500	2000	2500
Age of rock/millions of years	24	46	71	90	113

(a) Plot a graph to show how the age of rocks on the ocean floor varies with distance from the ridge. Draw a line of best fit. **(b)** Explain how these data support Wegener's theory of continental drift.

8   Explain what the theory of continental drift Is and why Wegener's ideas were not accepted when he published them but they are accepted today.

The atmosphere today

Learning objectives

- list the gases that are in the atmosphere
- know that the proportions of these gases in the atmosphere has been the same for the last 200 million years
- describe how the gases in air can be separated by the fractional distillation of liquefied air.

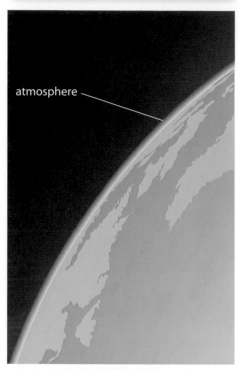

atmosphere

Figure 1 The atmosphere is a very thin layer compared with the diameter of the Earth.

The gases in the atmosphere

The **atmosphere** is a mixture of gases that surround the Earth. The atmosphere gets thinner, that is, there are fewer gas particles, as you go further from the Earth's surface. It is difficult to say exactly where the atmosphere stops as it gets gradually thinner.

Air is the mixture of gases in the lower part of the atmosphere. About 99% of the air is made up of nitrogen and oxygen. About four-fifths of the air is nitrogen and one-fifth is oxygen. This mixture has stayed fairly constant for the last 200 million years.

There are small amounts of other gases, which make up the other 1%. Some of these are noble gases, mainly argon. Noble gases are the unreactive gases in group 0 of the periodic table (see lesson C1 1.4). A small amount of carbon dioxide is also found in air. Water vapour is present in the air as well. The amount of water vapour changes due to changes in the weather and humidity. Because of this, water vapour is not included when giving the composition of air.

Air also contains very small or trace amounts of harmful gases, including sulfur dioxide, nitrogen oxide and carbon monoxide. These gases are formed in a number of ways: some are made by natural processes, such as volcanic activity, or from plants and animals decomposing; some are made by human activities such as burning fossil fuels.

Table 1 Percentage composition of gases in dry air.

Gas	Formula	Percentage in dry air
nitrogen	N_2	78
oxygen	O_2	21
argon	Ar	0.9
carbon dioxide	CO_2	0.04
other gases		traces

The proportion of these gases in the atmosphere has remained much the same for the last 200 million years (see lesson C1 7.5).

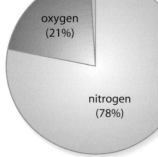

Figure 2 The gases in dry air.

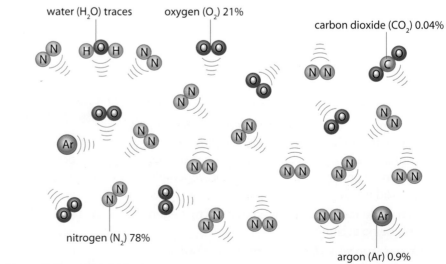

Figure 3 Gas molecules in air.

Separating the gases in air

The gases in air can be separated and used for different purposes. The most abundant gas, nitrogen, is very unreactive. It is used instead of air inside food packaging, such as crisp packets, because the oxygen in air would allow respiring bacteria to make the food go off faster. Oxygen is used to help patients breathe in hospitals. Argon is used as a non-reactive gas when welding and was used to fill filament light bulbs before low-energy light bulbs were introduced.

The gases in air are separated by **fractional distillation**. Air is cooled until it becomes a liquid, at about −200 °C. As it warms up again the separate substances boil at different temperatures and can be collected separately.

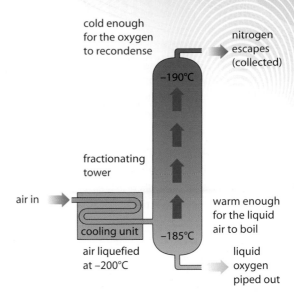

Figure 4 Simplified diagram of fractional distillation of liquefied air.

Practical

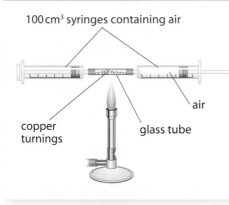

100 cm³ syringes containing air

air

copper turnings

glass tube

When air is passed over heated copper the oxygen in the air reacts with the copper. From this we can find the percentage of oxygen in air.

Figure 5 Apparatus for measuring percentage oxygen in air.

Nitrogen is the gas in crisp packets.

Questions

1   **(a)** Name the main two gases in air and give their proportions. **(b)** Give a use for each of these two gases. **(c)** How long has the atmosphere been like this?

2   **(a)** There are small amounts of noble gases in the air. What are noble gases? **(b)** Name the main noble gas found in the air. **(c)** Give a use for the main noble gas found in air.

3   **(a)** List the main three elements found in air. **(b)** List two compounds found in air.

4   Why is the proportion of gases in air given without any water?

5   Draw a bar chart to show the main gases in air.

6   Nitrogen is used in food packaging. Explain why.

7   There are trace amounts of harmful gases in the air. **(a)** Give one example of a natural process that produces trace amounts of a harmful gas. **(b)** Give one example of an artificial process that produces trace amounts of a harmful gas.

8   Air is a useful mixture of gases. Describe and explain how the gases can be separated.

Examiner feedback

The fractional distillation of air is based upon the same principles as the fractional distillation of crude oil, as both are mixtures containing substances with different boiling points.

The changing atmosphere

The early atmosphere

The Earth is thought to be about 4.5 billion years old. Scientists have evidence that the atmosphere we have today is very different to the one the Earth used to have. However, scientists are uncertain how the early atmosphere formed, which gases were in it, and how it changed to give the atmosphere we have today. They have several theories about the early atmosphere and the changes.

Where did the gases come from?

Some scientists think that the gases that formed the early atmosphere came from inside the Earth. During the first billion years of its existence, the Earth was much hotter and there was a lot of volcanic activity. Since volcanoes release gases into the atmosphere, some scientists believe that the early atmosphere was formed from these gases. However, there are other theories about where the gases in the early atmosphere came from. One theory is that the gases came from comets that collided with the Earth.

Which gases were in the early atmosphere?

Many scientists believe that carbon dioxide, water vapour, ammonia and methane were in the early atmosphere because volcanoes release them today.

The Earth lies between Venus and Mars in the Solar System. There are volcanoes on Venus and Mars and their atmospheres are over 95% carbon dioxide. This has led some scientists to think that Earth's early atmosphere was mainly carbon dioxide. Scientists have evidence that the evolution of life on Earth caused our atmosphere to change. There is no life on Mars or Venus, so their atmospheres have remained the same.

As the hot, young Earth cooled down, the water vapour in the air is thought to have condensed, contributing to the formation of oceans.

There is evidence that leads most scientists to believe that there was little or no oxygen in the early atmosphere. For example, oxygen is not released by volcanoes, and the iron compounds found in the oldest rocks are compounds that could only form in the absence of oxygen.

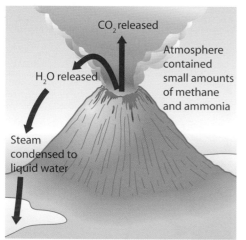

Figure 1 The gases in the early atmosphere came from volcanoes.

Life on Earth

Scientists do not know how life on Earth started, but there are many theories. Experiments by Miller and Urey showed how amino acids, molecules that are essential for life, could have been formed. They fired an electric spark, representing lightning, in a mixture of the gases methane, ammonia and hydrogen, representing some of the possible gases in the early atmosphere, plus water representing the oceans. A mixture of chemicals formed, including amino acids. This is sometimes called a 'primordial soup' meaning a rich mixture of chemicals essential for life. However, these experiments do not show how life itself could have started.

The formation of oxygen

Photosynthesising organisms evolved, including some bacteria and plants. They use up carbon dioxide and make oxygen. The oxygen we have in the atmosphere today is produced by photosynthesis. Over time, the amount

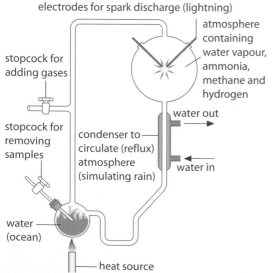

Figure 2 The Miller–Urey 'primordial soup' experiment.

of carbon dioxide decreased and the amount of oxygen increased. Most of the carbon from the carbon dioxide is now in fossil fuels or sedimentary rocks, mainly limestone (see lesson C1 7.5).

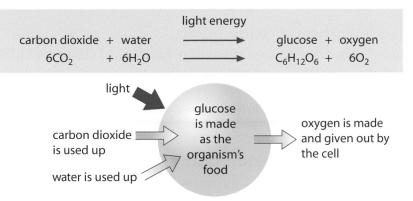

$$\text{carbon dioxide } + \text{ water } \xrightarrow{\text{light energy}} \text{ glucose } + \text{ oxygen}$$
$$6CO_2 \quad + \quad 6H_2O \quad \longrightarrow \quad C_6H_{12}O_6 \quad + \quad 6O_2$$

Figure 3 Photosynthesis carried out in a cell.

These structures are called stromatolites. Some are well over 2 billion years old and contain fossils of bacteria that photosynthesised.

The methane and ammonia in the early atmosphere would have been removed by reaction with the oxygen. The reaction of oxygen with ammonia would have produced some of the nitrogen in the atmosphere. However, there are several theories about where most of the nitrogen came from. Some scientists think it came from volcanoes, some from comets or from bacteria in the soil.

The proportions of the gases in the atmosphere are thought to have remained roughly constant for the last 200 million years (see lesson C1.7.5).

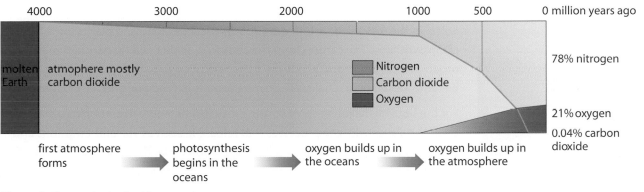

Figure 4 Changes in the Earth's atmosphere.

(see lesson C1 7.5)

Questions

1 Give two places that different scientists think the gases that formed the early atmosphere came from.

2 **(a)** Which gases are thought to have made up the early atmosphere? **(b)** Give one reason why many scientists think that the early atmosphere was mainly carbon dioxide.

3 How do some scientists think the oceans formed?

4 Give two reasons why scientists think there was little or no oxygen in the early atmosphere?

5 **(a)** Describe what was done in the Miller–Urey experiment. **(b)** What did this experiment show that

was significant? **(c)** Does this experiment explain how life may have started on Earth? Explain your answer.

6 **(a)** What happened on Earth that led to the formation of oxygen? **(b)** Describe how the oxygen in the atmosphere formed. **(c)** What process removed most of the carbon dioxide from the atmosphere?

7 What happened to the ammonia and the methane in the early atmosphere?

8 Oxygen makes up 21% of the air. Carbon dioxide makes up 0.04%. Describe the processes that have resulted in these gases reaching their present levels compared with their levels in the early atmosphere.

Carbon dioxide and global warming

Examiner feedback

Remember that the oceans still act as a reservoir of CO_2, but can no longer absorb as much as humans are producing.

Much of the Earth's carbon is locked up in sedimentary rocks like limestone.

Some of the Earth's carbon is locked up in fossil fuels.

Changes in carbon dioxide levels

The amount of carbon dioxide in the atmosphere has changed during the Earth's history. It fell from high levels in the early atmosphere to low levels, where it had remained for the last 200 million years. However, human activities are now raising the level of carbon dioxide in the atmosphere, and the increase is thought to be causing **global warming**.

Removal of carbon dioxide from the early atmosphere

Many scientists think that when the Earth was young, the main gas in the atmosphere was carbon dioxide. However, there is very little now. As the Earth cooled after its formation, water vapour in the air condensed to form the oceans. About half of the carbon dioxide from the atmosphere at that time is thought to have dissolved in the new oceans.

Some carbon dioxide reacted with substances in sea water, forming insoluble compounds, such as calcium carbonate, that sank to the bottom as sediment. Carbon dioxide also formed soluble compounds, such as calcium hydrogencarbonate, which was used by sea creatures to make calcium carbonate shells. As these creatures died, their shells fell into the sediment. Over time, these sediments formed sedimentary rocks, such as limestone, locking away the carbon in rocks.

Some of the dissolved carbon dioxide was used by algae and plants for photosynthesis. The carbon from the carbon dioxide became part of these organisms. As these organisms died, they decayed to form fossil fuels. The carbon became locked up in coal and as hydrocarbons in crude oil and natural gas.

For the last 200 million years or so, the processes that absorb carbon dioxide have been in balance with the processes that release it, such as volcanic eruptions and the decay of organisms – this is called the carbon cycle. Therefore the amount of carbon dioxide in the atmosphere remained stable at around 0.03% for that period.

Rising levels of carbon dioxide and global warming

In recent years carbon dioxide levels have started to rise, and are now nearer 0.04% than 0.03%. Scientists have evidence that much of this rise is due to burning large amounts of fossil fuels, which produce carbon dioxide. Large-scale deforestation is also thought to be contributing to the increase in carbon dioxide. Some of the extra carbon dioxide is being absorbed by the oceans, but not all of it.

Carbon dioxide is a **greenhouse gas**. This means that it stops some of the Earth's heat escaping into space and thus keeps the Earth warm (see lesson C1 4.4). However, there are signs that the mean temperature of the Earth is rising. Many scientists think that the increase in temperature is linked to the increased levels of carbon dioxide in the atmosphere. This global warming could lead to climate change and to rising sea levels, which could cause flooding. The world is seeking ways to reduce carbon dioxide emissions, with the two main suggestions being to increase energy efficiency and to burn less fossil fuel by using alternative sources of energy.

The effect of carbon dioxide levels on marine life

The oceans are slightly alkaline with a pH of about 8. When carbon dioxide dissolves in the oceans it forms carbonic acid. As the amount of carbon dioxide in the atmosphere changes, the pH of the oceans rises or falls. These changes affect marine life.

As the amount of carbon dioxide in the atmosphere is currently increasing, the pH of the oceans is falling. This lowers the amount of dissolved carbonates that many organisms need to make shells or skeletons. Some creatures may not be able to survive if the pH falls too far.

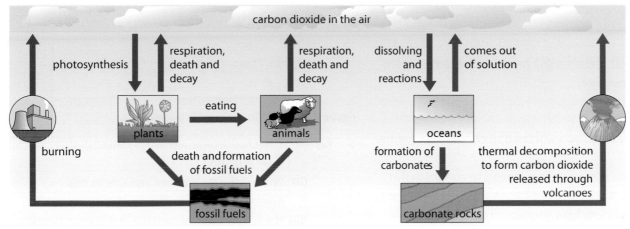

Figure 1 Processes that add and remove carbon dioxide from the air.

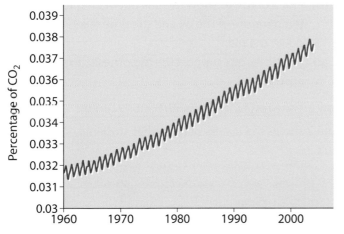

Figure 2 The amount of carbon dioxide in the air has increased over time.

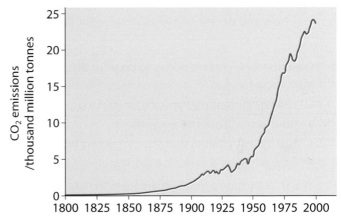

Figure 3 The amount of carbon dioxide released into the atmosphere from burning fossil fuels over time.

Questions

1. **(a)** What was the main gas in the air when the Earth was young? **(b)** Where is most of the carbon from the carbon dioxide now?

2. **(a)** Explain how plants remove carbon dioxide from the air. **(b)** Explain how this carbon ended up in fossil fuels.

3. **(a)** Explain how the oceans removed carbon dioxide from the air. **(b)** Explain how some of this carbon ended up in fossil fuels. **(c)** Explain how some of this carbon ended up in sedimentary rocks.

4. Give four ways in which carbon dioxide is released into the atmosphere.

5. Why has the amount of carbon dioxide in the atmosphere remained constant for the last 200 million years?

6. Carbon dioxide levels in the atmosphere are rising. Explain why.

7. Explain how and why the increasing level of carbon dioxide in the atmosphere is affecting marine life.

8. What effects could rising levels of carbon dioxide be having on the Earth and why? How could we lower levels of carbon dioxide?

ISA practice: the viscosity of oils

There are many different vegetable oils. The properties of these oils, including their **viscosity**, varies. Viscosity is a measure of how easily a liquid flows. The greater the viscosity of a liquid, the slower it flows.

You have been asked to measure the viscosity of a vegetable oil at different temperatures by timing how long drops of oil take to flow down a tile.

Hypothesis

It is suggested that there is a link between the temperature of an oil and its viscosity.

Section 1

1   In this investigation you will need to control some of the variables.

    (a) Name one variable you will need to control in this investigation. *(1 mark)*

    (b) Describe briefly how you would carry out a preliminary investigation to find a suitable value to use for this variable. Explain how the results will help you decide on the best value for this variable. *(2 marks)*

2   Describe how you are going to do your investigation. You should include:

- the equipment that you are going to use
- how you will use the equipment
- the measurements that you are going to make
- how you would make it a fair test.

    You may include a labelled diagram to help you to explain your method.

    In this question you will be assessed on using good English, organising information clearly and using specialist terms where appropriate. *(6 marks)*

3   Think about the possible hazards in your investigation.

    (a) Describe one hazard that you think may be present in your investigation. *(1 mark)*

    (b) Identify the risk associated with the hazard you have described, and say what control measures you could use to reduce the risk. *(2 marks)*

4   Design a table that will contain all the data that you are going to record during your investigation. *(2 marks)*

Total for Section 1: 14 marks

Section 2

Two students, Study Group 1, carried out an investigation into the hypothesis. Figure 1 shows the results of their investigation.

Time taken for the oil to flow down the tile
10°C: 2 min 5 sec
19°C: 60 sec
31°C: 32 sec
40°C: 21 sec
51°C: 18 sec

Figure 1 Results of Study Group 1's investigation.

5   **(a) (i)** What is the independent variable in this investigation?

        (ii) What is the dependent variable in this investigation?

        (iii) Name one control variable in this investigation. *(3 marks)*

    (b) Plot a graph to show the link between the temperature of the oil and the time taken to flow down the tile. *(4 marks)*

    (c) Do the results support the hypothesis? Explain your answer. *(3 marks)*

Below are the results of three other study groups.

Figure 2 shows the results of two other students – Study Group 2.

Time taken for the oil to flow down the tile
10°C: 115 sec
19°C: 50 sec
31°C: 22 sec
40°C: 11 sec
51°C: 8 sec

Figure 2 Study Group 2's results.

Table 1 shows the results from another two students – Study Group 3.

Table 1 Results from Study Group 3.

Temperature of oil/°C	Time for the oil to flow down the tile/s			
	Test 1	Test 2	Test 3	Mean
10	98	95	91	95
19	49	47	48	48
31	18	17	20	18
40	15	27	16	19
51	5	6	5	5

Table 2 gives the results of Study Group 4. Scientists in a lubricating oil laboratory measured the rate of lubricating oil flow in a vehicle engine at different temperatures. They repeated the tests with new oil, oil that had been used for 1000 hours and oil that had been used for 10 000 hours.

Table 2 Results from Study Group 4, scientists in a lubricating oil laboratory.

Temperature of engine/°C	Rate of oil flow/dm³/min		
	New oil	Oil after 500 hours' use	Oil after 1000 hours' use
150	7.4	8.5	12.6
200	9.2	10.4	12.8
250	11.1	12.4	12.4
300	13.7	15.1	12.5
350	16.5	18.0	13.0

6 Describe one way in which the results of Study Group 2 are similar to or different from the original results of Study Group 1, and give one reason for this similarity or difference. *(3 marks)*

7 (a) Draw a sketch graph of the results from Study Group 2. *(3 marks)*

(b) Does the data support the hypothesis being investigated? To gain full marks you should use all of the relevant data from Studies 1, 2 and 3 to explain whether or not the data supports the hypothesis. *(3 marks)*

(c) The data from the other groups only gives a limited amount of information. What other information or data would you need in order to be more certain as to whether or not the hypothesis is correct? Explain the reason for your answer. *(3 marks)*

(d) Use Studies 2, 3 and 4 to answer this question. What is the relationship between the temperature of an oil and its viscosity? How well does the data support your answer? *(3 marks)*

8 Look back at the method of investigation you described in answer to question 2. If you could repeat the investigation, suggest one change that you would make to the method, and give a reason for the change. *(3 marks)*

9 A company makes vehicle engines. How could the results of this investigation help the company decide on the length of time the oil should be used for before replacing it? *(3 marks)*

Total for Section 2: 31 marks
Total for the ISA: 45 marks

Assess yourself questions

1 The properties of some gases in air are shown. Name the gas with the properties given.

(a) This gas is essential for life as it is used for respiration. *(1 mark)*

(b) There are small amounts of this gas in the air but levels are rising and many scientists believe it may be causing global warming. *(1 mark)*

(c) This is the main gas in air. *(1 mark)*

(d) This gas makes up 0.9% of the air and is very unreactive. *(1 mark)*

(e) This gas is thought to have been the main gas in the air when the Earth was young. *(1 mark)*

(f) The gas that there was little or none of when the Earth was young that was produced by photosynthesis. *(1 mark)*

(g) The gas, of which there was a lot when the Earth was young, that was removed from the air by photosynthesis. *(1 mark)*

(h) The amount of this gas in the air depends on the humidity. *(1 mark)*

2 The amount of carbon dioxide in the air has remained constant, until recently, for about the last 200 million years.

(a) State two ways in which carbon dioxide is removed from the atmosphere. *(2 marks)*

(b) State two ways in which carbon dioxide is added to the atmosphere. *(2 marks)*

(c) The amount of carbon dioxide in the air has been increasing in recent years. Explain why. *(1 mark)*

(d) The blue line on the graph in Figure 1 shows how the amount of carbon dioxide in the air has changed in recent years. The red line on the graph shows how the Earth's temperature has changed.

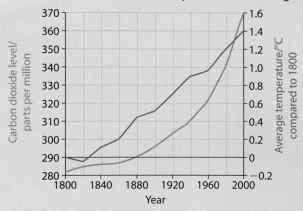

Figure 1 Graph showing changes in temperature and the amount of carbon dioxide in the air.

How strong a piece of evidence is this that increasing levels of carbon dioxide are causing global warming? *(2 marks)*

3 Table 1 shows the boiling points of the main gases in air.

Table 1 Boiling points of the main gases in air.

Gas	nitrogen	argon	oxygen
Boiling point/°C	−196	−186	−183

The gases in air are separated by fractional distillation of liquefied air. Which gas will boil first as the liquefied air is allowed to warm? *(1 mark)*

4 Figure 2 shows the structure of the Earth. Name each part shown. *(3 marks)*

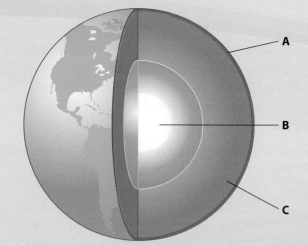

Figure 2 The structure of the Earth.

5 Figure 3 shows the tectonic plates on the Earth's surface.

(a) What are tectonic plates? *(2 marks)*

(b) Give the letters of two places where volcanoes and earthquakes are likely. *(1 mark)*

(c) Explain why volcanoes are not found in the UK today. *(1 mark)*

(d) What is happening to the distance between Africa and South America? Explain your answer. *(2 marks)*

(e) The Andes mountains are near D on the map.

  (i) Explain why mountains have formed in this area according to the theory of plate tectonics. *(2 marks)*

  (ii) Describe an older theory that explained how mountains formed as the Earth cooled. *(2 marks)*

(f) Scientists think that the movement of the plates is caused by convection currents.

  (i) Explain how these convection currents arise. *(3 marks)*

  (ii) About how far are the plates moving each year? *(1 mark)*

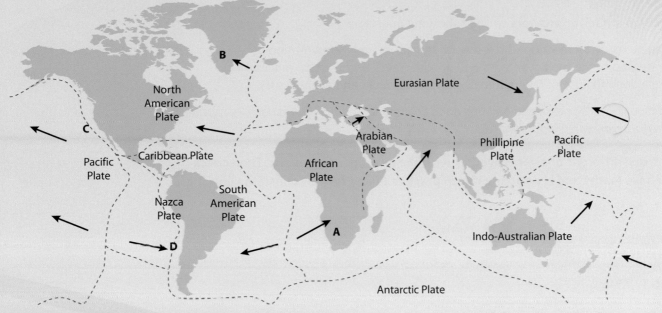

Figure 3 Tectonic plates

6 (a) Alfred Wegener proposed the theory of continental drift in 1915. Which one of the following ideas does not support his theory?

A Fossils of fish are found in different parts of the world.

B The shape of continents appear to fit together like a jigsaw.

C Rocks near the centre of the Atlantic Ocean are younger than those nearer the continents.

D Volcanoes and earthquakes appear in zones and are not spread evenly around the world. *(1 mark)*

(b) Fossils of the land reptile *lystrosaurus* have been found in Africa, India and Antarctica. Explain how this supports Wegener's theory of continental drift. *(2 marks)*

(c) Why were Wegener's ideas not accepted when he first put them forward? *(1 mark)*

7 Changes in water levels underground are a very strong indicator of earthquakes. There was a major earthquake in China in 1976 that killed about 250 000 people. Table 2 gives information about the water levels in a mine in the week before the earthquake. Water is pumped out of the mine each day and the more water pumped out, the higher the water level in the ground is.

Table 2 Water measurements.

Days before earthquake	7	6	5	4	3	2	1	0
Rate of water pumping out of the mine / m³/s	45	40	35	30	25	25	50	75

(a) Plot a graph to show how the pumping rate changed in the week before the earthquake. *(4 marks)*

(b) Describe the changes in water level in the week before the earthquake. *(2 marks)*

(c) Why is it important to monitor underground water levels in earthquake zones? *(2 marks)*

8 Carbon dioxide dissolves in the sea and then reacts to form some soluble compounds, such as calcium hydrogencarbonate, and some insoluble compounds, such as calcium carbonate. These compounds can all become part of sedimentary rocks, locking away carbon dioxide.

(a) Outline how insoluble compounds can become part of sedimentary rocks. *(2 marks)*

(b) Outline how soluble compounds can be used by marine organisms and become part of sedimentary rocks. *(3 marks)*

9 Explain and evaluate the effects of human activities on the atmosphere.

In this question you will be assessed on using good English, organising information clearly and using specialist terms where appropriate. *(6 marks)*

Here are three students' answers to the following question:

Ethanol can be made by reaction of steam with ethene or from fermentation of carbohydrates. Compare the advantages and disadvantages of these two processes in terms of reaction rate, use of raw materials, purity of product and energy costs.

In this question you will be assessed on using good English, organising information clearly and using specialist terms where appropriate. (6 marks)

Read the three different answers together with the examiner comments. Then check what you have learnt and try putting it into practice in any further questions you answer.

B Grade answer

Student 1

> The answer does not compare the two processes.

Making ethanol from ethene is fast and makes pure ethanol. Fermentation uses renewable raw materials.

> The answer does not mention energy costs.

Examiner comment

The information the student has given is all correct but would miss many of the marks. The student has not compared the two processes for any of the four areas asked about. For example, the student writes that production from ethene is fast but has not indicated if this is faster or slower than fermentation. Also, the student has not answered the whole question, only covering three of the four areas asked for, missing out the part about energy costs. The student has not indicated what raw materials are used.

A Grade answer

Student 2

> The answer does not compare the purity of ethanol produced by the two processes.

The production of ethanol from ethene is faster and makes pure ethanol. Fermentation uses renewable raw materials while ethene uses non-renewable raw materials. Fermentation has lower energy costs.

> The answer does not mention what the raw materials are.

Examiner comment

The student has compared the two processes in terms of reaction rate, use of raw materials and energy costs. For example, the student writes that production from ethene is 'faster'. However, there is no comparison about purity of product as the student has not stated if the ethanol made by fermentation is pure or not. The student has also not described what the raw materials are.

A* Grade answer

Student 3

The answer compares the two processes in all four areas.

The student has explained the difference in energy costs.

> The rate of reaction to produce ethanol from ethene is faster than from fermentation. It also produces purer ethanol. However, the production of ethanol from fermentation has lower energy costs because lower temperatures and pressures are used. Fermentation uses renewable raw materials using carbohydrates like sugar cane. By comparison, ethene is made from crude oil, which is non-renewable.

The identity and type of raw materials are given for both processes.

Examiner comment

This answer has covered all four areas asked for in the question. These are: reaction rate, use of raw materials, purity of product and energy costs. It provides a comparison between the two methods for each of these four areas. The answer goes into sufficient depth, providing an explanation of the difference in energy costs and including the identity of the raw materials for both processes, as well as whether they are renewable or non-renewable. The answer is structured in a logical sequence dealing with each of the four areas separately, one at a time. Clear sentences are used and spelling, punctuation and grammar are good.

- Read the question more than once. Read it the first time to get a general idea of what you are being asked, and then again to make sure you understand exactly what you are being asked to do.
- In questions that ask for several things, ensure that each part is answered. You could highlight each thing you are asked to do and tick them off as you cover them.
- In questions that ask for a comparison, make sure that you make relative comments that actually compare the things asked about.
- Present your answer in a clear, structured way. Think about the points you want to make to answer each part of the question and then answer them in turn.
- Do your best to use good spelling, punctuation and grammar throughout your answers.
- Try to use the correct scientific terminology.
- Avoid using words like 'it' and 'they'. Always say what you are referring to.

Examination-style questions

1 Here is a simplified periodic table showing the first 20 elements.

1	2	3	4	5	6	7	0
H hydrogen 1							**He** helium 2
Li lithium 3	**Be** berylium 4	**B** boron 5	**C** carbon 6	**N** nitrogen 7	**O** oxygen 8	**F** fluorine 9	**Ne** neon 10
Na sodium 11	**Mg** magnesium 12	**Al** aluminium 13	**Si** silicon 14	**P** phosphorus 15	**S** sulfur 16	**Cl** chlorine 17	**Ar** argon 18
K potassium 19	**Ca** calcium 20						

(a) **(i)** How many electrons do the elements neon and argon have in their outer shells? *(1 mark)*

(ii) What key property can you tell about fluorine from its position in the periodic table? *(1 mark)*

(b) Give the name, group number and period number for the element with electronic structure 2, 8, 4. *(3 marks)*

(c) What type of bonding would you expect to find in the following compounds? Explain your answer in each case.

(i) ammonia (NH_3) *(1 mark)*

(ii) sodium sulfide (Na_2S). *(1 mark)*

2 Dolomite (magnesian limestone) is a double carbonate of calcium and magnesium: $CaMg(CO_3)_2$.

(a) When heated strongly, carbon dioxide is driven off.

(i) What name is given to this type of reaction? *(1 mark)*

(ii) How could you show that the gas is carbon dioxide? *(1 mark)*

(iii) Complete and balance the equation for this reaction:

$$CaMg(CO_3)_2 \longrightarrow \text{_____} + MgO + CO_2$$ *(2 marks)*

(b) 2 g of pure dolomite, when heated to complete reaction, gives off 0.96 g of carbon dioxide. In a class experiment heating 2 g of dolomite chips each, four students ended up with the following amounts of mixed oxide residue:

student **A** 0.90 g, student **B** 1.06 g, student **C** 1.10 g, student **D** 1.4 g.

Comment on these results: what do they tell you about these four experiments? *(4 marks)*

3 Cobalt is a hard, silver-gray transition metal that is just a little less reactive than iron. It is used with iron to make hard-wearing steel alloys.

(a) Cobalt occurs in a natural mineral called cobaltite – a compound of cobalt, arsenic and sulfur. Metal ores are often converted to their oxides by roasting in air before the metal is extracted. What products would be formed in this way and why would it not be safe to do this in the science laboratory? *(3 marks)*

(b) Do you think that cobalt metal is extracted from its oxide by electrolysis or carbon reduction? Explain your answer. *(2 marks)*

(c) Why are metals such as iron and aluminium rarely used in their pure form? *(1 mark)*

4 Oil produced from plants can be refined and used to replace diesel for motor vehicles. What are the advantages and disadvantages of using these biofuels instead of fossil fuels? *(4 marks)*

5 Polymers can be made from unsaturated hydrocarbons called alkenes.

(a) Which two of the following five hydrocarbons are alkenes? *(1 mark)*

$$C_3H_8, C_2H_4, C_4H_{10}, C_2H_6, C_4H_8$$

(b) What are hydrocarbons? *(1 mark)*

(c) What is meant by the term *unsaturated* in this context? *(1 mark)*

(d) Bromine water can be used in a test to show that a compound is unsaturated. Describe what you would see in the reaction between bromine water and an unsaturated compound. *(1 mark)*

(e) Alkenes are made from the thermal decomposition of alkanes in a process called cracking. In this process, the alkanes are vaporised and passed over a hot catalyst. Balance this equation for the cracking of decane.

$$C_{10}H_{22} \longrightarrow C_6H_{14} + C_2H_4$$ *(1 mark)*

(f) The polymer poly(propene) can be made by the polymerisation of propene. The structure of propene is shown.

$$\begin{array}{c} CH_3 \quad H \\ | \qquad | \\ C = C \\ | \qquad | \\ H \quad\; H \end{array}$$

Draw the structure of poly(propene). *(1 mark)*

(g) Many plastic bottles are made from poly(propene). The amount of plastic bottles recycled has risen steadily over recent years. Some data are shown in the table below.

Year	1994	1996	1998	2000	2002	2004
Tonnes of plastic bottles recycled	2000	5000	9000	11 000	18 000	38 000

Plot a graph showing how the mass of plastic bottles recycled has changed in recent years. *(3 marks)*

(h) Some plastic bottles are recycled, some are burned in incinerators and some are buried in landfill. Describe some advantages and disadvantages of the different methods of disposal of plastic bottles.

In this question you will be assessed on using good English, organising information clearly and using specialist terms where appropriate. *(6 marks)*

6 **(a)** Vegetable oils are extracted from plants and have many uses. Rapeseed oil is a common vegetable oil used in the UK.

    (i) From which part of plants are vegetable oils extracted? *(1 mark)*

    (ii) Name one method, other than pressing, by which oil can be separated from the crushed plant material. *(1 mark)*

    (iii) For use in food manufacture, plant oils can be hardened in a reaction with heat and a nickel catalyst. Which chemical do the plant oils react with? *(1 mark)*

6 **(a) (iv)** Plant oils can also be made into a fuel. What is the name of the fuel? *(1 mark)*

  (b) Many foods are cooked in vegetable oil rather than water. Explain the advantage of cooking foods in vegetable oil rather than water. *(1 mark)*

  (c) Vegetable oils are unsaturated. The amount of unsaturation can be measured using the iodine number. The higher the iodine number, the more unsaturated the molecule.

    The table below shows the iodine number and melting point of some vegetable oils.

Vegetable oil	Iodine number	Melting point/°C
Coconut	10	25
Palm kernel	37	24
Olive	81	−6
Sunflower	125	−17
Linseed	178	−24

    (i) Plot a scattergraph with melting point on the vertical axis against iodine number on the horizontal axis. *(2 marks)*

    (ii) Describe the relationship between iodine number and melting point. *(1 mark)*

  (d) Vegetable oils do not dissolve in water. However, emulsions of water and vegetable oils can be made by adding an emulsifier such as lecithin from egg yolk.

    (i) Give one example of a food that is an emulsion. *(1 mark)*

    (ii) Give one advantage of an emulsion of vegetable oil and water over the separate oil or water. *(1 mark)*

    (iii) Describe key features of an emulsifier molecule. *(1 mark)*

7 **(a)** The outer part of the Earth is the atmosphere, which is a mixture of gases.
Match the gases below to their percentage in the atmosphere. *(1 mark)*

argon		oxygen		nitrogen

98%	78%	21%	1%

(b) The diagram below shows the way in which scientists believe the continents were arranged just over 200 million years ago. Alfred Wegener suggested this arrangement in 1915 but his ideas were not accepted for many years. Give some evidence that Alfred Wegener used to support his idea of continental drift and the main reason his ideas were not accepted at the time.

In this question you will be assessed on using good English, organising information clearly and using specialist terms where appropriate. *(6 marks)*

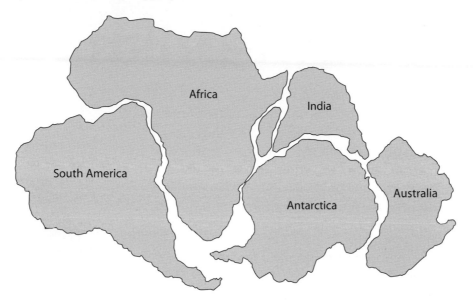

(c) The Earth is thought to be made up of three main layers: the core, the mantle and the crust. The upper mantle and crust are broken up into huge pieces called tectonic plates that are slowly moving. What is thought to be causing the plates to move? *(2 marks)*

The building blocks of chemistry

In this section you will look more deeply into the nature of matter, finding out how the arrangement and number of subatomic particles vary from element to element. In the first chapter you will see how the number and arrangement of electrons in an atom control the chemistry of the elements and the reactions they take part in. You will discover how non-metals can combine to form molecules, and how metals and non-metals combine to form salts and other ionic compounds.

In the second chapter you will learn how these chemical reactions at atomic level lead to the properties of the major groups of materials. Why do metals conduct electricity and why are they strong yet easy to shape? Why do salts have high melting points, yet can be brittle? Why are some non-metals, such as sulfur, soft and easy to melt, while others, such as diamond, are very hard with very high melting points? Why do different polymers have different properties? All these questions can be explained once you understand the processes of chemical bonding. You will also be introduced to nanotechnology.

Chapter 3 then looks at chemistry quantitatively. Why do different atoms have different masses? How can you use this knowledge to work out the formula of a chemical compound? You will learn how to calculate the yield of a product from a chemical reaction, once you know its balanced chemical equation. You will also learn about how substances are analysed using instruments such as mass spectrometers or techniques such as chromatography.

Test yourself

1. What controls the number of electrons in an atom and how are the electrons arranged around the nucleus?

2. What does the group number of an element in the periodic table tell you about its electron pattern?

3. How do metals and non-metals form compounds together?

4. How do non-metals form molecules?

5. How does a balanced chemical equation explain why the mass of products must equal the mass of reactants in a chemical reaction?

Objectives

By the end of this unit you should be able to:

- explain how metals and non-metals combine by ionic bonding
- explain how non-metals can combine with each other by covalent bonding
- explain the differences in the physical properties of non-metals, such as carbon and oxygen, in terms of their molecular structure
- explain the differences in the physical properties of metals and metallic compounds, such as salts, in terms of their ionic structure
- calculate formula masses and use these with balanced chemical equations to calculate yield
- explain how substances can be analysed by a variety of techniques.

Understanding compounds

John Dalton – a great experimenter and a great chemist.

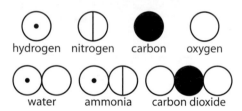

hydrogen nitrogen carbon oxygen

water ammonia carbon dioxide

Figure 1 Some of Dalton's elements and compounds.

Chemistry born from alchemy

A thousand years ago alchemists tried to make gold (rare and expensive) from 'base metal' such as lead (cheap and available). They failed, because they didn't really understand what happens when chemicals react. However, they did develop some interesting techniques such as distillation, and they spotted one or two useful patterns that helped later chemists. They also blew themselves up and poisoned themselves quite often.

John Dalton, who lived 200 years ago, was determined to make sense of all these reactions. He did some very careful experiments, weighing the reactants and products, and noticed that the same chemicals always reacted in the same relative amounts. This convinced him that everything must be made up of tiny particles – **atoms** – and that these atoms always combine in the same proportions for a given compound. In particular he concluded that:

- every **element** is made of its own distinctive atoms of a particular mass
- other chemicals are made from atoms that have joined together in some way.

Dalton gave his elements special symbols and drew pictures of some simple **compounds**. He may not have got everything quite right, but he certainly got chemistry moving in the right direction.

Elements, mixtures and compounds

We now know that there are just over 100 different elements. An element is a substance made from only one type of atom. It cannot be broken down into simpler substances. All the elements are listed in the periodic table (see lesson C2 1.5).

Hydrogen and oxygen are both elements. Hydrogen is a colourless gas that is flammable. Oxygen is a colourless gas in which other substances burn well. If you mix hydrogen and oxygen, then you still have the same substances: hydrogen and oxygen.

If a flame is put near the mixture of hydrogen and oxygen, a chemical reaction takes place producing the compound water. This is very different to both hydrogen and oxygen. It is a colourless liquid that can put fires out.

In a mixture of hydrogen and oxygen, atoms of oxygen are not joined to atoms of hydrogen; they are just mixed together. In the compound water, the hydrogen and oxygen atoms are joined to each other.

The equation for this chemical reaction is:

$$\text{hydrogen} + \text{oxygen} \longrightarrow \text{water}$$
$$2H_2 + O_2 \longrightarrow 2H_2O$$

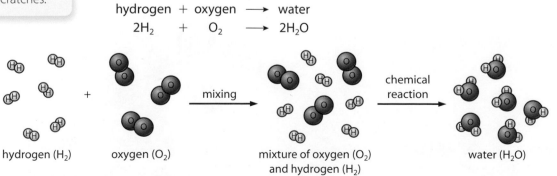

hydrogen (H$_2$)   oxygen (O$_2$)   mixture of oxygen (O$_2$) and hydrogen (H$_2$)   water (H$_2$O)

Figure 2 The reaction between hydrogen and oxygen.

More about compounds

A compound is a substance made from different elements chemically joined together. Although there are only just over 100 elements, there are millions of known compounds.

The properties of each substance in a mixture are the same as before they were mixed. However, the properties of a compound are different to those of the elements from which it is made. For example, sodium chloride (common salt) is made from sodium and chlorine. Sodium and chlorine are both very reactive, dangerous elements. Sodium chloride is completely different to both sodium and chlorine. Sodium chloride is stable and safe enough to eat.

The reaction between sodium and chlorine.

The equation for this chemical reaction is:

sodium + chlorine ⟶ sodium chloride

$$2Na + Cl_2 \longrightarrow 2NaCl$$

Compounds can be broken down (decomposed) into simpler substances. This can often be achieved by heating (thermal decomposition) or using electricity (electrolysis). For example, passing an electric current through molten aluminium oxide breaks it down into aluminium and oxygen.

Questions

1 (a) What was Dalton's big (or small?) idea about the nature of materials? (b) How did he check to see if his ideas were correct?

2 How can the elements in the compound aluminium oxide be separated?

3 Which element do all of these acids have in common: sulfuric acid (H_2SO_4), phosphoric acid (H_3PO_4), nitric acid (HNO_3), and hydrochloric acid (HCl)?

4 What is the difference between a mixture of hydrogen and oxygen and the compound water? Explain in terms of how the atoms are arranged.

5 Here is a list of substances. Which are elements and which are compounds? Mg, SO_2, Co, CO, Br_2, KBr, $CaCO_3$, K_2O.

6 Look at the diagrams in Figure 3 and say for each whether it shows an element, a compound or a mixture.

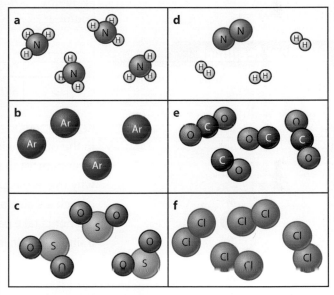

Figure 3 Element, mixture or compound?

7 Magnesium burns in oxygen to form magnesium oxide. What pattern does this table of masses show?

Mass of magnesium/g	Mass of oxygen reacting/g
0.3	0.2
0.36	0.24
0.24	0.16

8 Carbon dioxide is a compound made from the elements carbon and oxygen. Use these three substances to explain the difference between elements and compounds.

Ionic bonding

Happiness is a stable electron structure

Helium, neon and argon belong to a family of elements found in **group 0** of the periodic table called the noble gases. They have this name because they keep to themselves and do not join in chemical reactions at all. What is it that makes them so stable and unreactive?

Science in action

Noble gases are often used to fill light bulbs with heated filaments (such as xenon headlights) because they do not react with the hot wire.

In the first chemistry module (lesson C1 1.3), you saw how, for the first 20 elements, **electrons** stack up around the **nucleus** in **energy levels** (**electron shells**): up to two in the first, eight in the second, and eight in the third. As you learned in lesson C1 1.4, noble gas atoms have eight electrons in the outer shell. After the first 20 elements, things get a bit more complicated.

Table 1 Energy levels.

Element number	Name	Number of electrons	Electron configuration
2	helium	2	2
10	neon	10	2,8
18	argon	18	2,8,8

Krypton, xenon and radon have eight electrons in the outer shell. This appears to be a very stable arrangement that is not easy to disrupt. That is why the noble gases are so unreactive. In fact, this is such a stable arrangement that atoms of other elements take part in chemical reactions to achieve a similar result. In the process, they often have to team up.

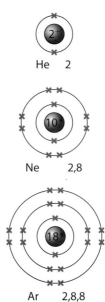

He 2

Ne 2,8

Ar 2,8,8

Figure 1 The electronic structure of the noble gases makes them very unreactive.

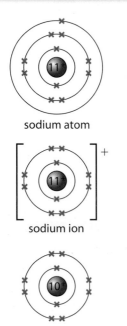

sodium atom

sodium ion

neon atom

Figure 2 Alkali metal ions have the stable electronic structure of a noble gas, but have a positive electrical charge.

Fast and loose: the single outer electron

Group 1 of the periodic table contains the alkali metals, such as sodium (see lesson C1 1.4). These elements have just one electron in their outer energy level. It is quite easy to lose this electron as not much energy is needed to pull it away from its atom. If this happens, the atom is no longer neutral. It has one more positive **proton** in the nucleus than it has negative electrons. This means that overall the particle now has a single positive charge. Charged particles like this are called **ions**.

A sodium ion is smaller than a sodium atom as it has one electron shell fewer. Ions like this have the electronic structure of a noble gas, so they are very stable.

On the scrounge for electrons

Group 7 of the periodic table contains non-metallic elements called the halogens. Group 7 elements, such as chlorine, all have seven electrons in their outer energy level. Group 7 atoms do not easily *lose* electrons, but they can *capture* an extra electron to form a stable ion that has the electronic structure of a noble gas. The ion that is formed has one extra electron, so it has a single negative charge. Such ions are called halide ions.

Metals and halogens: made for each other

All metal atoms have easily removable electrons in their outer energy levels. When they lose these electrons they form stable positive ions. Group 2 metals form 2+ ions, group 3 metals form 3+ ions, and so on.

Non-metal atoms with five, six or seven electrons in their outer energy levels gain extra electrons to get a stable structure. Group 5 non-metals have to gain three electrons to form 3– ions; group 6 non-metals have to gain two electrons to form 2– ions. Forming such ions enables the atoms to attain the stable electronic structure we see in the atoms of the noble gases (group 0).

Getting hitched: the ionic bond

You may have spotted the obvious connection. Metallic atoms become stable positive ions by losing electrons. Non-metallic atoms become stable negative ions by gaining electrons. Together, these oppositely charged ions form a compound, an entirely new substance, that does not necessarily resemble either of the elements that have come together to form it. Often, the compound is a crystal, a solid that consists of an enormous **lattice** of ions held together by the forces of attraction between them (see lesson C1 1.2). This method of joining is called an **ionic bond**. The compound is called an **ionic compound**.

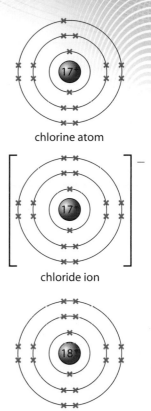

chlorine atom

chloride ion

argon atom

Figure 3 Halide ions have the stable electronic structure of a noble gas, but with a negative electrical charge.

Route to A*

With a clear understanding of the logic behind ionic bonding, you will be able to work out any combination of metal/non-metal – rather than simply remembering the ones you have seen before.

Science skills

These electronic structure diagrams are a good example of how scientists use simple models to help understand complex problems. Atoms are, of course, three-dimensional and far more complex than shown, but these simple diagrams are very powerful, and help us to understand and predict the properties of the elements.

Questions

1. Atoms of an unknown element X have eight electrons in their outer shell. What predictions can you make about the properties of this element X? Explain your answer.

2. What ions will be formed by the alkali metals lithium (Li), potassium (K) and rubidium (Rb)?

3. Draw electron configuration diagrams to show how potassium (K – element 19) becomes an ion.

4. Aluminium is in group 3. What is the charge of an aluminium ion and why?

5. Atoms of iodine have seven electrons in their outer energy shells. Is iodine a metal or non-metal and what ion does it form?

6. What do sodium ions (Na^+), chloride ions (Cl^-) and neon atoms all have in common?

7. Suggest a reason why metals are able to conduct electricity.

8. Sodium is in group 1, calcium is in group 2 and chlorine is in group 7. Explain why sodium chloride is NaCl, but calcium chloride is $CaCl_2$.

Rules for ionic bonding

Reactive sodium burns in toxic chlorine.

Could you tell that common salt is a compound of the dangerously reactive elements sodium and chlorine?

Examiner feedback

Remember that there is no such thing as an NaCl particle. This is just shorthand for the simplest ratio of the ions.

The salt on your table

As we saw in lesson C2 1.1, sodium is a soft and dangerously reactive metal in group 1 of the periodic table. Chlorine is a green poisonous gas in group 7. If sodium is burnt in chlorine the violent reaction leaves white crystals of a new compound: sodium chloride. This is the common salt you put on your chips.

The sodium and chlorine have combined chemically to form a new compound with new properties. Each sodium atom has lost an electron to become a positive sodium ion (Na^+). Each chlorine atom has gained an electron to form a negative chloride ion (Cl^-). The oppositely charged particles are attracted to form an ionic bond. The charges balance out so the compound is neutral. We can write equations for this.

$$\text{sodium} + \text{chlorine} \longrightarrow \text{sodium chloride}$$
$$2Na + Cl_2 \longrightarrow 2NaCl$$

The balanced chemical equation represents the proportion of the different ions present in the crystal, in this case 1:1. The ions do not actually pair up in this way. They stack up in a giant structure called an ionic lattice (see lesson C2 2.2). This regular stacking pattern gives rise to the typical cubic shape of salt crystals.

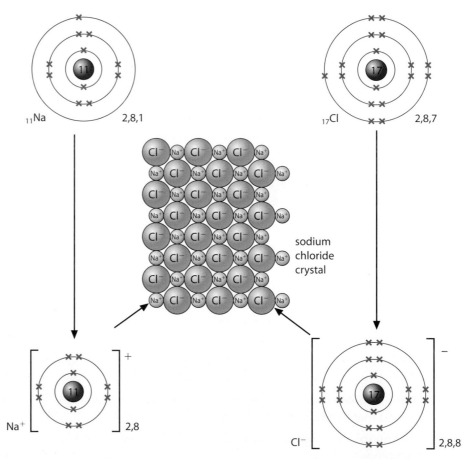

Figure 1 Common salt contains ions stacked in an ionic lattice.

Any group 1 metal will react with any group 7 halogen in the same way. For example, potassium and bromine will react to form the stable salt potassium bromide (KBr), which is made from K^+ and Br^- ions.

Keeping it in balance

Group 2 metals such as calcium have two electrons in their outer shell. If they lose these to form the stable noble gas structure, they become ions with a 2+ charge. If calcium reacts with chlorine, for example, you need two chloride (Cl^-) ions to balance the charge of one calcium (Ca^{2+}) ion and make a neutral compound. The simple chemical formula for calcium chloride is therefore $CaCl_2$, and the balanced equation is:

$$Ca + Cl_2 \longrightarrow CaCl_2$$

Group 6 non-metals such as oxygen have six electrons in their outer shell. They need to gain two electrons to achieve the stable noble gas structure, becoming ions with a 2− charge. So if oxygen reacts with sodium, for example, you need two sodium (Na^+) ions to balance the charge of one oxide (O^{2-}) ion. The simple chemical formula for sodium oxide is therefore Na_2O, and the balanced equation is:

$$4Na + O_2 \longrightarrow 2Na_2O$$

Of course, if calcium reacts with oxygen, the charges on the Ca^{2+} and O^{2-} ions cancel out at a simple 1:1 ratio again:

$$2Ca + O_2 \longrightarrow 2CaO$$

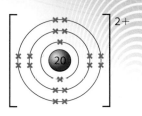

Figure 2 The calcium Ca^{2+} ion has the electron structure of argon: 2, 8, 8.

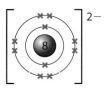

Figure 3 The oxide O^{2-} ion has the electron structure of neon: 2, 8.

Finding the formulae of ionic compounds

You can work out the formulae of ionic compounds using Table 1. This includes some common compound ions that are made up of several atoms joined together. For example, the nitrate ion (NO_3^-) has one nitrogen atom joined to three oxygen atoms and has a charge of −1.

To work out the formula (Table 2), you must make the total charge of the positive and negative ions balance. If the formula contains more than one of a compound ion, the compound ion is put in brackets for clarity.

Table 1 Ion charges.

Positive ions		Negative ions	
aluminium	Al^{3+}	bromide	Br^-
calcium	Ca^{2+}	chloride	Cl^-
copper(II)	Cu^{2+}	fluoride	F^-
hydrogen	H^+	iodide	I^-
iron(II)	Fe^{2+}	oxide	O^{2-}
iron(III)	Fe^{3+}	sulfide	S^{2-}
lithium	Li^+	carbonate	CO_3^{2-}
magnesium	Mg^{2+}	hydrogen carbonate	HCO_3^-
potassium	K^+		
sodium	Na^+	hydroxide	OH^-
zinc	Zn^{2+}	nitrate	NO_3^-
ammonium	NH_4^+	sulfate	SO_4^{2-}

Table 2 Working out the formulae of some ionic compounds.

	Potassium oxide	Calcium sulfide	Aluminium oxide	Copper carbonate	Calcium hydroxide	Ammonium sulfate
Positive ions	K^+ K^+	Ca^{2+}	Al^{3+} Al^{3+}	Cu^{2+}	Ca^{2+}	NH_4^+ NH_4^+
Negative ions	O^{2-}	S^{2-}	O^{2-} O^{2-} O^{2-}	CO_3^{2-}	OH^- OH^-	SO_4^{2-}
Formula	K_2O	CaS	Al_2O_3	$CuCO_3$	$Ca(OH)_2$	$(NH_4)_2SO_4$

Questions

1. Aluminium is in group 3. Explain why aluminium forms an Al^{3+} ion.

2. Calcium loses two electrons to form a 2+ ion. Why is calcium chloride $CaCl_2$?

3. Why is the formula for sodium oxide Na_2O?

4. Calcium and chloride ions both have the electron configuration 2,8,8. What is different about them?

5. Magnesium oxide has the formula MgO. Explain why this is.

6. What is unusual about the ammonium ion compared with all of the other compound ions shown?

7. Use the table of ion charges to write the formula of:
   (a) potassium bromide, **(b)** iron(III) oxide,
   (c) aluminium hydroxide, **(d)** calcium nitrate,
   (e) zinc sulfate, and **(f)** calcium hydrogen carbonate.

8. Some transition elements such as iron can vary the number of electrons in their outer energy level. Iron forms two different sulfates, for example, $FeSO_4$ and $Fe_2(SO_4)_3$. For each case, how many electrons have been lost from the outer energy level of the iron atom? Explain your answer.

Covalent bonding

Learning objectives

- explain how non-metals form molecules
- identify the bonds formed between atoms
- understand that some elements form molecules on their own, but most molecules are compounds
- use different models to show covalent bonds for different purposes.

No electrons to spare?

Non-metals cannot form ionic compounds on their own. Their atoms need extra electrons to get a stable 'noble gas' electron configuration, but without metal atoms where can they get the electrons?

The answer is to share. Non-metal atoms can join their outer energy levels together and share one or more pairs of electrons. The shared electrons form very strong **covalent bonds**. When atoms share electrons, they stay together. We can imagine them looking like soap bubbles stuck together. These arrangements are called **molecules**. Because no electrons have been gained or lost, molecules carry no electrical charge. This arrangement can be very stable.

Choosing the right model

Atoms and molecules are far too small to see, so we can't draw accurate pictures of them. Instead we use models and symbols to help us visualise what is going on in chemical reactions. As we may want to focus on different aspects of the reaction we use different models for different purposes. Don't let this confuse you. Just use the best model for what you want to think about (see Figure 2).

- Molecular models help us to visualise the shape of molecules.
- Electron energy level diagrams show how covalent bonds are formed.
- Structural formulae show covalent bonds clearly.
- Simple formulae show the atoms involved to help us balance reaction equations.

The elements that won't go out alone

Chlorine atoms are one electron short of that stable 'noble gas' arrangement, so they need to share just one pair of electrons to form a Cl_2 molecule. You can show it in an electron shell diagram. Draw the outer shells of the two atoms slightly overlapping, and arrange the electrons in pairs around the circles. Use dots for electrons on one, and crosses for electrons on the other. You will end up with one dot and one cross in the overlap where the two atoms have joined together, and you will see that each chlorine atom now has eight electrons in its outer shell, like argon. The shared pair of electrons makes the covalent bond. The two electrons are exactly the same; using a dot and a cross just helps you to see how the shared pair is formed. A simpler version of this diagram removes the circles altogether and just shows the dots and crosses.

Similarly, oxygen atoms are two electrons short of a stable shell. In the O_2 molecule they share two electron pairs to make a double covalent bond.

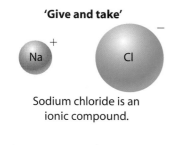

'Give and take'

Sodium chloride is an ionic compound.

'Sharing'

Water is a covalent compound.

Figure 1 Ions stay separate, but covalent bonds make atoms stick together like bubbles.

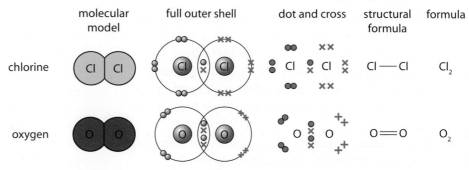

Figure 2 Some different ways to show covalent bonds.

Compounds made through sharing

Atoms of different non-metallic elements can also join together to make covalent compounds. Chlorine can form a single covalent bond with hydrogen to make hydrogen chloride (HCl). Oxygen needs to share two electrons and so can join with two hydrogen atoms to make hydrogen oxide, or water (H_2O). Nitrogen is three electrons short of a stable shell. It can share a pair of electrons with each of three hydrogen atoms to make an ammonia molecule (NH_3). Carbon is four electrons short of a stable shell. It can share a pair of electrons with each of four hydrogen atoms to make a methane molecule (CH_4).

Taking it further

The nitrogen atom in ammonia has a 'lone pair' of electrons not involved in covalent bonding. A passing hydrogen ion can be captured here, the nitrogen providing both electrons for a strong covalent bond. This forms the ammonium ion (NH_4^+), which carries the positive charge from the **hydrogen ion** and can form salts just like an alkali metal. An example is the **fertiliser** salt ammonium nitrate, NH_4NO_3.

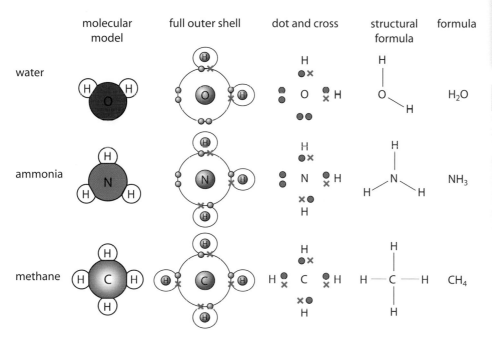

Figure 3 Common compounds with different numbers of covalent bonds.

Science in action

As every single atom in a diamond is attached to its neighbours by strong covalent bonds, diamonds are very hard indeed. In the oil industry, diamond-studded drills are used to cut through rock.

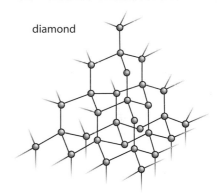

Figure 4 Diamond is a giant covalent structure built from carbon atoms.

Giant molecules

Some elements and compounds, such as carbon and silicon dioxide, exist as giant structures (lattices) in which every atom is linked to its neighbours by strong covalent bonds (see lesson C2 2.1). These are called **giant covalent structures**.

Questions

1 Many gases, such as hydrogen and oxygen, exist as molecules (H_2, O_2). What kind of bonding is there within these molecules?

2 Why do the noble gases not need to form covalent bonds?

3 Draw an electronic structure diagram for an H_2 molecule. (Remember, the first energy level can only take two electrons.)

4 Explain in simple terms the difference between ionic and covalent bonding.

5 Draw an electronic structure diagram for an HCl molecule.

6 Suggest a reason why there are not any strong forces between simple molecules.

7 Sand is made from silicon dioxide (SiO_2). What is it about the structure of this compound that makes sand hard?

8 Explain carefully how a carbon atom combines with two oxygen atoms to form carbon dioxide. Draw dot-and-cross and structural formulae diagrams for a CO_2 molecule.

Metals

Useful metals

Just over three-quarters of the elements are metals. Humans have depended on metals since the Bronze Age. Today, we make much use of metals such as iron, aluminium and copper.

Metals all have electrons in their outer shells that are easily lost, giving stable positive ions. These 'loose' outer electrons give metals their special properties (see lesson C1 3.5).

Science skills

Table 1
The atomic radii of the group 4 elements.

Element	Atomic radius/10^{-12}m
C	70
Si	110
Ge	125
Sn	145
Pb	154

Use this to explain why group 4 elements change from metals to non-metals down the group.

Figure 1 All the elements shaded blue are metals.

The structure of metals

In the solid, the metal atoms are packed close together in a regular **metallic structure**. The outer shell electrons are lost from each atom and become free to move throughout the metal. This leaves a giant structure (lattice) of positive metal ions surrounded by **delocalised** electrons.

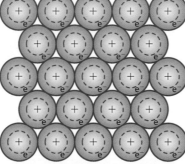

metal atoms packed together

Outer energy level (shell) electrons are lost from atoms leaving a lattice of positive metal ions surrounded by delocalised electrons. The shaded area represents the delocalised electrons.

Figure 2 The structure of metals.

The structure is held together by the electrostatic attraction between the positive metal ions and the negative delocalised electrons. This attraction is strong, and so metals have high melting and boiling points.

Metals conduct heat

Metals conduct energy, in the form of heat, because the delocalised electrons are free to move and can carry energy with them.

- Metals feel cold to the touch because they conduct the energy given out away from your hand, which cools down.
- Computer microprocessors use metal fins to carry energy given out away from the processor and stop it from overheating.
- Pans are often made from metals because they conduct heat from the flame or hotplate to your food.

Delocalised electrons transfer energy and cool the computer chip.

Metals conduct electricity

Delocalised electrons also allow metals to conduct electricity very well. Electric cables are usually made of copper, which is an excellent conductor. Any impurities in the copper can cause irregularities in the metallic structure and impede the flow of electrons, so copper for cables has to be at least 99.9% pure.

Even the vibrations of the metal ions can affect the flow of electricity. As a metal gets hot, the ions vibrate more and the current is reduced. If you cool the metal down, it conducts electricity much better. At very low temperatures, certain materials become superconductors. They have no electrical resistance and can carry enormous currents. Supercooled superconducting metal coils are used for the powerful electromagnets in **magnetic resonance imaging (MRI) scanners** and in particle accelerators such as the **large hadron collider**.

Metals are strong and easy to shape

Metals are strong enough to build machines and bridges with, but they are also easy to shape as they can be bent and hammered into shape. This is because the layers of metal ions can slide over each other while keeping the strong attraction between the positive metal ions and the delocalised electrons.

Supercooled metals are used for powerful electromagnets like those in the large hadron collider.

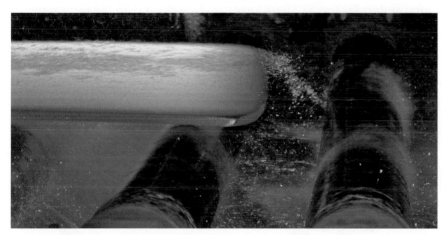

Hot steel can be squeezed out to make girders.

Science in action

The powerful magnetic fields needed to contain the superheated core of a nuclear fusion reactor are produced by the enormous electric currents that can flow through supercooled, superconducting electromagnets.

Questions

1   **(a)** Describe the structure of a metal. **(b)** Describe the bonding in a metal.

2   Why do metals have high melting points?

3   Explain why metals can be used to make: **(a)** pans; **(b)** electrical wires.

4   Explain why metals can be bent and shaped.

5   Suggest a reason why metals are usually heated before being beaten or pressed into shape.

6   Copper is used to make electrical wires. **(a)** Explain why copper is used to make electrical wires.

(b) Draw a diagram to show the structure of copper. **(c)** Use your diagram to explain why copper conducts electricity. **(d)** Use your diagram to explain why copper can be stretched into wires. **(e)** Why does copper have to be 99.9% pure to conduct electricity efficiently?

7   Gold is a better conductor than copper. Suggest a reason why it isn't used for household wiring.

8   Explain carefully how temperature affects the way electricity can flow though a metal. Give an example of how this effect has been put to good use.

Simple molecular substances

Molecules

Many substances are made up of molecules. This includes some elements such as oxygen (O_2) and hydrogen (H_2), and many compounds such as water (H_2O) and glucose ($C_6H_{12}O_6$). Many elements that are non-metals are made of molecules. Compounds made from non-metals are also made of molecules.

A molecule is two or more atoms joined by covalent bonds. In a covalent bond, two electrons are shared between two atoms. This shared electron pair joins the atoms together.

The formula of a simple molecular substance tells us how many atoms of each type are in one molecule. For example, the formula CH_4 (methane) tells us that a methane molecule is made up of one carbon atom and four hydrogen atoms.

Intermolecular forces

Covalent bonds within molecules are very difficult to break. If they are broken, this constitutes a chemical change as different substances are formed. For example, if the covalent bonds in water are broken, hydrogen and oxygen are formed.

Between the molecules there are no bonds. However, there are weak forces called **intermolecular forces**. These forces are far weaker than the three types of bonding (ionic, covalent and metallic).

Melting and boiling points

When simple molecular substances melt or boil, it is the weak forces between molecules that are overcome. The covalent bonds do not break. For example, the molecules in water as a solid (ice), liquid (water) and gas (steam) are all H_2O molecules. The molecules in methane as a solid, liquid and gas are all CH_4 molecules.

This photo shows water as ice, water and steam. All three states are made up of H_2O molecules.

Table 1 The structure of the simple molecular substance methane as a solid, a liquid and a gas.

State	Solid	Liquid	Gas
Space-filling diagrams (better represent what molecules look like)			
Stick diagrams			

Simple molecular substances have low melting and boiling points because the forces between the molecules are weak. This means that many simple molecular substances are gases, liquids, or solids with low melting points.

Table 2 Melting and boiling points of some simple molecular substances.

Substance	Ethanol	Iodine	Methane	Naphthalene	Oxygen	Water
Formula	C_2H_5OH	I_2	CH_4	$C_{10}H_8$	O_2	H_2O
Melting point/°C	−114	114	−183	80	−219	0
Boiling point/°C	78	184	−162	218	−183	100
State at room temperature	liquid	solid	gas	solid	gas	liquid

Electrical conductivity

An electric current is the flow of electrically charged particles. Molecules have no electric charge — they are neutral. This means that simple molecular substances do not conduct electricity in any state.

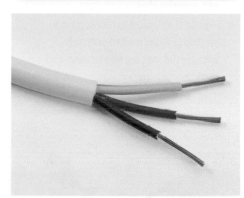

The molecules in the plastic coating on this electrical cable do not conduct electricity. This protects us from electric shocks.

Questions

1 (a) What is a molecule? (b) What is a covalent bond?

2 (a) Name three elements made from molecules. (b) Name three compounds made from molecules.

3 Which of the following compounds are made from molecules? Calcium oxide (CaO), hydrogen sulfide (H_2S), methanol (CH_3OH), copper sulfate ($CuSO_4$), sodium chloride (NaCl), silane (SiH_4), ammonia (NH_3).

4 (a) The formula of glucose is $C_6H_{12}O_6$. What does this formula tell us about glucose? (b) How are the atoms held together in a molecule of glucose?

5 Bromine molecules have the formula Br_2. They contain two bromine atoms joined by a single covalent bond. Bromine is a liquid at room temperature that easily turns into a gas due to a low boiling point. Explain why bromine has a low boiling point.

6 Simple molecular substances, including pure water, do not conduct electricity. Explain why.

7 Look at the data in Table 3 and decide which of the substances are made of simple molecules.

Table 3 Properties of simple molecules.

Substance	A	B	C	D	E
Melting point/°C	−85	808	39	41	3550
Boiling point/°C	−20	1465	701	182	4827
Conductivity as solid	insulator	insulator	conductor	insulator	insulator

8 Describe and explain the properties of simple molecular substances by a discussion of their structure and bonding.

Table 4 Data on some simple molecular substances.

Substance	Relative mass of molecule (M_r)	Boiling point/°C
methane (CH_4)	16	−162
ethane (C_2H_6)	30	−89
propane (C_3H_8)	44	−42
butane (C_4H_{10})	58	0
pentane (C_5H_{12})	72	36
hexane (C_6H_{14})	86	69

a Plot a graph of boiling point against relative mass of molecules and draw a line of best fit.

b Describe the relationship between the mass of the molecule and the boiling point.

Ionic substances

Ions

Many compounds are made up of ions. Ions are particles that are electrically charged because they contain different numbers of protons and electrons. Ionic substances are compounds made from both metals and non-metals. Simple examples include sodium chloride (common salt), copper sulfate and calcium carbonate.

Ionic lattice

All ionic compounds are solids at room temperature. Inside the compound there are billions of ions from one edge of the solid right across to the other in all directions. The ions are packed together in an ordered, regular structure. This is a giant lattice.

Each ion is surrounded by several ions of opposite charge. Opposite electrical charges attract each other. So each ion is attracted by strong **electrostatic attraction** to all the ions surrounding it. This electrostatic attraction between positive and negative ions is known as ionic bonding.

The formula of an ionic compound tells us the ratio of the ions in the compound. For example, the formula NaCl tells us that there is one sodium ion (Na^+) for every chloride ion (Cl^-) in the structure. The formula Al_2O_3 means that there are two aluminium ions (Al^{3+}) for every three oxide ions (O^{2-}) in the structure.

Melting and boiling points

The attraction between positive and negative ions is strong. In order to melt and boil ionic compounds lots of these strong attractive forces between ions of opposite charges have to be overcome. This takes a lot of energy and so ionic compounds have high melting and boiling points. This means that all ionic compounds are solids at room temperature.

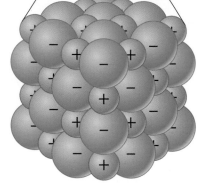

Figure 1 Sodium chloride is made of a giant lattice of positive and negative ions.

Table 1 The melting and boiling points of some ionic compounds.

Substance	Sodium chloride	Magnesium bromide	Aluminium oxide	Potassium carbonate
Formula	NaCl	$MgBr_2$	Al_2O_3	K_2CO_3
Melting point/°C	808	711	2040	896
State at room temperature	solid	solid	solid	solid

Table 2 The structure of an ionic compound as a solid, a liquid and a gas.

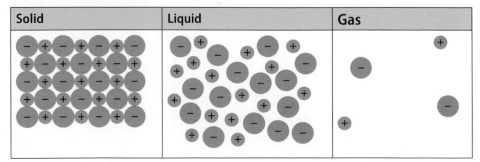

Solid	Liquid	Gas

Electrical conductivity

Ions are electrically charged particles. An electric current is the flow of electrically charged particles. As a solid, the ions are vibrating in fixed positions and cannot move around. When the ionic compound is melted, the ions can move around and so can conduct electricity.

Many ionic compounds dissolve in water. When they dissolve the ions are free to move around in the solution. This means that ionic compounds that are soluble in water also conduct electricity when dissolved.

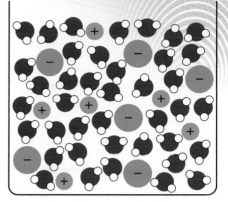

Figure 2 When sodium chloride dissolves in water the positive and negative ions separate and mix in with the water molecules.

Science in action

When ionic compounds are in solution they break down into simpler substances and can conduct electricity. This is called **electrolysis**. Electrolysis has many uses. The electrolysis of sodium chloride solution is a major industrial process producing hydrogen, chlorine and sodium hydroxide from the sodium chloride solution. Each of these products has many uses.

Examiner feedback

When ionic compounds conduct electricity, it is the ions that are moving to carry the current, not electrons.

Questions

1   What are ions?

2   What is ionic bonding?

3   Which of the following compounds have ionic structures? Zinc oxide (ZnO), butane (C_4H_{10}), nickel sulfate ($NiSO_4$), calcium chloride ($CaCl_2$), fluorine (F_2) and hydrazine (N_2H_4).

4   Calcium oxide is an ionic compound with the formula CaO and a melting point of 2600 °C. **(a)** Explain what this formula means. **(b)** Explain why calcium oxide has a high melting point.

5   Give the formula of the following ionic compounds (see lesson C2 1.3 for ion charges). **(a)** Potassium oxide. **(b)** Magnesium carbonate. **(c)** Aluminium sulfate.

6   Potassium iodide, KI, is an ionic compound. **(a)** Explain why it does not conduct electricity as a solid. **(b)** Explain why it does conduct electricity when it is melted. **(c)** Explain why it does conduct electricity when it is dissolved in water.

7   Look at the data in Table 3 and decide which of the substances are ionic compounds.

Table 3 Compounds and their properties.

Substance	A	B	C	D	E
Melting point/°C	763	3550	1453	1520	358
Boiling point/°C	1452	4827	2785	1680	684
Conductivity as solid	insulator	insulator	conductor	insulator	insulator
Conductivity as liquid	conductor	insulator	conductor	conductor	conductor

8   Describe and explain the properties of ionic substances with a discussion of their structure and bonding.

Taking it further

The melting point of an ionic compound is a good indication of the strength of ionic bonding. The stronger the attraction between the ions, the higher the melting point. Two factors that affect the strength of ionic bonding are the size and charge of ions.

From the data below, how do you think the size and charge of an ion affect bond strength?

1 picometre (pm) $= 1 \times 10^{-12}$ m.

Substance	Formula	Melting point (°C)
Calcium oxide	CaO	2572
Magnesium oxide	MgO	2852
Sodium oxide	Na_2O	1132

Ion	Formula	Radius of ion (pm)
Calcium	Ca^{2+}	100
Magnesium	Mg^{2+}	72
Sodium	Na^+	102

Covalent structures

Different forms of carbon

Diamond and graphite are forms of carbon. Both have very high melting points, but diamond is hard and an insulator while graphite is soft and conducts electricity. An understanding of their structure is needed to explain the similarities and differences.

Giant covalent structures

Atoms can join together by covalent bonding to make molecules. In a simple molecular substance, there are millions of separate but identical molecules. For example, water is made up of water molecules (H_2O) in which two hydrogen atoms are joined to one oxygen atom. The molecules themselves are not joined together.

Atoms can also be joined by covalent bonding to form a giant covalent structure. In such a structure all the atoms are covalently bonded in a massive network. An example is diamond. In diamond, the carbon atoms are linked together in a giant lattice, which extends all through the material. Giant covalent substances are sometimes called **macromolecular** substances, but they are not molecules.

Other examples of giant covalent substances include graphite (another form of carbon), and silicon dioxide (silica, SiO_2). The formula of silicon dioxide shows that the ratio of silicon to oxygen atoms in the structure is 1 : 2.

Melting and boiling points

In order to melt or boil a giant covalent substance, lots of covalent bonds have to be broken. These bonds are very strong and need a lot of energy in the form of heat to break them. For this reason, giant covalent structures have very high melting points. Diamond, for example, melts at over 3500 °C.

Diamond and graphite

Although diamond and graphite are both forms of carbon and have giant covalent structures, they do not have the same properties.

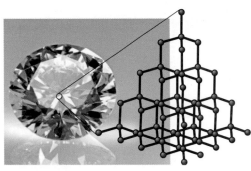

Figure 1 Diamond (a type of carbon) forms a giant covalent structure.

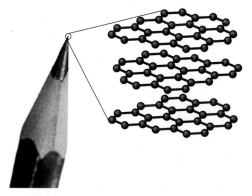

Figure 2 The lead in a pencil is made of graphite, another form of carbon. In graphite, the carbon atoms form layers.

Table 1 Diamond and graphite.

	Diamond	Graphite
Structure	Each C atom is joined to four others by covalent bonds.	Each C atom is joined to three others by covalent bonds. This forms layers, which are free to slide over each other. The layers are not bonded to each other.
Melting point	Very high, because lots of strong covalent bonds have to be broken.	Very high, because lots of strong covalent bonds have to be broken.
Hardness	Very hard, because the atoms are bonded in a rigid network.	Soft and slippery, because the layers can slide over each other (there are weak forces between the layers).
Electrical conductivity	Insulator, because there are no electrons free to move around.	Conductor, because there is one electron from each carbon atom free to move along the layers (it has delocalised electrons).

Fullerenes

In the 1980s a third form of carbon was discovered that forms large molecules, but not giant covalent structures. **Fullerenes** are molecules made up of linked carbon rings. The first to be discovered was a molecule containing 60 carbon atoms in a spherical shape. It was called buckminsterfullerene, after the American architect Buckminster Fuller who built domes with a similar structure.

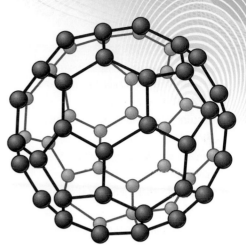

Figure 3 Fullerenes are made up of carbon rings joined together. This is buckminsterfullerene (C_{60}).

The domes at the Eden Project have a hexagonal structure similar to that of fullerenes.

Since the original discovery, more fullerenes have been made in different sizes and shapes. They are proving very useful, for example for drug delivery into the body, in lubricants and as catalysts. **Nanotubes** (see lesson C2 2.6) are used for reinforcing materials such as tennis rackets.

(see lesson C2 2.6)

Questions

1. **(a)** What is a covalent bond? **(b)** Which two types of structures contain atoms joined by covalent bonds?
2. Give three examples of substances that have a giant covalent structure.
3. Explain why all giant covalent substances have very high melting points.
4. Explain clearly why diamond is hard but graphite is soft and slippery.
5. Explain clearly why diamond is an insulator but graphite is a conductor.
6. Look at the data in Table 2 and decide which of the substances are giant covalent substances.

Table 2 Properties of substances.

Substance	A	B	C	D	E
Melting point/°C	3550	−36	564	3727	850
Conductivity as solid	insulator	insulator	conductor	conductor	insulator

7. **(a)** What are fullerenes? **(b)** Give some uses for fullerenes.
8. Carbon exists in different forms. Some are giant covalent structures, such as diamond and graphite, while fullerenes are simple molecular structures. Explain the difference between simple molecular and giant covalent structures.

Examiner feedback

It is very important to realise that giant covalent structures are not made of molecules. Some students think they are big molecules, but they are not. In a substance made of molecules, each molecule has the same number of atoms in them, whereas in a giant covalent substance there is one continuous network of atoms linked by covalent bonds.

Route to A*

For diamond and graphite, it is important to be able to explain the similarities in (e.g. melting point) and differences between (e.g. hardness, conductivity) their properties by an understanding of their structures.

Science in action

Some drill bits are tipped with diamond. This is because diamond is the hardest known substance and so the drill bits can be used to drill through very hard substances.

Metals and alloys

Thermal and electrical conductivity

There are over 100 elements, most of which are metals. As you learned in lesson C2 1.5, one very important property is that metals conduct heat and electricity. Saucepans are often made from metal because this allows heat to pass easily into the food. Electrical wires are made of metals, often copper, because metals are good electrical conductors.

Metals all have the same type of structure in which the outer shell electrons are delocalised, which means that the electrons can move through the metal. This is why metals conduct heat and electricity so well.

Alloys

Metals can be bent and hammered into shape. This is because the layers of metal atoms can slide over each other while maintaining the metallic bonding that holds the atoms together. However, pure metals are often too soft to be useful, because the layers slide over each other too easily.

Metals can be made harder by turning them into alloys. An **alloy** is a mixture of a metal with small amounts of other elements, usually other metals or carbon. The atoms of the added elements are a different size from the metal atoms. They 'jam up' the metal structure, stopping the layers of atoms from sliding past each other so easily. This makes alloys harder.

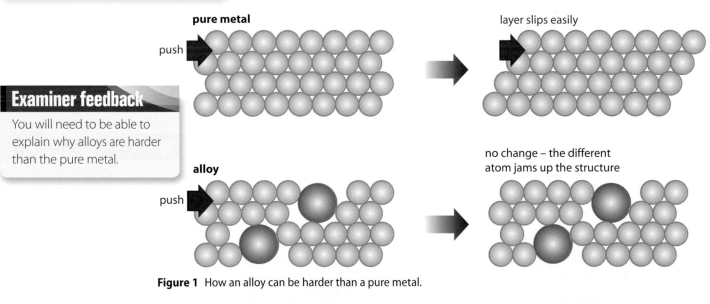

Figure 1 How an alloy can be harder than a pure metal.

The purity of gold is measured in carats. Pure gold is 24 carats. The purest gold used for most jewellery is 22 carats.

Steel is the most common alloy. It is a mixture of iron with small amounts of carbon or other elements. There are many different types of steel, containing different amounts of various alloying elements. Stainless steels, for example, contain chromium for rust resistance, while tool steels have added tungsten for hardness.

Pure gold is too soft to make practical jewellery with. Other metals, such as copper and silver, are added to make a harder and stronger alloy.

Shape memory alloys

A **shape memory alloy** is an alloy that can return to its original shape after being deformed. One example is **nitinol**, which is an alloy of nickel and titanium. Objects made from nitinol are cold-forged in a particular shape. If the object is bent or deformed in some way, then warming or heating it will return it to its original shape.

One common use for nitinol is in the wires in dental braces. In the warmth of the mouth, the wires try to return to their original size and shape. This pulls or pushes the teeth into position. Unlike stainless steel, nitinol braces do not have to be frequently replaced or tightened. The teeth are corrected faster and the braces are more comfortable.

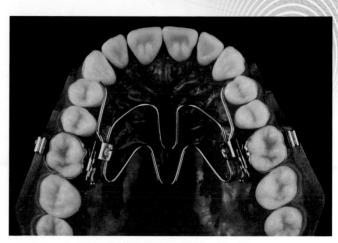

One use of the shape memory alloy nitinol is for dental braces.

Science in action

Another use of nitinol is to repair collapsed arteries. A squashed tube made of nitinol mesh, called a stent, is slid into the artery. As it warms up in the body it expands to its original size and opens the artery up.

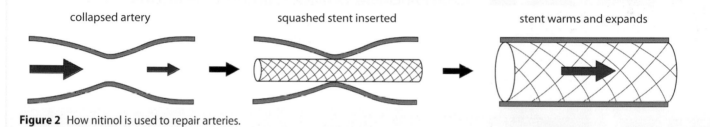

collapsed artery   squashed stent inserted   stent warms and expands

Figure 2 How nitinol is used to repair arteries.

Questions

1 Explain why metals conduct heat.

2 Explain why metals conduct electricity.

3 Explain why metals can be bent and hammered into shape.

4 Pure iron is too soft to be useful. Explain why pure iron is soft.

5 **(a)** The properties of iron are improved by making it into alloys called steel. What are alloys? **(b)** Why are alloys harder than pure metals?

6 Stainless steel is a very useful alloy of iron. Why is this alloy so useful compared with iron and other steels?

7 Tungsten is used to make steels which are used to make tools such as hammers. What property is needed for tool steels?

8 Nitinol is a shape memory alloy. What is a shape memory alloy? Give one example of a use of nitinol. Explain why nitinol is better in this use than a traditional metal or alloy.

A*

Science skills

Here are some data about the hardness of some gold alloys. The hardness is measured by the Vickers Hardness scale (HV), where the higher the value the greater the hardness.

Gold alloy	% gold	Hardness (HV)
24 carat	100	55
22 carat	91.7	138
21 carat	87.5	190
18 carat	75.0	212

a Plot a graph of hardness against percentage gold in the alloy and draw a line of best fit.

b Describe the relationship shown.

Polymers with different properties

Figure 1 The structure of poly(ethenol), also known as polyvinyl alcohol or PVA.

One major use of poly(ethenol) is as an adhesive.

Types of polymer

There are many different **polymers**, with a wide range of uses. These range from simple plastic bags and bottles to non-stick coatings, breathable clothing fibres, water-absorbing hydrogels and shape memory polymers. The properties of polymers are determined by their structure. This can depend on the monomers the polymer is made from, or on the way the polymer is made.

Different starting materials

Different monomers produce polymers with different properties. For example, plastic shopping bags are often made from poly(ethene), common name polythene, made from ethene $CH_2 = CH_2$. A different polymer, poly(ethenol) or PVA, is made from ethenol $CH_2 = CHOH$. Poly(ethenol), unlike poly(ethene), dissolves in water. Some hospital laundry bags are made from poly(ethenol) so that the bag dissolves in the wash.

Changing the polymerisation conditions

Different polymers with different properties and uses can also be made from the same monomer. This is done by changing the reaction conditions when the polymer is made. For example, there are two forms of poly(ethene): high-density poly(ethene) (HDPE) and low-density poly(ethene) (LDPE). Different conditions of temperature and pressure and a different catalyst are used to make HDPE and LDPE.

Table 1 LDPE and HDPE.

Polymer	Reaction conditions	Difference in structure	Properties	Uses
low-density poly(ethene) LDPE	temperature: 200 °C pressure: 2000 atm catalyst: trace of oxygen	molecules are highly branched and therefore loosely packed	flexible, soft	bags, cling film
high-density poly(ethene) HDPE	temperature: 60 °C pressure: 2 atm catalyst: Ziegler–Natta (titanium based catalysts)	molecules are less branched and therefore tightly packed (this makes the polymer more rigid)	stiffer, harder	buckets, bottles

Thermosoftening and thermosetting polymers

In most polymers, the long polymer chains are not joined to each other. These polymers are called **thermosoftening** polymers (thermoplastics). They soften and melt on heating because there are weak forces between the polymer chains. This means that such polymers can be recycled, as they can be melted and remoulded.

In **thermosetting** polymers (thermosets) the polymer chains are joined to each other by strong covalent bonds called **cross-links**. These polymers do not soften or melt on heating and so cannot be recycled. Thermosetting polymers are usually hard and rigid because of the cross-links, whereas thermosoftening polymers are more flexible.

Table 2 Thermosoftening and thermosetting polymers.

Type of polymer	Thermosoftening	Thermosetting
Examples	poly(ethene), poly(propene), poly(chloroethene)	melanine, bakelite
Effect of heating	softens and melts	does not soften/melt (chars or decomposes if hot enough)
Structure	no cross-links between polymer chains (the chains are often tangled up)	cross-links between polymer chains

Practical

A type of slime can be made by reacting a solution of poly(ethenol) with borax. The borax forms cross-links between the poly(ethenol) chains, making it more viscous. The amount of borax used can be changed and the viscosity measured.

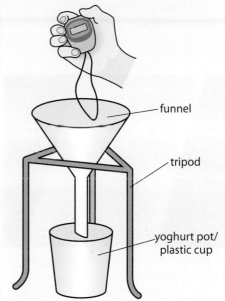

Figure 2 This apparatus can be used to measure the viscosity of slime by timing how long the slime takes to pass through the funnel.

Questions

1. What two things can be changed to make polymers with different properties?

2. **(a)** Why are plastic buckets not made out of low-density poly(ethene)?
   (b) Why is cling film not made out of high-density poly(ethene)?

3. High- and low-density poly(ethene) are both made from ethene. What is done differently to make these two polymers with different properties?

4. Decide whether each of the following polymers is a thermosoftening or a thermosetting polymer. **(a)** Cyanoacrylate glue (superglue) sets to form a hard adhesive that does not soften or melt on heating. **(b)** Vulcanised rubber is used to make car tyres. It does not soften as the tyres become hot, so the tyres are long-wearing. **(c)** Polycarbonates are used instead of glass in many greenhouses. Polycarbonate melts at 267 °C.

5. Poly(propene) is a thermosoftening polymer. Melamine is a thermosetting polymer. Explain why poly(propene) can be recycled but melamine cannot.

6. How could you test a polymer to see if it is thermosoftening or thermosetting?

7. When poly(ethenol) reacts with borax, slime is formed. Explain why slime becomes more viscous if more borax is used.

8. Thermosoftening polymers and thermosetting polymers have different properties. Describe and explain these differences.

Route to A*

It is important to be able to explain the difference in structure and properties of thermosoftening and thermosetting polymers.

Examiner feedback

To explain the difference in properties between thermosoftening and thermosetting polymers, it is essential to discuss the interactions between polymer chains. In thermosoftening polymers there are weak forces between the chains, whereas in thermosetting polymers there are strong covalent bonds between the chains.

Nanoscience

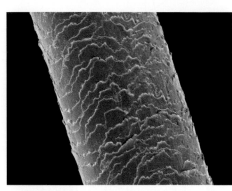

This human hair is 800 times thicker than the biggest nanoparticles.

The beads on this cotton cloth are nanoparticles of a water-repellent polymer, creating waterproof cotton.

What is nanoscience?

Nanoscience is a relatively new area of science. There are many potential benefits of nanoscience but also some concerns. You may already be using products containing nanoparticles.

Nanoscience is the study of **nanoparticles**. A **nanometre** (nm) is one millionth of a millimetre. A nanoparticle is a particle between 1 and 100 nm in size. These nanoparticles contain a few hundred atoms. They are too small to be seen, even with the most powerful light microscope. For comparison, a human hair is about 80 000 nm thick, many times thicker than a nanoparticle.

Nanoparticles can have very different properties from the same materials in bulk. For example, nanoparticles:

- may be a different colour
- may have a different strength
- may react differently
- may have different electrical/thermal conductivity.

The main reason why nanoparticles have different properties from large pieces of the same material (in bulk) is that the nanoparticles have a much larger surface area to volume ratio. This means that a much higher fraction of the atoms are on the surface.

Uses of nanoparticles

Sunblock lotions contain titanium dioxide to block out harmful ultraviolet (UV) rays from the sun. In traditional sunblocks, the large particles of titanium dioxide used also reflected light, giving the sunblock a white appearance. Now many sunblocks use nanoparticles of titanium dioxide. They still block out UV rays, but the sunblock is colourless. Titanium dioxide and other nanoparticles are also starting to be used in cosmetics.

Bulk silver metal is very unreactive. However, nanoparticles of silver can kill bacteria. Clothes manufacturers incorporate silver nanoparticles into some clothes, to kill bacteria and prevent smells. Some deodorant sprays may soon contain silver nanoparticles.

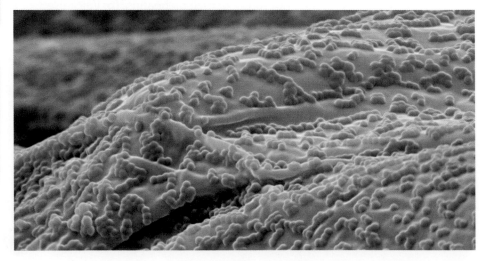

Nanoparticles are also being used as coatings. Self-cleaning windows have a coating of titanium dioxide nanoparticles, which causes rain to wash off dirt. Some new refrigerators have a coating of silver nanoparticles on the inside to kill bacteria.

Nanoparticles of gold appear red rather than the yellow colour of the bulk material. Gold is unreactive as a bulk material, but gold nanoparticles can act as a catalyst in some chemical reactions. A catalyst is a substance that speeds up a reaction without getting used up. Many other new catalysts are being made from nanoparticles.

Nanoparticles also have potential use as sensors. For example, gold nanoparticles have been used to detect toxic lead ions. These sensors can be very selective, only detecting a specific substance.

Carbon nanotubes are examples of fullerenes (see lesson C2 2.3) and have many potential uses. One property of these nanotubes is that they can be used as semiconductors. They have the potential to replace silicon in microchips, making computers faster and more powerful. They may also be useful in the treatment of cancer. Carbon nanotubes are very strong and are being used to make stronger but lighter construction materials. They have been used in tennis rackets and golf clubs, and could also be used to strengthen steel and concrete.

Concerns about nanoparticles

There are some concerns about the effects of nanoparticles on people. Given that nanoparticles have different properties to the bulk material, it is possible that some nanoparticles may be toxic even if the bulk material is not. For example, nanoparticles are usually more reactive than the bulk material and can pass through the skin and cell membranes. Further research is needed to find out what the effects might be.

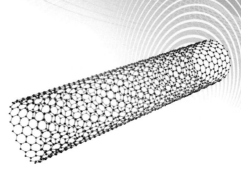

Figure 1 The structure of a single-walled carbon nanotube. Some types have a double wall.

Science in action

You can now buy pyjamas containing silver nanoparticles from high street shops. Some of the marketing for these pyjamas has been for hospital patients, with some evidence that they may help to protect patients from infection by the MRSA bacterium.

Science in action

Self-cleaning windows have a surface coating of nanoparticles. When there is daylight on the windows, energy from the Sun allows the nanoparticles to break down the dirt. The nanoparticles also affect rain so that rain droplets are spread out in a layer across the surface. This layer of water washes away the broken-down dirt.

Questions

1 Explain what the term nanoparticle means.

2 Explain what the term nanoscience means.

3 List some ways in which the properties of nanoparticles can differ from the bulk material.

4 Why do nanoparticles of a substance have different properties to the bulk material?

5 Carbon nanotubes are very strong and are used in making some golf clubs and tennis racquets. Carbon nanotubes are fullerenes. What are fullerenes?

6 Why are some clothes being made that contain nanoparticles of silver?

7 Give one example of the use of nanoparticles, and explain the benefit of using them, in each of the following areas: **(a)** construction materials **(b)** coatings **(c)** cosmetics such as sunblocks **(d)** catalysts **(e)** sensors **(f)** computers.

8 Describe some of the potential advantages and disadvantages of using nanoparticles.

Assess yourself questions

1 Which of the following substances are:

(a) elements? (b) compounds? *(2 marks)*

Mg, CO, Co, S_8, Br_2, H_2S, C_4H_{10}, Ar, C_2H_6O, silver nitrate, chromium, sulfur dioxide

2 Which of the following substances have:
- an ionic structure
- a simple molecular structure
- a metallic structure
- a giant covalent structure?

(a) magnesium oxide (MgO)

(b) lead (Pb)

(c) diamond (C)

(d) ammonia (NH_3)

(e) carbon disulfide (CS_2)

(f) nickel (Ni)

(g) silicon dioxide (SiO_2)

(h) buckminsterfullerene (C_{60})

(i) calcium bromide ($CaBr_2$)

(j) iron sulfide (FeS) *(10 marks)*

3 Calcium chloride ($CaCl_2$) is an ionic compound. It has a high melting point of 772 °C. It does not conduct electricity as a solid, but does conduct electricity when melted or dissolved in water.

(a) Explain why calcium chloride has a high melting point. *(2 marks)*

(b) Explain why calcium chloride does not conduct electricity as a solid but does when melted or dissolved in water. *(3 marks)*

(c) Copy and complete Figure 1 to give the electronic structure of the calcium ions in calcium chloride. *(1 mark)*

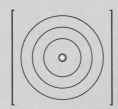

Figure 1 Electron shells.

(d) Copy and complete Figure 1 again to give the electronic structure of the chloride ions in calcium chloride. *(1 mark)*

4 Write the formula of the following ionic compounds.

(a) potassium oxide

(b) aluminium sulfide

(c) magnesium bromide

(d) sodium carbonate

(e) magnesium hydroxide

(f) silver nitrate

(g) iron(III) sulfate

(h) ammonium carbonate *(8 marks)*

5 (a) The electronic structure of the Na^+ ion is 2,8. Give the electronic structure of the following ions:

Mg^{2+}, F^-, S^{2-}, Al^{3+}, K^+, Cl^- *(6 marks)*

(b) What do the electronic structures of all common ions have in common? *(1 mark)*

6 Potassium reacts with iodine to form an ionic compound. Potassium is a group 1 element. Iodine is a group 7 element.

(a) What name are the elements in group 1 known by? Give the charge on the potassium ion and explain why it has this charge. *(2 marks)*

(b) What name are the elements in group 7 known by? Give the charge on the iodide ion and explain why it has this charge. *(2 marks)*

(c) Give the formula of potassium iodide. *(1 mark)*

7 (a) Draw 'stick' diagrams of the following molecules.

(i) H_2O (ii) CO_2 (iii) NH_3 (iv) N_2

(b) A 'stick' diagram is shown of each of the following molecules in Figures 2, 3 and 4. Draw a 'dot-cross' diagram for each to show the outer shell electrons.

(i) 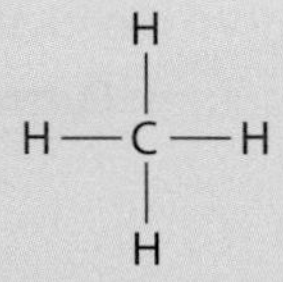
```
        H
        |
    H — C — H
        |
        H
```
Figure 2 Methane.

(ii) Cl — Cl

Figure 3 Chlorine.

(iii) O ═ O

Figure 4 Oxygen.

8 Use the data in Table 1 to decide whether the following substances have simple molecular, ionic, metallic or giant covalent structures.

Table 1 Properties of substances.

	Melting point/°C	Boiling point/°C	Electrical conductivity as solid	Electrical conductivity as liquid	Electrical conductivity as solution
A	735	1435	✗	✓	✓
B	1610	2230	✗	✗	insoluble
C	7	81	✗	✗	insoluble
D	370	914	✗	✓	insoluble
E	114	183	✗	✗	✗
F	1455	2837	✓	✓	insoluble

(7 marks)

9 Graphite, diamond and buckminsterfullerene are all forms of the element carbon. Diamond and graphite have a giant covalent structure whereas buckminsterfullerene is a simple molecular substance with the formula C_{60}.

(a) Diamond and graphite both have very high melting points. Explain why. *(2 marks)*

(b) Graphite conducts electricity but diamond does not. Explain this difference. *(2 marks)*

(c) Diamond is very hard while graphite is soft. Explain this difference. *(3 marks)*

(d) Buckminsterfullerene has a lower melting point than diamond. Explain why. *(2 marks)*

10 (a) Iron is a metal. Metals have high melting points, conduct electricity as solids and can be bent and shaped.

  (i) Explain why iron conducts electricity. *(2 marks)*

  (ii) Explain why iron has a high melting point. *(2 marks)*

  (iii) Explain why iron can be bent and shaped. *(2 marks)*

(b) Steels are alloys of iron. Alloys are harder than pure metals.

  (i) What is an alloy? *(2 marks)*

  (ii) Why are alloys harder than pure metals? *(2 marks)*

(c) The data in Table 2 shows how the strength of steel varies with the percentage of carbon in the steel. Strength is measured here as tensile strength which is where the steel is stretched until it breaks. The units are meganewtons (MN), where 1 MN = 1 000 000 N.

Table 2 Carbon in steel.

Percentage C in steel	0.15	0.3	0.6	0.8
Tensile strength/MN	15	39	69	92

  (i) Plot a graph of strength against percentage of carbon. Draw a line of best fit. *(4 marks)*

  (ii) Describe the relationship between the percentage of carbon and the strength of the steel. *(1 mark)*

  (iii) The tensile strength of steel containing 3% carbon is 20 MN. Comment on this value in the light of the data on your graph. *(2 marks)*

(d) Shape memory alloys (SMAs) are examples of smart alloys. These alloys will return to their original shape when heated. A good example is nitinol, an alloy of nickel and titanium. Give a use of SMAs and describe how the special properties of SMAs allow this use. *(2 marks)*

11 Some clothes are now made containing nanoparticles of silver. Silver as a bulk material is an unreactive metal, but nanoparticles of silver have different properties and are added to clothes to kill bacteria. Some people are concerned that silver nanoparticles might be harmful even though bulk silver is safe.

(a) What are nanoparticles? *(1 mark)*

(b) Why do nanoparticles have different properties to bulk materials? *(1 mark)*

(c) Why might nanoparticles of silver be harmful even if bulk silver is safe? *(1 mark)*

12 Table 3 Some information about four polymers.

Polymer	low-density poly(ethene)	high-density (poly)ethene	isotactic poly(propene)
Monomer	ethene	ethene	propene
Conditions under which polymer is made	temperature: 200°C pressure: 2000 atm catalyst: trace of oxygen	temperature: 60°C pressure: 2 atm catalyst: Ziegler-Natta	temperature: 60°C pressure: 2 atm catalyst: Ziegler-Natta
Properties	flexible, soft	hard, strong	strong
Uses	bags	buckets, bottles	rope, carpet

(a) High-density poly(ethene) and low-density poly(ethene) have different properties and uses. High-density poly(ethene) is harder and stronger. In what way is their production different that leads to this difference in properties? *(1 mark)*

(b) High-density poly(ethene) and isotactic poly(propene) have different properties and uses. Isotactic poly(propene) is more flexible. In what way is their production different that leads to this difference in properties? *(1 mark)*

(c) The three polymers in Table 3 are thermosoftening. These polymers soften and melt on heating, and can be remoulded. Some other polymers are thermosetting, and do not soften or melt on heating and cannot be remoulded. Explain this difference in terms of structure and bonding.
In this question you will be assessed on using good English, organising information clearly and using specialist terms where appropriate. *(6 marks)*

13 (a) Describe some simple tests to show that a substance is ionic.

(b) Describe some simple tests to show that a substance is simple molecular.

The structure and mass of atoms

The structure of atoms

Atoms are made of three smaller subatomic particles called **protons**, **neutrons** and **electrons**. Table 1 shows the relative mass and electric charge of these particles. Both mass and charge are measured relative to a proton, which is assigned a mass of 1 and a charge of +1.

Table 1 The mass and charge of the main subatomic particles.

Subatomic particle	Relative charge	Relative mass
proton	+1	1
neutron	0 (neutral)	1
electron	−1	0.0005 (negligible)

At the centre of the atom is a tiny **nucleus**. The nucleus contains all the atom's protons and neutrons, so most of the mass of the atom is concentrated there. The electrons move around the nucleus occupying energy levels (shells).

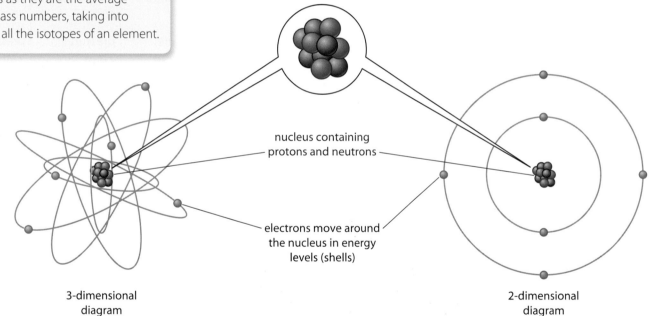

nucleus containing protons and neutrons

electrons move around the nucleus in energy levels (shells)

3-dimensional diagram

2-dimensional diagram

Figure 1 Close-up of an atom. Note that the sizes of the protons, neutrons and electrons are not to scale.

Atomic mass

Different atoms have different masses because they contain different numbers of subatomic particles. The mass of atoms is important because it allows scientists to work out what mass of chemicals to use in reactions.

Atoms are described by their **mass number** and **atomic (proton) number**, which can be shown as follows.

Mass number = number of protons + number of neutrons ⟶ 23

Atomic number (proton number) = number of protons ⟶ 11

Na

Figure 2 Definitions of mass number and atomic number.

The mass number and atomic number can be used to work out how many protons, neutrons and electrons there are in an atom. Remember that atoms are neutral, so the number of electrons always equals the number of protons.

For example, in $_{11}^{23}$Na, the number of:

protons   = atomic number   = 11

neutrons = mass number − atomic number = 23 − 11 = 12

electrons = atomic number   = 11

Isotopes

The number of protons in an atom determines which element it is. Atoms of different elements have different numbers of protons. For example, all atoms with 17 protons are chlorine atoms.

Isotopes are atoms of the same element with different mass numbers. In other words, they are atoms with the same number of protons but different numbers of neutrons. For example, there are two isotopes of chlorine, ^{35}Cl and ^{37}Cl.

Relative atomic mass (A_r)

The mass number of atoms tells us the relative mass of an individual atom. The mass of an atom is measured relative to one atom of the main carbon isotope, ^{12}C, which has a mass of exactly 12.

Most elements consist of a mixture of isotopes, so the relative mass of an element is an average figure based on the abundance of each isotope. It is called the **relative atomic mass** (A_r). For example, about three-quarters of chlorine atoms are ^{35}Cl with a relative mass of 35, while one-quarter of chlorine atoms are ^{37}Cl atoms with a relative mass of 37. The average (mean) mass of all chlorine atoms, the relative atomic mass (A_r), is 35.5.

Taking it further

Scientists now know that protons and neutrons are made up of even smaller particles. These smaller particles are called quarks.

Table 2 The two isotopes of chlorine.

Isotope	$_{17}^{35}$Cl	$_{17}^{37}$Cl
Atomic number	17	17
Mass number	35	37
Protons	17	17
Neutrons	18	20
Electrons	17	17

Science in action

Ideas about the structure of the atom have changed a great deal over time. The model changed as first electrons, then protons and the nucleus, energy levels and neutrons were discovered in turn. The current model is still developing as scientists learn more.

Questions

1   Explain what the mass number and the atomic number of an atom represent.

2   Explain why all atoms are neutral.

3   Give the number of protons, neutrons and electrons in the following atoms.

   $_{9}^{19}$F $_{18}^{40}$Ar $_{19}^{39}$K

4   $_{35}^{79}$Br and $_{35}^{81}$Br are isotopes of bromine. **(a)** By considering the number of subatomic particles in each atom, explain why these two atoms are isotopes. **(b)** Explain why both atoms are bromine atoms. **(c)** About half of all bromine atoms are ^{79}Br and half are ^{81}Br. What is the relative atomic mass of bromine?

5   Identify the element: **(a)** whose atoms contain 16 protons; **(b)** whose atoms contain 19 electrons; and **(c)** with some atoms having atomic number 6 and mass number 14.

6   Explain why the mass number of an atom is a whole number but relative atomic mass is not.

7   Explain why the mass of electrons is regarded as being negligible when considering the mass of atoms.

8   Describe the structure of atoms in detail.

A*

Taking it further

Isotopes of the same element have the same chemical properties. This is because they have the same electronic structure. For example, ^{35}Cl and ^{37}Cl atoms have the same chemical properties as they both have the electronic structure 2,8,7.

The mole

Relative formula mass (M_r)

The **relative formula mass** (M_r) of a substance is the sum of the relative atomic masses (A_r) of all the atoms shown in its formula. For example, H_2O contains two hydrogen atoms ($A_r = 1$) and one oxygen atom ($A_r = 16$) and so has a relative formula mass of 18.

Percentage by mass of elements in compounds

Water has a relative formula mass of 18. Most of this mass is from the oxygen, which has a relative mass of 16. This means that 16 18ths, which is 88.9%, of the mass of the water molecule, is oxygen.

We can work out the percentage by mass of any element in a compound using the following equation:

$$\text{Percentage by mass of an element in a compound} = 100 \times \frac{\text{relative mass of all the atoms of that element}}{M_r}$$

e.g. percentage of O in $H_2O = 100 \times \frac{16}{18} = 88.9\%$

e.g. percentage O in $Ca(NO_3)_2 = 100 \times \frac{(6 \times 16)}{164} = 58.5\%$

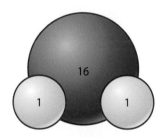

Calculate the M_r of water, H_2O.
(Relative atomic masses: H = 1, O = 16.)
$M_r = (2 \times 1) + 16 = 18$

Figure 1 Calculating the relative formula mass of water.

One mole of a substance

A pair is two of something. A dozen is twelve of something. And in science, a **mole** is 602 204 500 000 000 000 000 000 of something! Scientists count particles using moles. A mole of particles of many substances fits well into a boiling tube or beaker. For example, a mole of water molecules has a mass of 18 g. There are about 15 moles of water in a glass of water.

The number 602 204 500 000 000 000 000 000 was carefully chosen so that the mass of that number of particles equals the relative formula mass (M_r) in grams. For example, the M_r of water is 18 and so the mass of 1 mole of water molecules is 18 g. The M_r of carbon dioxide is 44, so the mass of 1 mole of carbon dioxide molecules is 44 g.

If 1 mole of water molecules has the mass 18 g, then 2 moles has the mass 36 g (2 × 18). The general equation is shown below. This can be remembered by thinking of 'Mr Moles'.

$$\text{Mass (g)} = M_r \times \text{moles}$$

When calculating the number of moles, the mass must be measured in grams. Some common conversion factors are shown in Table 1.

This man is drinking about 15 moles of water molecules.

Table 1 Conversion factors.

1 mg	= 0.001 g
1 kg	= 1000 g
1 tonne	= 1 000 000 g

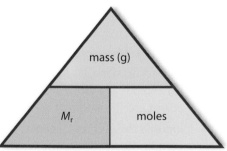

Figure 2 This triangle relates mass, M_r and moles. By covering the quantity you need to calculate you can find the equation to use.

Table 2 shows some example calculations using this equation.

Table 2 Calculations using mass (g) = M_r × moles.

Substance	Formula	M_r	Mass	Moles
Water	H_2O	18	36 g	$= \dfrac{\text{mass (g)}}{M_r} = \dfrac{36}{18} = 2$
Calcium carbonate	$CaCO_3$	100	1 tonne (1 million g)	$= \dfrac{\text{mass (g)}}{M_r} = \dfrac{1\,000\,000}{100} = 10\,000$
Carbon dioxide	CO_2	44	$= M_r \times \text{moles}$ $= 44 \times 0.25$ $= 11\,g$	0.25
Ammonium nitrate	NH_4NO_3	80	$= M_r \times \text{moles}$ $= 80 \times 150$ $= 12\,000\,g$	150
Unknown gas	unknown	$= \dfrac{\text{mass (g)}}{\text{moles}} = \dfrac{3.2}{0.10} = 32$	3.2 g	0.10
Unknown solid	unknown	$= \dfrac{\text{mass (g)}}{\text{moles}} = \dfrac{0.61}{0.005} = 122$	0.61 g	0.005

Questions

1 Write the following masses in grams: **(a)** 20 kg **(b)** 5 tonnes **(c)** 50 mg.

2 Calculate the relative formula mass (M_r) of the following substances. (A_r values: H = 1, N = 14, O = 16, Na = 23, Mg = 24, Al = 27, S = 32, Cl = 35.5, K = 39, Ca = 40, Fe = 56, Cu = 63.5.)

   (a) O_2 **(b)** Na **(c)** S_8 **(d)** NH_3 **(e)** $FeCl_3$ **(f)** $Ca(OH)_2$
   (g) $Mg(NO_3)_2$ **(h)** NH_4NO_3 **(i)** $Al_2(SO_4)_3$ **(j)** $CuSO_4.5H_2O$

3 Calculate the percentage by mass of the element shown in the following: (A_r values: H = 1, N = 14, O = 16, Mg = 24, S = 32, K = 39, Ca = 40, Fe = 56, Cu = 63.5, Br = 80.)

   (a) Mg in $MgBr_2$ **(b)** Fe in Fe_2O_3 **(c)** N in NH_4NO_3
   (d) O in $Ca(NO_3)_2$ **(e)** O in $CuSO_4.5H_2O$

4 Calculate the mass of 1 mole of the following substances (A_r values: H = 1, N = 14, O = 16, Na = 23, Mg = 24, Al = 27, S = 32, Cl = 35.5.)

   (a) Mg **(b)** Cl_2 **(c)** Na_2O **(d)** $Al(OH)_3$ **(e)** $(NH_4)_2SO_4$

5 Rewrite the equation linking mass, moles and M_r to show: **(a)** moles as a subject **(b)** M_r as the subject.

6 How many moles are there in each of the following substances? (A_r values: C = 12, O = 16, Na = 23, S = 32, Fe = 56.)

   (a) 8 g of SO_3 **(b)** 371 g of Na_2CO_3 **(c)** 1 kg of Fe_2O_3

7 What mass would the following quantities have? (Relative atomic masses: H = 1, N = 14, O = 16, Mg = 24, Cl = 35.5.)

   (a) 2.5 moles of Cl_2 **(b)** 20 moles of NH_4OH **(c)** 0.02 moles of $Mg(NO_3)_2$

8 What is the M_r of the following substances? **(a)** Cyclohexane, for which 0.05 moles has a mass of 4.2 g. **(b)** Aspirin, for which 0.001 moles has a mass of 0.18 g.

Examiner feedback

In calculations, a good general rule in chemistry is to work to 3 significant figures.

Science in action

The relative formula mass (M_r) of a substance can be measured by mass spectroscopy. This is an instrumental method of analysis (see lesson C2 3.6).

Examiner feedback

In all chemical calculations, most of the marks are for the method, rather than the final answer. Always show full working and write correct mathematical statements. If you make a slip in a calculation and get the wrong answer, you will get no marks if you have not shown your working, whereas you may only lose 1 mark if you have shown full working.

Reacting-mass calculations

What a chemical equation tells us

It is important that scientists know what masses of chemicals to use in a reaction to make a certain amount of product. These masses can be calculated.

The equation below shows the reaction between hydrogen and nitrogen to make ammonia. It tells you how many particles of each substance are involved in the reaction: one mole of nitrogen (N_2) molecules reacts with three moles of hydrogen (H_2) molecules to make two moles of ammonia (NH_3) molecules.

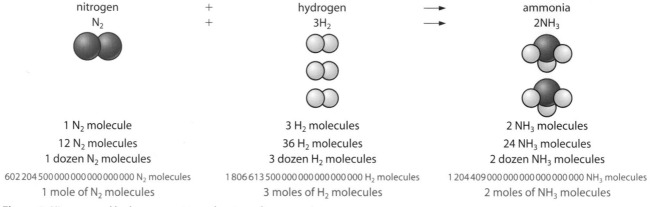

| nitrogen | + | hydrogen | → | ammonia |
| N_2 | + | $3H_2$ | | $2NH_3$ |

1 N_2 molecule	3 H_2 molecules	2 NH_3 molecules
12 N_2 molecules	36 H_2 molecules	24 NH_3 molecules
1 dozen N_2 molecules	3 dozen H_2 molecules	2 dozen NH_3 molecules
602 204 500 000 000 000 000 000 N_2 molecules	1 806 613 500 000 000 000 000 000 H_2 molecules	1 204 409 000 000 000 000 000 000 NH_3 molecules
1 mole of N_2 molecules	3 moles of H_2 molecules	2 moles of NH_3 molecules

Figure 1 Nitrogen and hydrogen react together to make ammonia.

Calculating reacting masses

There are two general ways to calculate the mass of chemicals that react together or are produced in a reaction. One method uses ratios based on relative formula masses and the other method calculates the number of moles taking part in the reaction.

Table 1 Calculating masses.

Using ratios	Using moles
• Calculate the M_r of the substance whose mass you are given and the substance whose mass you are calculating.	• Calculate the number of moles of the substance whose mass is given (moles = mass (g) ÷ M_r).
• Multiply these M_r values by the balancing numbers in the chemical equation to find the mass ratio.	• Use the chemical equation to work out how many moles of the substance asked about are used or made.
• Work out the mass that would react with/be made from 1 g of the substance whose mass you are given using the mass ratio.	• Calculate the mass of the substance asked for (mass (g) = M_r × moles).
• Scale this up from 1 g to the mass you are given.	

Example 1

Iron is made when aluminium reacts with iron oxide. This reaction is used to weld railway lines together. What mass of aluminium is needed to react with 640 g of iron oxide? (Relative atomic masses: O = 16, Al = 27, Fe = 56.)

$$Fe_2O_3 + 2Al \longrightarrow 2Fe + Al_2O_3$$

M_r Fe_2O_3 = 160, Al = 27	Moles of $Fe_2O_3 = \dfrac{mass\ (g)}{M_r} = \dfrac{640}{160} = 4$
Fe_2O_3 reacts with 2 Al	Moles of Al = moles of Fe_2O_3 × 2 = 4 × 2 = 8
160 g of Fe_2O_3 reacts with 54 g (2 × 27) of Al	Mass of Al = M_r × moles = 27 × 8 = 216 g
1 g of Fe_2O_3 reacts with $\dfrac{54}{160}$ g of Al	
640 g of Fe_2O_3 reacts with 640 × $\dfrac{54}{160}$ g of Al = 216 g	

Example 2

Calcium hydroxide (slaked lime) is used by farmers to **neutralise** acidic soil. Calcium hydroxide is made by adding water to calcium oxide (quicklime). What mass of calcium hydroxide is made from 14 kg of calcium oxide? (Relative atomic masses: H = 1, O = 16, Ca = 40.)

$$CaO + H_2O \longrightarrow Ca(OH)_2$$

M_r CaO = 56, Ca(OH)$_2$ = 74 CaO makes Ca(OH)$_2$ 56 g of CaO makes 74 g of Ca(OH)$_2$ 1 g of CaO makes $\frac{74}{56}$ g of Ca(OH)$_2$ 14 kg of CaO makes $14 \times \frac{74}{56}$ kg of Ca(OH)$_2$ = 18.5 kg	Moles of CaO = $\frac{\text{mass (g)}}{M_r} = \frac{14\,000}{56} = 250$ Moles of Ca(OH)$_2$ = moles of CaO = 250 Mass of Ca(OH)$_2$ = $M_r \times$ moles = 74 × 250 = 18 500 g = 18.5 kg

Example 3

Titanium is a metal. One of its uses is to make replacement hip joints. Titanium can be made by reacting titanium chloride with sodium. What mass of titanium chloride reacts with 460 g of sodium? (Relative atomic masses: Na = 23, Cl = 35.5, Ti = 48.)

$$TiCl_4 + 4\,Na \longrightarrow Ti + 4\,NaCl$$

M_r Na = 23, TiCl$_4$ = 190 4Na reacts with TiCl$_4$ 92 g (4 × 23) of Na reacts with 190 g of TiCl$_4$ 1 g of Na reacts with $\frac{190}{92}$ g of TiCl$_4$ 460 g of Na reacts with $460 \times \frac{190}{92}$ g of TiCl$_4$ = 950 g	Moles of Na = $\frac{\text{mass (g)}}{M_r} = \frac{460}{23} = 20$ Moles of TiCl$_4$ = moles of Na ÷ 4 = 20 ÷ 4 = 5 Mass of TiCl$_4$ = $M_r \times$ moles = 190 × 5 = 950 g

Questions

1. Describe in words what this balanced equation means: $2H_2 + O_2 \longrightarrow 2H_2O$

2. Calcium oxide is formed when calcium carbonate is heated. What mass of calcium oxide is formed from 50 g of calcium carbonate? (Relative atomic masses: C = 12, O = 16, Ca = 40.)
$$CaCO_3 \longrightarrow CaO + CO_2$$

3. What mass of hydrogen is produced when 96 g of magnesium reacts with hydrochloric acid? (Relative atomic masses: H = 1, Mg = 24.)
$$Mg + 2HCl \longrightarrow MgCl_2 + H_2$$

4. What mass of oxygen reacts with 46 g of sodium? (Relative atomic masses: O = 16, Na = 23.)
$$4Na + O_2 \longrightarrow 2Na_2O$$

5. What mass of water is formed when 1 kg of methane burns? (Relative atomic masses: H = 1, C = 12, O = 16.)
$$CH_4 + 2O_2 \longrightarrow CO_2 + 2H_2O$$

6. Propane (C_3H_8) is often used as the fuel in gas fires and barbecues. What mass of oxygen is needed to burn 110 g of propane? (Relative atomic masses: H = 1, C = 12, O = 16.)
$$C_3H_8 + 5O_2 \longrightarrow 3CO_2 + 4H_2O$$

7. What mass of aluminium reacts with 10.65 g of chlorine to make aluminium chloride? (Relative atomic masses: Al = 27, Cl = 35.5.)
$$2Al + 3Cl_2 \longrightarrow 2AlCl_3$$

8. What mass of aspirin, $C_6H_4(OCOCH_3)COOH$, can be made from 1 g of salicylic acid, $C_6H_4(OH)COOH$? (Relative atomic masses: H = 1, C = 12, O = 16.)
$$C_6H_4(OH)COOH + (CH_3CO)_2O$$
$$\longrightarrow C_6H_4(OCOCH_3)COOH + CH_3COOH$$

Reaction yields

Why don't reactions produce as much as expected?

If you react 4 g (2 moles) of hydrogen with 32 g (1 mole) of oxygen, you should in theory make 36 g (2 moles) of water. In a reaction, the atoms in the **reactants** are rearranged to make the **products**, so there should be the same mass of products as of reactants. However, if you do this reaction you will almost certainly end up with less than 36 g of water.

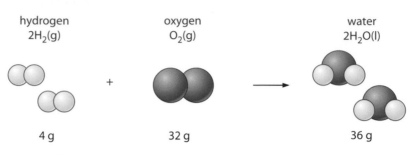

hydrogen $2H_2(g)$		oxygen $O_2(g)$		water $2H_2O(l)$
4 g	+	32 g	→	36 g

Figure 1 Chemical reaction to form water.

There are a number of reasons why you do not get as much product as you might expect.

1 When you carry out a reaction in the laboratory, some chemicals are lost along the way. For example, some material may get stuck to the sides of a test tube or flask.

2 Sometimes other reactions can take place as well the reaction you want. For example, when methane is burned not all of it reacts to form carbon dioxide because other reactions make carbon monoxide or carbon (soot) instead.

3 Some reactions are **reversible** and the products can turn back into the reactants.

Percentage yield

The amount of product obtained from a reaction is known as the yield. The **percentage yield** compares the amount that is actually produced with the amount expected in theory.

$$\% \text{ yield} = \frac{\text{mass of product obtained}}{\text{maximum theoretical mass of product}} \times 100$$

For example, if you only obtain 27 g of water from the reaction in Figure 1, where the maximum theoretical yield is 36 g, then the percentage yield is 75%. This means that only 75% of the water that could have been produced has been obtained.

$$\text{Percentage yield} = \frac{27}{36} \times 100 = 75\%$$

The higher the percentage yield of a reaction, the better. A high yield means that fewer raw materials are used to make the same amount of product. This means that the process is more sustainable. Scientists work hard to improve their processes and techniques to obtain the highest possible yield.

When you bake a cake, some of the ingredients get left in the bowl or on the cake tin. The same sort of thing happens in a chemical reaction.

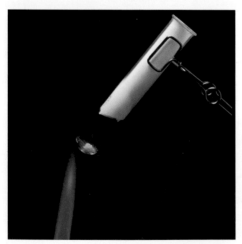

As the white ammonium chloride is heated at the bottom of the tube it breaks apart into ammonia and hydrogen, which react together to re-form ammonium chloride in the cooler parts of the tube.

Reversible reactions

Some chemical reactions are reversible. This means that both the forward and reverse reactions can take place. For example, the gases ammonia and hydrogen chloride react together to form the white solid ammonium chloride. On heating, ammonium chloride breaks up into ammonia and hydrogen chloride.

ammonia + hydrogen chloride $\rightleftharpoons$ ammonium chloride

$$NH_3 \quad | \quad HCl \quad \rightleftharpoons \quad NH_4Cl$$

Route to A*

Some questions may ask you to work out the theoretical yield before the actual yield. The calculation of a theoretical yield is a reacting mass calculation like in lesson C2 3.3. It is important to work carefully to perform this reacting mass calculation before calculating the percentage yield.

Questions

1 Why is it desirable to have a high yield in a chemical reaction?

2 Calculate the percentage yield in each of the following.

	Theoretical maximum mass of product	Mass of product obtained
(a)	10 g	4 g
(b)	2 g	1.6 g
(c)	50 g	39 g

3 When 6 kg of hydrogen reacts with 28 kg of ammonium, you might expect to produce 34 kg of ammonia. In practice, only about 10 kg is formed. **(a)** Explain why you would expect to produce 34 kg of ammonium in this reaction. **(b)** Give three possible reasons why less than 34 kg of product is formed. **(c)** Calculate the percentage yield for this reaction.

4 Quicklime (CaO) is made by the thermal decomposition of calcium carbonate:
$$CaCO_3 \longrightarrow CaO + CO_2$$
(a) Calculate the theoretical mass of quicklime that can be formed from 200 g of calcium carbonate. (Relative atomic masses: C = 12, O = 16, Ca = 40.) **(b)** In a reaction, 108 g of quicklime was obtained from 200 g of calcium carbonate. Calculate the percentage yield.

5 Aluminium is extracted from aluminium oxide, which comes from the ore bauxite, by electrolysis:
$$2Al_2O_3 \longrightarrow 4Al + 3O_2$$
(a) Calculate the theoretical mass of aluminium that can be formed from 1 kg of aluminium oxide. (Relative atomic masses: O = 16, Al = 27.) **(b)** In a reaction, 480 g of aluminium was obtained from 1 kg of aluminium oxide. Calculate the percentage yield.

6 Hydrogen is made by the reaction of methane with steam:
$$CH_4 + H_2O \longrightarrow 3H_2 + CO_2$$
(a) Calculate the theoretical mass of hydrogen that can be formed from 1 kg of methane. (Relative atomic masses: H = 1, C = 12.) **(b)** In a reaction, 250 g of hydrogen was obtained from 1 kg of methane. Calculate the percentage yield.

7 Sulfur trioxide is formed by reaction of sulfur dioxide with oxygen:
$$2SO_2 + O_2 \rightleftharpoons 2SO_3$$
(a) Calculate the theoretical mass of sulfur trioxide that can be formed from 100 g of sulfur dioxide. (Relative atomic masses: O = 16, S = 32.) **(b)** In a reaction, 105 g of sulfur trioxide was formed from 100 g of sulfur dioxide. Calculate the percentage yield. **(c)** Look at the equation for this reaction and give the main reason why the yield is less than 100%.

8 The production of aspirin has a 75% yield. Explain clearly what this means and suggest reasons why it is less than 100%.

C2 3.5

Empirical formulae

Learning objectives

- explain what the empirical formula of a compound represents
- calculate the empirical formula of a compound from the mass or percentage of the elements in the compound.

What is an empirical formula?

All substances have an **empirical formula**. This represents the simplest whole number ratio of atoms (or ions) of each element in a substance. For example, the empirical formula of silicon dioxide (silica) is SiO_2. This means that the ratio of silicon (Si) atoms to oxygen (O) atoms is 1:2; there are twice as many oxygen atoms as silicon atoms.

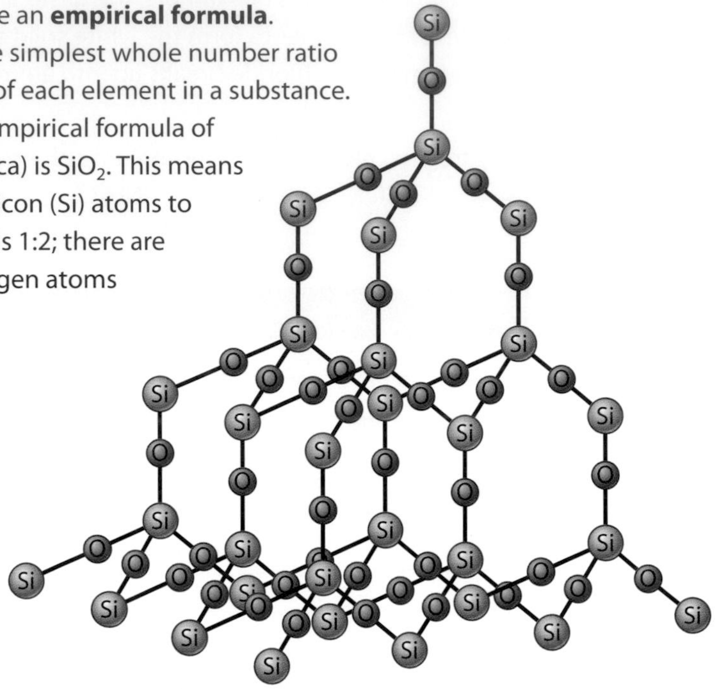

Figure 1 Silicon dioxide (empirical formula = SiO_2).

Examiner feedback

Only substances made of molecules can have a molecular formula. Substances like sodium chloride (NaCl), which is ionic, and copper (Cu), which is metallic, do not have a molecular formula as they are not made of molecules.

Substances that are made of molecules also have a **molecular formula**. This represents the number of atoms of each element in one molecule. For example, ethene has the molecular formula C_2H_4 – this means that there are two carbon (C) and four hydrogen (H) atoms in one molecule. It has the empirical formula CH_2, meaning that the simplest ratio of carbon to hydrogen atoms is 1:2.

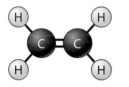

Figure 2 Ethene (molecular formula = C_2H_4, empirical formula = CH_2).

For some molecules, the molecular formula is the same as the empirical formula. For example, the molecular formula of water is H_2O; there are two hydrogen (H) atoms and one oxygen (O) atom in each molecule. The empirical formula is also H_2O, meaning that the simplest ratio of hydrogen to oxygen atoms is 2:1.

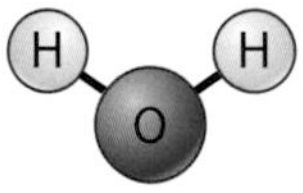

Figure 3 Water (molecular formula = H_2O, empirical formula = H_2O).

Calculating an empirical formula

Substances can be analysed to find which elements they are made from. This analysis gives the mass (or percentage by mass) of each element in the substance. The empirical formula can be calculated from this information as follows.

(a) Make a column for each element.

(b) Divide the mass (or percentage) of each element by its relative atomic mass (A_r).

(c) Simplify this ratio by dividing all the answers by the smallest answer.

(d) Find the simplest whole number ratio. The numbers are from real experiments so may not be exact whole numbers. However, you may need to multiply the answers by 2, 3 or 4 to get close to whole numbers.

(e) Write the empirical formula.

Science in action

Compounds can be analysed by a method called instrumental microanalysis. This provides information about which elements a compound contains and the percentage by mass of each element. This information is used to calculate the empirical formula.

Example 1

Analysis of a compound found in an iron **ore** determined that it contained 70% iron and 30% oxygen by mass. Find the empirical formula of the iron oxide in this ore. (A_r: Fe = 56, O = 16.)

		Fe	O
(a)	Make a column for each element.	Fe	O
(b)	Divide each percentage by the A_r.	$\frac{70}{56} = 1.25$	$\frac{30}{16} = 1.875$
(c)	Simplify this ratio.	$\frac{1.25}{1.25} = 1$	$\frac{1.875}{1.25} = 1.5$
(d)	Find the simplest whole number ratio.	$1 \times 2 = 2$	$1.5 \times 2 = 3$
(e)	Write the empirical formula.	Fe_2O_3	

Example 2

Analysis of a compound found that it contained 2.4 g of carbon, 0.4 g of hydrogen and 3.2 g of oxygen. Find the empirical formula of this compound. (A_r: C = 12, H = 1, O = 16.)

		C	H	O
(a)	Make a column for each element.	C	H	O
(b)	Divide each mass by the A_r.	$\frac{2.4}{12} = 0.2$	$\frac{0.4}{1} = 0.4$	$\frac{3.2}{16} = 0.2$
(c)	Simplify this ratio.	$\frac{0.2}{0.2} = 1$	$\frac{0.4}{0.2} = 2$	$\frac{0.2}{0.2} = 1$
(d)	Find the simplest whole number ratio.	1	2	1
(e)	Write the empirical formula.	CH_2O		

What is the empirical formula of the iron oxide in this ore?

Questions

1. **(a)** Propene has the molecular formula C_3H_6. Explain what this means.
   (b) Propene has the empirical formula CH_2. Explain what this means.

2. A compound is found to contain 40% sulfur and 60% oxygen by mass. Find the empirical formula of the compound. (Relative atomic masses: O = 16, S = 32.)

3. A compound is found to contain 8.9 g of lead and 6.1 g of chlorine. Find the empirical formula of the compound. (Relative atomic masses: Cl = 35.5, Pb = 207.)

4. A compound is found to contain 18.2% potassium, 59.4% iodine and 22.4% oxygen by mass. Find the empirical formula of the compound. (Relative atomic masses: O = 16, K = 39, I = 127.)

5. A compound is found to contain 2.61 g of carbon, 0.65 g of hydrogen and 1.74 g of oxygen. Find the empirical formula of the compound. (Relative atomic masses: H = 1, C = 12, O = 16.)

6. A hydrocarbon is found to contain 81.8% carbon by mass. Find the empirical formula of the hydrocarbon. (Relative atomic masses: H = 1, C = 12.)

7. 1.0 g of aluminium reacts with chlorine to form 4.94 g of aluminium chloride. Calculate the empirical formula of aluminium chloride. (Relative atomic masses: Al = 27, Cl = 35.5.)

8. Some iron wool is placed in a crucible and heated until the mass stops increasing. Data is shown below. (Relative atomic masses: O = 16, Fe = 56.)
   Mass of empty crucible = 25.27 g
   Mass of crucible plus iron wool = 25.67 g
   Mass of crucible and contents at the end = 25.84 g
   (a) Why does the mass of the crucible and its contents increase?
   (b) Why was the crucible heated until the mass stopped increasing?
   (c) Calculate the empirical formula of the iron oxide formed.

Ⓐ*

Practical

Magnesium oxide can be made by heating magnesium ribbon in a limited oxygen supply, using the apparatus in Figure 4. If the reactant and product are weighed, the empirical formula for magnesium oxide can be calculated.

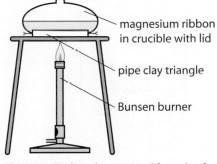

magnesium ribbon in crucible with lid

pipe clay triangle

Bunsen burner

Figure 4 Finding the empirical formula of magnesium oxide.

Taking it further

For substances made from molecules, the molecular formula can be worked out using the relative formula mass and the empirical formula.

Analysis

Why analyse?

It is very important that we can analyse substances to identify what they contain and to measure quantities. Substances that are commonly tested include samples of air, water, food and medicines. Air and water may be tested to check levels of pollutants, while food and medicines are analysed to make sure they do not contain anything that could be harmful.

There are many different methods of analysis. They include different types of chromatography.

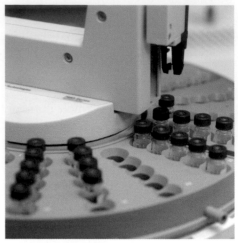

Instrumental analysis.

Paper chromatography

Paper chromatography has been used for many years to analyse coloured substances. Chromatography separates the different dyes or **pigments** in a coloured substance.

Artificial colourings added to foods can be analysed in this way. A sample of the colour from the food is placed on a piece of chromatography paper along with samples of known dyes, then a solvent is added. The solvent soaks up the paper, taking the dyes with it. The more soluble a colouring is, the further up the paper it travels. Different colourings move different distances up the paper.

Instrumental methods

Many **instrumental techniques** have been developed by scientists to analyse substances. In an instrumental method, a sample is placed inside a device that performs some sort of analysis. These methods are fast, accurate and very sensitive. They can be used to test very small samples.

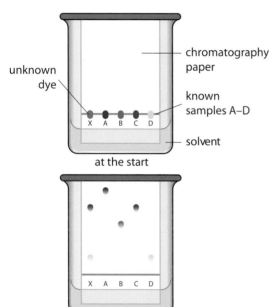

Figure 1 Analysis of unknown dye X by paper chromatography. The analysis shows that X contains two separate dyes, the known substances C and D.

Gas chromatography

A very common instrumental method is **gas chromatography**. The sample is injected into the machine and vaporised. The sample passes through a long column packed with a solid that is wound into a coil. An inert (unreactive) gas, such as nitrogen, is passed through the column to move the sample through. Different substances in the mixture travel through the column at different speeds. This means that they reach the end of the column at different times and so are separated.

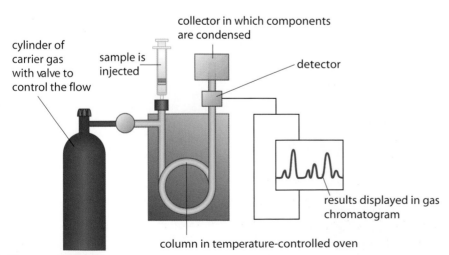

Figure 2 Gas chromatography.

The time taken for a substance to reach the detector at the end of the column is called its **retention time**. A gas chromatogram is produced that shows how many compounds are in the sample and the retention time of each one. The retention time can help to identify the substance. In the example in Figure 3, three compounds have been separated and detected.

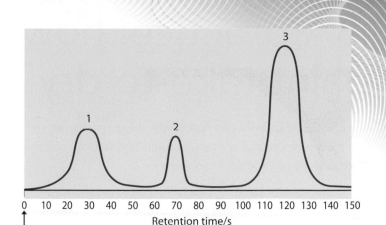

Figure 3 This sample contains three substances.

Mass spectroscopy

Another instrumental technique is **mass spectroscopy**. A mass spectrometer can identify tiny amounts of a substance very quickly and accurately. It does this by measuring the mass of the particles in the substance.

In mass spectroscopy the molecules lose an electron to form a **molecular ion**. The relative formula mass of the substance equals the mass of the molecular ion. Often, the molecular ion breaks apart and other lighter fragments are detected as well.

A mass spectrometer is often attached to a gas chromatography machine. After a sample has been separated by chromatography, the mass spectrum of each compound can be recorded, allowing it to be identified.

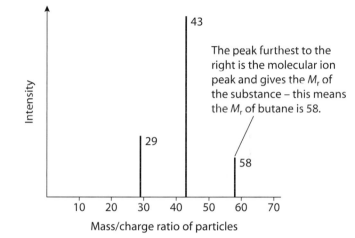

The peak furthest to the right is the molecular ion peak and gives the M_r of the substance – this means the M_r of butane is 58.

Figure 4 An example of a mass spectrum. This is butane.

Questions

1 Look at Figure 5. On the left it shows the chromatogram of the artificial colourings used to colour some sweets. On the right it shows the chromatogram of the sweets themselves. **(a)** Which sweets contain one colouring? **(b)** Which sweet contains the most colourings and which ones does it contain? **(c)** Which colourings are in the yellow sweets?

Figure 5 Chromatograms of sweets.

2 How does paper chromatography separate the substances in a mixture?

3 How does gas chromatography separate the substances in a mixture?

4 Look at the gas chromatogram in Figure 3. **(a)** Record the retention time of each substance. **(b)** Use the data table opposite to identify which substances are in the sample. **(c)** Which substance is there most of in the sample?

Substance	P	Q	R	S	T
Retention time / s	70	135	56	118	30

5 In mass spectroscopy, a molecular ion is formed. What is a molecular ion?

6 What key piece of information do we find from mass spectroscopy in order to help identify a compound?

7 How can mass spectroscopy be used with gas chromatography to help identify substances?

8 What is instrumental analysis, what is it used for and why is it so useful?

ISA practice: how long does paint take to dry?

The paint used to paint cars reacts with oxygen in the air, so that it becomes solid. Car manufacturers and repairers often place a newly painted car in an oven.

A manufacturer of paint for cars has asked some students to investigate the effect of temperature on the time it takes for paint to become solid.

Section 1

1   Write a hypothesis about how you think temperature will affect the rate of reaction. Use information from your knowledge of rates of reaction to explain why you made this hypothesis. *(3 marks)*

2   Describe how you could carry out an investigation into this factor.

    You should include:

    • the equipment that you would use
    • how you would use the equipment
    • the measurements that you would make
    • how you would make it a fair test.

    You may include a labelled diagram to help you explain the method.

    In this question you will be assessed on using good English, organising information clearly and using specialist terms where appropriate. *(6 marks)*

3   Think about the possible hazards in the investigation.

    (a) Describe *one* hazard that you think may be present in the investigation. *(1 mark)*

    (b) Identify the risk associated with the hazard that you have described, and say what control measures you could use to reduce the risk. *(2 marks)*

4   Design a table that you could use to record all the data you would obtain during the planned investigation.
    (2 marks)

    Total for Section 1: 14 marks

Section 2

Study Group 1 was two students, who carried out an investigation into the hypothesis. They used car spray paint and decided the paint was solid when a match would no longer scratch the surface. Figure 1 shows their results.

Paint samples dried at different temperatures
20°C: paint solidified after 120 min
50°C: paint solidified after 50 min
70°C: paint solidified after 34 min
90°C: paint solidified after 26 min
110°C: paint solidified after 20 min

Figure 1 Study Group 1's results.

5   **(a)** Plot a graph of these results. *(4 marks)*

    (b) What conclusion can you make from the investigation about a link between the temperature and the rate of reaction?
    You should use any pattern that you can see in the results to support your conclusion. *(3 marks)*

    (c) Look at your hypothesis, the answer to question 1. Do the results support your hypothesis?

    Explain your answer. You should quote some figures from the data in your explanation. *(3 marks)*

Below are the results from three more study groups.

Study Group 2 was another group of two students. Figure 2 shows their results.

Paint samples dried at different temperatures
20°C: paint solidified after 115 min
40°C: paint solidified after 58 min
60°C: paint solidified after 28 min
80°C: paint solidified after 25 min
100°C: paint solidified after 25 min

Figure 2 Study Group 2's results.

Study Group 3 was a third group of students. Their results are given in Table 1.

Table 1 Results from Study Group 3.

Temperature at which paint is solidified/°C	Time for paint to solidify/min			
	Test 1	Test 2	Test 3	Mean of tests
15	145	142	138	142
30	95	98	92	95
45	75	102	73	83
60	38	36	34	36
75	29	26	27	27

Study Group 4 is a group of researchers who looked on the internet and found a graph showing how temperature affects the rate of reaction between a metal and sulfuric acid (Figure 3).

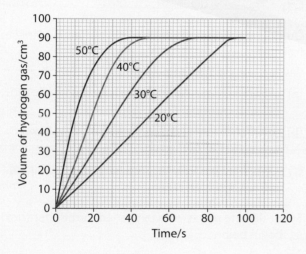

Figure 3 Graph found by Study Group 4.

6 **(a)** Draw a sketch graph of the results from Study Group 1. *(3 marks)*

(b) Look at the results from Study Groups 2 and 3. Does the data support the conclusion you reached about the investigation in question 5(a)? Give reasons for your answer. *(3 marks)*

(c) The data contain only a limited amount of information. What other information or data would you need in order to be more certain whether the hypothesis is correct or not?
Explain the reason for your answer. *(3 marks)*

(d) Look at Study Group 4's results. Compare them with the data from Study Group 1. Explain how far the data supports or does not support your answer to question 5(b). You should use examples from Study Group 4's results and from Study Group 1. *(3 marks)*

7 **(a)** Compare the results of Study Group 1 with Study Group 2. Do you think that the results for Study Group 1 are *reproducible*?
Explain the reason for your answer. *(3 marks)*

(b) Explain how Study Group 1 could use results from other groups in the class to obtain a more *accurate* answer. *(3 marks)*

8 Applying the results of the investigation to a context.

Suggest how ideas from the original investigation and the other studies could be used by the manufacturers to decide on the best temperature at which to operate the car paint oven. *(3 marks)*

Total for Section 2: 31 marks

Total for the ISA: 45 marks

Assess yourself questions

1 Atoms of fluorine have an atomic number of 9 and a mass number of 19. Explain what the terms *atomic number* and *mass number* mean. *(2 marks)*

2 An atom of phosphorus can be represented as

$$^{31}_{15}P$$

(a) What is the atomic number of this atom? What is the mass number? *(2 marks)*

(b) How many protons, neutrons and electrons are there in this atom? *(3 marks)*

3 Magnesium atoms consist of a mixture of isotopes. The main isotopes are ^{24}Mg, ^{25}Mg and ^{26}Mg.

(a) How many protons do these three isotopes contain? How many neutrons do these contain? *(2 marks)*

(b) What are isotopes? *(2 marks)*

(c) (i) The relative atomic mass of magnesium is 24.3. Explain why this is not a whole number. *(1 mark)*

(ii) The masses of all atoms are measured relative to another atom. Which atom is this? *(1 mark)*

4 Identify the following atoms by giving the symbol with mass number (e.g. ^{24}Mg). You will need to use the periodic table to help.

(a) An atom with 13 protons and 14 neutrons.

(b) An atom with 53 protons and 74 neutrons.

(c) An atom with 35 protons and 44 neutrons.

(d) An atom with atomic number 18 and mass number 40. *(4 marks)*

5 Using the periodic table calculate the relative formula mass, M_r, of the following substances.

(a) CO_2

(b) H_2SO_4

(c) Br_2

(d) $Al(OH)_3$

(e) $Fe_2(SO_4)_3$ *(5 marks)*

6 Calculate the percentage by mass of:

(a) Mg in $MgBr_2$

(b) Fe in Fe_2O_3

(c) N in NH_4NO_3

(d) O in $Fe(OH)_3$

(e) O in $Ca(NO_3)_2$ *(5 marks)*

7 Calculate the mass of 1 mole of:

(a) Na

(b) Na_2CO_3

(c) $(NH_4)_2SO_4$ *(3 marks)*

8 Propane (C_3H_8) is often used as the fuel in gas fires and barbecues. What mass of oxygen is needed to burn 110 g of propane? *(3 marks)*

$$C_3H_8 + 5O_2 \longrightarrow 3CO_2 + 4H_2O$$

9 Ammonium sulfate is a very good fertiliser. It is made by the reaction of ammonia with sulfuric acid. How much ammonium sulfate fertiliser can be made from 1 kg of ammonia? *(3 marks)*

$$2NH_3 + H_2SO_4 \longrightarrow (NH_4)_2SO_4$$

10 Aluminium is made by the electrolysis of aluminium oxide. What mass of aluminium is formed from 100 g of aluminium oxide? *(3 marks)*

$$Al_2O_3 \longrightarrow 2Al + 3O_2$$

11 A common way to make hydrogen is to react methane with steam. What mass of hydrogen can be formed from 40 g of methane? *(3 marks)*

$$CH_4 + H_2O \longrightarrow CO + 3H_2$$

12 Quicklime (calcium oxide) is made by heating limestone (calcium carbonate) in a lime kiln.

$$CaCO_3 \longrightarrow CaO + CO_2$$

(a) What mass of calcium oxide is formed from 150 g of calcium carbonate? *(3 marks)*

(b) In practice, only 80 g of calcium oxide is formed. Calculate the percentage yield. *(1 mark)*

(c) Give *three* possible reasons why the percentage yield is less than 100%. *(3 marks)*

13 Sulfur dioxide reacts with oxygen to form sulfur trioxide as shown below.

$$2SO_2 + O_2 \longrightarrow 2SO_3$$

(a) What mass of sulfur trioxide is formed from 32 g of sulfur dioxide? *(3 marks)*

(b) In practice, only 35 g of sulfur trioxide is formed. Calculate the percentage yield. *(1 mark)*

14 (a) The molecular formula of glucose is $C_6H_{12}O_6$. The empirical formula of glucose is CH_2O. What is an empirical formula? *(2 marks)*

(b) The molecular formulae of some substances are shown. Give the empirical formula of these substances. *(5 marks)*

(i) benzene = C_6H_6

(ii) hydrazine = N_2H_4

(iii) propane = C_3H_8

(iv) ethanoic acid = $C_2H_4O_2$

(v) butanol = $C_4H_{10}O$

15 Use the data to calculate the empirical formula of the following compounds.

    (a) N = 30.4%, O = 69.6% *(3 marks)*

    (b) Fe = 1.75 g, O = 0.75 g *(3 marks)*

    (c) C = 82.5%, H = 17.2% *(3 marks)*

    (d) C = 38.7%, H = 16.1%, N = 45.2% *(3 marks)*

    (e) P = 4.75 g, H = 0.46 g *(3 marks)*

16 A sample of gallium was burned in oxygen to form gallium oxide. 1.00 g of gallium burned to form 1.34 g of gallium oxide. Calculate the empirical formula of gallium oxide. *(4 marks)*

17 An organic compound containing carbon, hydrogen and oxygen only was found to contain 48.6% carbon and 8.1% hydrogen. Find the empirical formula of this compound. *(2 marks)*

18 Paper chromatography using ethanol as solvent was carried out to find the pigments in a paint. The results are shown in Figure 1. Which pigments does the paint contain? *(1 mark)*

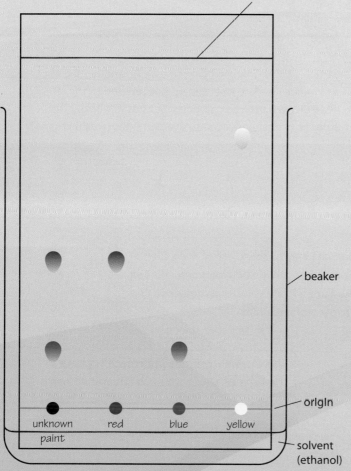

Figure 1 Chromatogram of the pigments in a paint.

19 Figure 2 shows a typical chromatogram for corn oil, a vegetable oil. It shows the fatty acids from the oil. Use the chromatogram to answer the questions that follow.

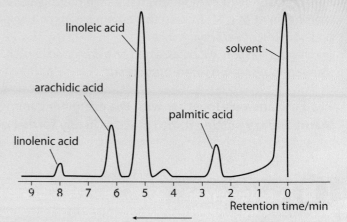

Figure 2 Gas chromatogram of corn oil.

    (a) Which fatty acid travelled through the gas–liquid column fastest? *(1 mark)*

    (b) How many fatty acids were found in the mixture? Which fatty acid was present in the largest amount? *(2 marks)*

    (c) Gas–liquid chromatography is an instrumental method of analysis. Give three advantages of instrumental methods compared with chemical analysis. *(3 marks)*

    (d) Name another instrumental method that can be used to analyse the substances separated by gas–liquid chromatography. *(1 mark)*

20 Mass spectroscopy can be used to find the relative formula mass (M_r) of compounds using the peak for the molecular ion. The mass spectrum of methyl ethanoate is shown. Methyl ethanoate is a solvent used in glues.

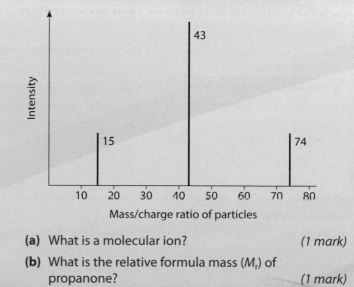

    (a) What is a molecular ion? *(1 mark)*

    (b) What is the relative formula mass (M_r) of propanone? *(1 mark)*

Here are three students' answers to the following question:

Sodium chloride (common salt) has a high melting point of 801 °C. Water has a low melting point of 0 °C. Explain this difference in melting points by discussing the structure and bonding of each substance.

In this question you will be assessed on using good English, organising information clearly and using specialist terms where appropriate. (6 marks)

Read the answers together with the examiner comments. Then check what you have learnt and try putting it into practice in any further questions you answer.

B Grade answer

Student 1

> Sodium chloride has a high melting point because there are strong forces between the particles. The melting point of water is low because there are weak forces between the particles.

The answer does not say what type of particle they are.

Examiner comment

This answer contains correct basic information but lacks detail. This question is about structure and bonding but the student has not mentioned the types of structure and bonding that sodium chloride and water have. The student only refers to particles rather than being more specific and writing about ions and molecules. Also, the student has referred to the strength of attractions between particles, but has not linked this to the amount of energy needed to overcome those attractions.

A Grade answer

Student 2

This answer is better because it specifies which type of particles the substances contain.

> Sodium chloride has an ionic structure. There are strong forces between the ions, which mean it has a high melting point. Water has a simple molecular structure with weak forces between the molecules, meaning it has a low melting point.

This answer includes information about the type of structure each substance has.

Examiner comment

This answer is better because it describes the type of structure that each substance has. It also specifies the type of particles that the forces act between. However, the answer does not link the strength of these forces to the energy required to overcome them.

A* Grade answer

Student 3

> The structure of sodium chloride is ionic. There are strong attractive forces between the positive and negative ions. It takes a lot of energy to overcome these strong attractions so sodium chloride has a high melting point. The structure of water is simple molecular. There are strong covalent bonds between the atoms in each molecule, but weak forces of attraction between the molecules. When melted, it is the weak forces between the molecules that are overcome, which does not take much energy and so it has a low melting point.

> This answer links the strength of the forces between particles to the amount of energy needed to overcome them.

Examiner comment

This answer is excellent. It refers to the type of structure and bonding that each substance has. For sodium chloride, it refers to the 'attractive forces between positive and negative ions' rather than just 'forces between ions'. For water, the answer clearly distinguishes between the strong covalent bonds within the molecule, which are not overcome on melting, and the weak attractive forces between molecules, which are overcome. It also links the strength of these forces to the amount of energy needed to overcome them and so the melting point.

The answer is organised well and makes points in a logical order. It is also well written with good spelling, punctuation and grammar.

- Read the whole question carefully.
- In chemistry, the use of correct terminology can be very important. For example, at GCSE level, rather than referring to particles you should use the terms atoms, molecules or ions as appropriate.
- In questions worth a lot of marks where you are asked to explain something, you must give depth and detail to your answers.
- For longer answer questions, plan your answer before you start writing. Think through and note down the key points you want to make on the exam paper, perhaps by the question, before starting to write your answer.
- Present your answer in a clear, structured way, listing your arguments in a logical order.
- Do your best to use good spelling, punctuation and grammar throughout your answers.
- Avoid using words like 'it' and 'they'. Always be specific by saying what you are referring to.

Rates, salts and electrolysis

In this section, you will find out about how to measure and control the rates of chemical reactions. Temperature, concentration, pressure and surface area are all important factors. Substances called catalysts affect the rates of reactions. They can help to reduce the cost of industrial processes.

You will find out about energy transfers by chemical reactions. Exothermic reactions release energy to the surroundings and are useful for self-heating cans. Endothermic reactions absorb energy from the surroundings and are useful for sports injury packs.

Different salts are useful for different purposes, like ammonium nitrate for fertilisers and copper sulfate for fungicides. You will find out how to make them, and learn how to work out the best way to make a particular salt.

Ionic compounds break down to form simpler substances when electricity is passed through them, if they are molten or dissolved in water. This process of electrolysis is useful for making chemicals from salt and aluminium from aluminium ore. You will find out how to predict what happens when electricity is passed through a particular ionic substance.

Test yourself

1 Describe the structure of an atom.

2 What happens when elements react together?

3 Describe what happens when calcium carbonate reacts with acids.

4 Describe what happens during the complete combustion of methane, a hydrocarbon fuel. Include an equation in your answer.

5 Name the process used to extract aluminium from aluminium oxide, and to purify copper.

Objectives

By the end of this unit you should be able to:

- describe and explain the factors that affect the rate of chemical reactions
- describe the development, advantages and disadvantages of using catalysts in industrial processes
- interpret graphs in terms of the rate of a reaction
- describe and explain the differences between exothermic and endothermic reactions
- evaluate everyday uses of exothermic and endothermic reactions
- describe how different soluble and insoluble salts may be produced
- explain neutralisation in terms of the ions produced by acids and alkalis
- explain the use of electrolysis to manufacture aluminium and chemicals from salt, and to electroplate objects.

Reaction rates

Following the reaction

Chemical reactions may be very slow, like the slow rusting of an old car, or very fast, like the petrol/air explosions that drive the car's engine. Many, of course, lie somewhere between the two extremes.

Either way, reaction rates are very important in industry.

- If a reaction runs too slowly, it will make the process very inefficient and raise production costs.
- If a reaction runs too quickly, it might get out of control and cause an explosion.

First, we must measure...

Reaction rates need to be controlled, but before you can do that you need to be able to measure them. In practical terms, you need to choose something to measure that indicates the progress of the reaction. Then you keep track of that quantity over a sensible time interval. Examples of useful things to measure are:

- the mass of a reactant used up during a reaction
- the volume of a gas (product) produced during a reaction.

The rate of reaction is 'what happens' divided by 'how long it takes', that is:

$$\frac{\text{amount of reactant used}}{\text{time}} \quad \text{or} \quad \frac{\text{amount of product formed}}{\text{time}}$$

Rusting is a relatively slow reaction.

Speed it up

Often in industry you are trying to make a reaction go faster. This can make the industrial process more efficient and so help to make more money for your company. The key factors that affect the rate of a reaction are:

- temperature
- **concentration** (in liquids) or pressure (in gases)
- surface area
- the use of a catalyst.

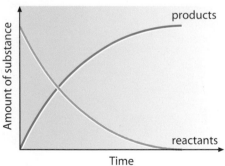

Figure 1 This graph shows how the reactants are used up and the products are formed during a reaction.

Practical

There are several ways of measuring the rate of a reaction. If the reaction produces a gas there are two simple methods we can use. For example, when calcium carbonate reacts with hydrochloric acid to form carbon dioxide gas:

calcium carbonate + hydrochloric acid $\longrightarrow$ calcium chloride + water + carbon dioxide

$CaCO_3$ + $2HCl$ $\longrightarrow$ $CaCl_2$ + H_2O + CO_2

One method is to measure the volume of gas given off over time using a gas syringe. If you plot a graph of gas volume against time, it is steepest at the start. The rate of the reaction is the greatest here, as there are the most acid and marble chips available to react. As the reactants are used up there are less of them. The reaction gets slower and the graph levels off. When the graph is totally flat, the reaction has finished because one or both of the reactants has been used up.

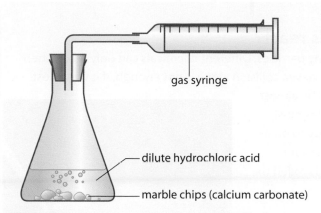

Figure 2 Measuring the volume of gas produced in the reaction between hydrochloric acid and marble chips, a natural form of calcium carbonate.

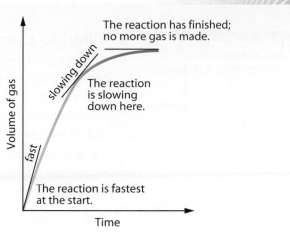

Figure 3 A typical volume/time graph.

A second method is to follow the reaction by measuring how the mass changes as the carbon dioxide is lost into the air. If the experiment is carried out on a balance, the loss of mass can be measured every minute until the reaction finishes. The loss of mass from the flask is the mass of escaping carbon dioxide. A graph of the mass of the carbon dioxide produced over time would look very similar to the volume/time graph, for the same reasons.

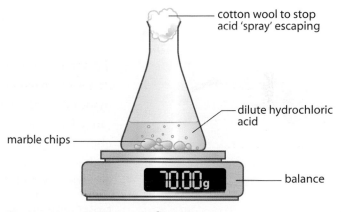

Figure 4 Measuring the mass of escaping gas.

Questions

1 What problems can be caused in industry if chemical reactions are **(a)** too fast or **(b)** too slow?

2 Magnesium ribbon reacts with excess acid. How could you tell if the reaction has finished?

3 Why would it be difficult to measure the rate of reaction for something like the rusting of iron?

4 Gas can be collected in an upside-down measuring cylinder in a bowl of water. Suggest why it is better to use a gas syringe.

5 Copper dissolves in hot nitric acid, giving off nitrogen dioxide gas. How could you measure the rate of this reaction?

6 For the marble chips/acid reaction:
 (a) How do you know that the reaction has finished?
 (b) Marble chips are usually left in the flask at the end of this reaction. Which reactant has been used up? **(c)** Sketch the shape of graph you would get if you plotted the *mass of the apparatus and reactants against time*.
 Explain your graph.

7 Magnesium dissolves in sulfuric acid to give magnesium sulfate, water and hydrogen gas.
 (a) List the reactants and products. **(b)** Describe how you could follow this reaction using a gas syringe. **(c)** Sketch the shape of the graph you would expect to get from this. **(d)** Suggest why it might be more difficult to follow this reaction by mass loss on a balance than the marble/acid reaction.

8 Malachite (copper carbonate) chips dissolve in sulfuric acid to give copper sulfate, water and carbon dioxide. Describe a simple experiment to follow the rate of this reaction using an open beaker and a balance.

The effect of temperature

Why do chemicals react?

All chemicals consist of tiny particles. Different chemicals can only react when these particles collide. However, collision alone is not enough. If it was, most chemical reactions would be almost instantaneous. The particles must collide with enough energy to break the existing chemical bonds and reform in a new way. This is called the **collision theory**.

The minimum energy required for any given reaction is called its **activation energy**. If the particles do not have this activation energy, they cannot react. Many reactions need a 'kick start' of energy to get started.

Matches need a kick start of energy before they catch fire: you have to strike them.

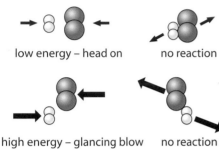

Figure 1 You need the right kind of collision before you get a reaction.

low energy – head on   no reaction

high energy – glancing blow   no reaction

high energy – head on   reaction

Temperature matters

At any given temperature, collisions of many different types occur. They will range from maximum-energy, 'head on' impacts to gentler glancing blows. The higher the temperature, the higher the average speed of the particles.

Faster particles carry more kinetic energy. Collisions that generate enough energy to break the bonds occur more often, so a larger proportion of the particles have a chance to react. This is why raising the temperature speeds up reactions, and cooling slows them down.

Milk turning sour is a chemical reaction.

In the marble chip/acid experiment you could plot a graph of the time it takes to produce the gas against temperature, but it is better to convert the figures to show the rate of production, and plot a graph of that instead, as shown below. The rate of gas production is found by:

$$\text{rate} = \frac{\text{volume of gas/cm}^3}{\text{time taken/s}}$$

Table 1 Results of the experiment.

Temperature/°C	20	32	44	56	68	80
Time to produce 100 cm³/s	400	200	100	50	25	12.5
Rate/cm³/°C		0.25	0.5	1	2	4	8

You may not be surprised to find that the gas is produced more quickly when you raise the temperature.

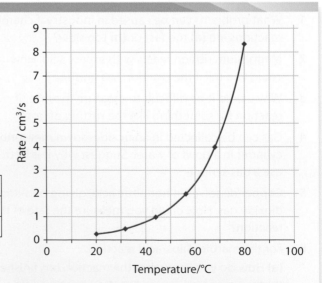

Figure 2 Graph showing rate of reaction plotted against temperature.

Monitoring the reaction

The marble chips (calcium carbonate) in acid experiment (see lesson C2 4.1) can be run at different temperatures by putting the apparatus in a water bath. You could use a gas syringe and time how long it takes to produce $100\,cm^3$ of gas, for example, at different temperatures.

The pace hots up

Looking at a rate/temperature graph you might expect all industrial reactions to be run at as high a temperature as possible. High temperatures are often used, but there are limits, as the products may themselves break down if heated too much.

Cost is also important. You need a lot of energy to reach very high temperatures, and energy costs money. Industrial chemists are always on the lookout for the most cost-effective ways to run their chemical reactions.

Practical

A common lab experiment uses the reaction between sodium thiosulfate and acid to investigate reaction rates. This reaction produces sulfur, which makes the solution turn cloudy. The end-point for this reaction is when the cross disappears, as in Figure 3.

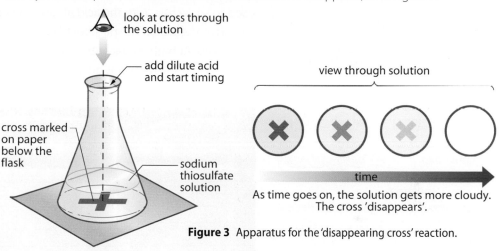

look at cross through the solution

add dilute acid and start timing

cross marked on paper below the flask

sodium thiosulfate solution

view through solution

time

As time goes on, the solution gets more cloudy. The cross 'disappears'.

Figure 3 Apparatus for the 'disappearing cross' reaction.

Questions

1 The glasses of milk shown in the photograph on the previous page are the same age. Which do you think was left in a warm room, and which was left in the fridge?

2 **(a)** Why doesn't a Bunsen burner light as soon as you turn on the gas? **(b)** What does a spark provide that sets off this reaction?

3 Why should you never turn on a light switch if you can smell a gas leak?

4 Why does food cook faster in a 'deep fat fryer' than in boiling water?

5 Explain why raising the temperature increases the rate of reaction.

6 In the calcium carbonate/acid reaction shown in lesson C2 4.1: **(a)** What factors, other than temperature, might affect how fast the carbon dioxide is produced? **(b)** How would you ensure that these other factors were not affecting the results?

7 **(a)** From the graph of reaction rate vs temperature shown in Figure 2, approximately what rise in temperature doubles the reaction rate?

(b) Sketch the graph from Figure 2. Now draw two new sketch graphs on the same axes, one for an initially faster reaction and the other for an initially slower reaction.

8 Plot a graph of 'time taken' against temperature using the data in Table 2.

Table 2 Results for the thiosulfate 'disappearing cross' experiment.

Temperature/°C	Time for cross to disappear/s
20	360
30	180
40	91
50	46
60	23

Explain why your graph looks different to the reaction rate versus temperature graph (Figure 2) and from your graph predict how long the reaction would take at 45 °C.

Opportunities for interaction

Close encounters

The concentration of a solution tells us how much **solute** is dissolved in the water. If there is a large amount of solute in a small amount of water, the solution is **concentrated**. If there is a lot of water and little solute, the solution is **dilute**. Before two particles can react, they must meet. This is more likely to happen in a concentrated solution than in a dilute one, so we should expect reaction rates to increase with the concentration of the solution.

The reaction of acid on marble (calcium carbonate) (see lesson C2 4.1) can be used to show this. With the acid at low concentrations, the particles are widely spread in the water. The number of collisions between them and the marble will be limited. At higher concentrations, the chance of a collision between the acid particles and the marble is greatly increased. The marble chips fizz much faster in concentrated acid.

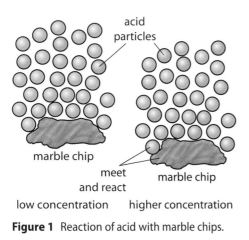

low concentration   higher concentration

Figure 1 Reaction of acid with marble chips.

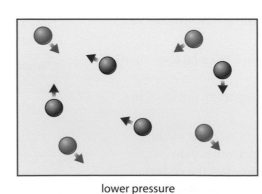

dilute   medium   concentrated

The more concentrated the acid, the more the fizz.

Gas pressure

Similar arguments apply to gases, but here **pressure** takes the place of concentration. All else being equal, the greater the pressure, the greater the number of particles of gas in a given space. If you double the pressure you will squash the particles into half the volume.

lower pressure   higher pressure

Figure 2 Higher pressure brings gas particles closer together.

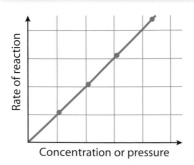

Figure 3 Typical graph of reaction rate against concentration or pressure.

The closer the particles are together, the more chance they have of colliding. If you double the pressure, you will double the rate of reaction. Graphs drawn for rate against concentration and rate against pressure look very similar.

Practical

Magnesium ribbon dissolves in sulfuric acid to give hydrogen gas. The gas can be collected in a gas syringe.

magnesium + sulfuric acid $\longrightarrow$ magnesium sulfate + hydrogen

$$Mg + H_2SO_4 \longrightarrow MgSO_4 + H_2$$

A student dissolved 2-cm strips of magnesium ribbon in different concentrations of acid and timed how long it took to collect $20\ cm^3$ of hydrogen in each case.

Look at Table 1.

a Which solution is the most concentrated?

b Which solution gives the fastest reaction?

c Why didn't the student use a solution containing no acid?

d Plot a graph of volume of acid used (cm^3) against time to collect $20\ cm^3$ of gas. Plot volume of acid on the x-axis, time taken on the y-axis.

e Use your graph to find the time needed to collect $20\ cm^3$ of gas if the volume of acid used was: **i** $15\ cm^3$ **ii** $45\ cm^3$.

f Which variables should be kept constant during this reaction?

g Add an extra column to your table labelled 'Rate of CO_2 production in cm^3/s'. Work out this rate for A to E.

h Plot a new graph of 'volume of acid' against your newly calculated rate of reaction. How does this compare with the graph from **d**? Explain the difference in appearance.

Table 1 The student's results.

Solution	Volume of acid	Volume of water	Time to collect $20\ cm^3$ of gas
A	10	40	200
B	20	30	100
C	30	20	67
D	40	10	50
E	50	0	40

Questions

1 A student made two different solutions of sodium chloride:
solution A: 2 g dissolved in $200\ cm^3$ water
solution B: 4 g dissolved in $500\ cm^3$ water
(a) Calculate the concentrations of these solutions in grams per cm^3. **(b)** Which solution has the greater concentration?

2 Why is dilute acid not as dangerous as concentrated acid?

3 Bleach can be used straight from the bottle to kill germs in the toilet. Why must it be diluted down before being used to bleach clothes?

4 A concentrated lemon drink says 'dilute with five parts water before drinking'. What would you notice if you diluted it with 10 parts of water instead? Explain this effect.

5 In the marble chip/acid reaction, what would happen to the rate of reaction if you doubled the concentration of the acid?

6 Explain how increasing the concentration of reactants increases the rate of a reaction. Sketch a graph to illustrate your answer.

7 Sulfur dioxide gas reacts with oxygen gas to form gaseous sulfur trioxide: $2SO_2 + O_2 \longrightarrow 2SO_3$
(a) How would increasing the pressure change the rate of the reaction? **(b)** Explain your answer in terms of the collision theory.

8 The graph shows the volume of carbon dioxide produced when marble chips were dissolved in excess hydrochloric acid.

Calculate the rate of this reaction over the first 20 seconds, showing your working. Describe what happens to the rate of reaction over the next minute, explaining your answer. The experiment was repeated with all variables the same except that the acid was twice as strong. Make a copy of this graph and sketch the line you would expect to get for this second reaction.

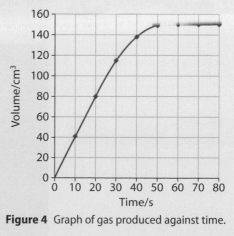

Figure 4 Graph of gas produced against time.

Where reactions happen

Free up some surface

If you cut a potato into small pieces it will cook much faster than a whole potato. The smaller the pieces of potato, the faster they cook. This is because more surface is exposed to the boiling water. It is the same with chemical reactions.

You have probably seen this idea in action when you try to light a campfire. The fire is much easier if you start with small twigs and splinters. You cannot set a large log on fire with just a match. Smaller twigs have a larger surface area than a big log, so they catch fire more easily.

The heart of the matter

Every time you break up a solid, you expose an extra bit of material to react. If you have the same amount of material in pieces of half the width, you will have twice the surface area (for the same amount of material), which will double the reaction rate.

1 cube of 4 cm　　= 8 cubes of 2 cm　　= 64 cubes of 1 cm

volume = 64 cm³

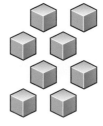
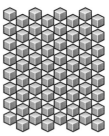

… but the total surface area =

$1 \times 6 \times 16 = 96\,cm^2$　　$8 \times 6 \times 4 = 192\,cm^2$　　$64 \times 6 \times 1 = 384\,cm^2$

Figure 1 Smaller cubes, larger surface area.

Sparklers are made from tiny iron filings glued onto an iron wire. Iron filings burn because they have such a large surface area.

A chip off the block

One large lump of marble (calcium carbonate) dropped into a beaker of acid fizzes steadily. But the same amount of marble in powdered form foams up out of the beaker. Collision theory can be used to explain this. The acid particles can only collide with the carbonate particles exposed at the surface. Crushing a marble chip increases the surface area several thousand times. The reaction rate goes up accordingly.

Which beaker had the powdered marble?

Lost to the air

If this reaction takes place in an open beaker, the carbon dioxide escapes into the air. If this is done on a balance, the total mass will go down as the carbon dioxide is lost. If you measure how much gas is lost in a given time, you can work out the rate of the reaction.

Example

In an experiment adding acid to 1 mm marble chips, the mass fell from 50 g to 45 g in 50 seconds.

Mass lost (= amount of carbon dioxide produced) = 5 g. Time taken = 50 s.

So: rate $= \dfrac{\text{mass lost}}{\text{time taken}} = \dfrac{5\,g}{50\,s} = 0.1\,g/s$

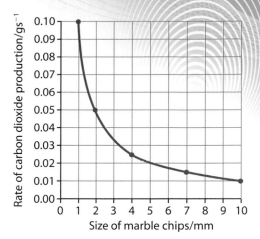

Figure 2 This graph shows how the reaction rate speeds up when you use smaller marble chips.

Practical

A student carried out an experiment to compare the rate of reaction of 0.1 g magnesium ribbon and 0.1 g magnesium powder when reacting with excess sulfuric acid. She measured the volume of hydrogen produced using a gas syringe.

Table 1 The student's results.

Time/s	Volume of hydrogen from Mg ribbon/cm³	Volume of hydrogen from Mg powder/cm³
0	0	0
10	22	52
20	40	68
30	56	77
40	68	82
50	76	84
60	82	86
70	85	86
80	86	86
90	86	86

a Plot a graph showing these results, with time as the x-axis and volume as the y-axis.

b Why is the final volume of hydrogen the same in both experiments?

c Why do the graphs flatten off near the end of the reaction?

d Which reaction produced the most hydrogen in the first 30 seconds?

e What is the average rate of production of hydrogen in cm³ per second for each reaction during this time?

f Explain the difference in the results.

Questions

1 Why does chewing your food well help you to digest it?

2 Why do small twigs catch fire more quickly than a big log?

3 Pythons swallow their food whole, and then take days to digest it. Explain why.

4 Flour mills and biscuit factories can easily produce clouds of flammable dust. **(a)** Why is this dangerous? **(b)** What could they do to overcome it?

5 Using the graph of reaction rate against marble chip size: **(a)** Find the mass loss per second for the 2 mm chips. **(b)** How long would it take the mass to drop by 1 g for these chips? **(c)** How much faster did the 2 mm marble chips react compared with the 4 mm chips?

Don't attempt this

6 If you broke a 1 cm cube up into 1 mm cubes, by what factor would the surface area increase?

7 Why is it that delicate carvings on the outside of churches seem to suffer more from the effects of acid rain than large blocks made from the same stone?

8 Explain carefully why iron filings on sparklers burn when heated, yet an iron poker can be left red hot in a fire for hours with no obvious reaction.

Catalysts

Learning objectives

- explain what a catalyst is
- evaluate the advantages and disadvantages of using a catalyst in industrial processes.

Science in action

Living things rely on biological catalysts called enzymes. Many industrial processes now use these enzymes to make their products. For example, sugar syrup is made from corn starch using enzymes.

Everlasting activity

We are going to look at one final way of speeding up reactions. Ever helped others get things done just by being there with them and smiling? Some chemicals have the same effect on reactions. Just by being there while the reaction is taking place, they make things go much faster. Here's an example: the breakdown of hydrogen peroxide, an unstable compound of hydrogen and oxygen. Left on its own, hydrogen peroxide will slowly break down into water and oxygen gas:

hydrogen peroxide $\longrightarrow$ water + oxygen

Drop a spatula of powdered manganese dioxide into hydrogen peroxide and the mixture starts to fizz rapidly as oxygen is given off. There is nothing particularly surprising in that, you might think — just another chemical reaction in progress.

However, if you filter the mixture after the reaction has finished, you get back the same amount of manganese dioxide as you put in. The oxygen has escaped and the water is left behind, but the manganese dioxide is unchanged. It can be used all over again.

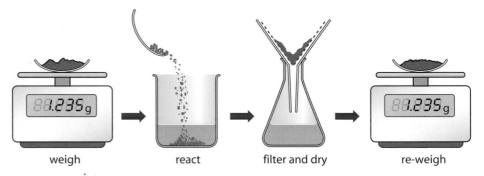

weigh   react   filter and dry   re-weigh

Figure 1 Manganese dioxide speeds up the breakdown of hydrogen peroxide, but remains unchanged itself.

Weakening the bonds

The reaction that occurs is the simple breakdown of hydrogen peroxide to water and oxygen, which would have occurred slowly on its own. The manganese dioxide has simply speeded up this reaction without itself being altered. It has acted as a **catalyst**.

$$2H_2O_2 \xrightarrow{\text{manganese dioxide catalyst}} 2H_2O + O_2$$

Catalysts work by lowering the activation energy needed for the reaction. The process is complex but a simple model is that the existing bonds are weakened on the catalyst surface. This makes it easier for them to be broken during collisions so that new bonds can form. The overall result is that the reaction happens more easily.

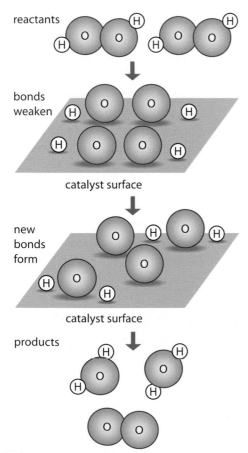

Figure 2 A catalyst works by weakening bonds, which lowers the activation energy.

Science in action

Many manufacturers are developing fuel cells as replacements for petrol and diesel engines. In a **fuel cell**, hydrogen combines with oxygen to form water and also produce electricity. The reaction is catalysed by a platinum catalyst.

Surface area matters

This kind of reaction can only happen on the surface of the catalyst. So the catalyst has to be in the smallest pieces possible. Breaking it up into a fine powder may not help, because the particles may just clump together. The answer is to attach small particles to tiny wire grids just millimetres apart. This allows the reactants to get to as large a surface as possible.

Nanotechnology (see lesson C2 2.6) could take this approach to a new level. Create particles of catalyst just a thousandth of a millimetre across and their surface area goes up a thousand times for the same amount of material. This makes the catalyst a thousand times more effective.

The catalyst for the job

In the simple breakdown of hydrogen peroxide, iron oxide or copper oxide will work too, but manganese dioxide gives the fastest reaction. It is important to find the right catalyst for the particular reaction you want to speed up.

Transition metals or their oxides are often used as catalysts:

- Iron is used in the production of ammonia.
- Nickel is used to turn oils into fats for margarine or chocolate.
- Platinum and rhodium are used in the production of nitric acid.
- Vanadium oxide is used in the production of sulfuric acid.

Disadvantages of some catalysts include:

- They may become poisoned by impure reactants; they then need purifying before being re-used.
- Some, such as the platinum used in fuel cells, are very expensive.

Overall, catalysts are very important in industry. Without them some reactions would be far too slow to be useful so much more fuel would need to be burnt to raise the reaction temperature and speed things up. This would cost the company money, reducing their profitability or making their products more expensive. Burning that extra fuel would also waste valuable resources and increase pollution, giving out much more carbon dioxide, which is linked to global warming.

The catalyst in a car's catalytic converter is held in a 'millimetre-sized' grid. Nanocatalysts could one day make cars completely pollution-free.

Making chocolate like this needs a nickel catalyst.

Examiner feedback

Many different catalysts are used in industry. You do not need to know them all, but you should be able to explain what is happening if you are given the details of a particular reaction.

Questions

1 Use the diagrams in Figure 1 to explain how you could show that manganese dioxide is not taking part directly in the hydrogen peroxide reaction.

2 How do catalysts allow reactions to run at lower temperatures?

3 Why is it not good enough to simply grind a catalyst into a fine powder?

4 How could 'nanocatalysts' help revolutionise the way we tackle pollution?

5 Platinum is very expensive. Why might this not matter so much for a catalyst?

6 Exhaust gases flow out from an engine very rapidly and so could escape into the air before they have time to react to form less harmful products. How does a catalytic converter overcome this problem?

7 How could you compare the effectiveness of different transition metal oxides as catalysts for the hydrogen peroxide reaction?

8 In a particular industrial process, using a catalyst means that the reaction can run efficiently at 100 °C instead of 300 °C. What are the potential benefits of using a catalyst in this reaction?

Reactions with energy to spare

Learning objectives

- explain how energy transfers are associated with chemical reactions
- describe how exothermic reactions transfer energy to the surroundings
- explain how this process can be used to our advantage.

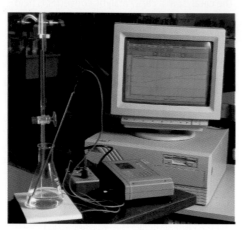

You could follow the rise in temperature during neutralisation with a temperature sensor connected to a computer.

Enough energy is released in the thermit reaction to melt the iron.

Calcium (left) and magnesium (right) both fizz in acid. The test tubes get hot.

Energy on the loose

It is not always easy to tell when a chemical reaction has occurred. For example, if you mix cold, dilute hydrochloric acid and sodium hydroxide solution, you will not see any obvious change. But touch the tube carefully and you will *feel* that something has happened. The tube has been warmed by the energy given out as the reaction takes place.

Chemical changes are often accompanied by changes in temperature, as energy is transferred to or from the surroundings. In the case above, energy is given out during neutralisation, which heats the surroudings. Chemical reactions that give out energy like this are called **exothermic** reactions.

$$acid + alkali \longrightarrow salt + water \quad \boxed{energy\ given\ out}$$

$$e.g.\ HCl + NaOH \longrightarrow NaCl + H_2O \quad \boxed{energy\ given\ out}$$

Energy release runs in families

Many chemical reactions are exothermic like this, particularly reactions that start reacting as soon as the reactants are mixed. For example, metals reacting with acids:

$$metal + acid \longrightarrow salt + hydrogen \quad \boxed{energy\ given\ out}$$

$$e.g.\ Mg + H_2SO_4 \longrightarrow MgSO_4 + H_2 \quad \boxed{energy\ given\ out}$$

Displacement reactions are also exothermic. An example is when aluminium displaces iron in the spectacular thermit reaction:

$$aluminium + iron\ oxide \longrightarrow aluminium\ oxide + iron \quad \boxed{energy\ given\ out}$$

$$2Al + Fe_2O_3 \longrightarrow Al_2O_3 + 2Fe \quad \boxed{energy\ given\ out}$$

Oxidation reactions, such as calcium metal turnings tarnishing as they react with the oxygen in air, also give out energy:

$$calcium + oxygen \longrightarrow calcium\ oxide \quad \boxed{energy\ given\ out}$$

$$2Ca + O_2 \longrightarrow 2CaO \quad \boxed{energy\ given\ out}$$

A spark gets things going

Combustion reactions, such as burning fuels, are exothermic. For these reactions it is usually the energy we want, not the products. We burn methane to heat our homes and petrol to run our cars.

$$fuel\ (e.g.\ oil,\ gas\ or\ wood) + oxygen \longrightarrow carbon\ dioxide + water \quad \boxed{energy\ given\ out}$$

You have to light the fuel with a spark or match to start the reaction going. This provides the activation energy (see lesson C2 4.2) needed for the reaction to begin. After that, the reaction generates enough energy of its own to keep the process going.

Energy out of control

Once a fuel starts to burn, the exothermic release of energy can easily get out of control. Fuels such as petrol are highly flammable and have to be handled with special care. Hazard symbols are used to warn of the dangers when transporting or storing fuels.

Highly flammable
These substances easily catch fire.

Explosive
These substances cause an explosion.

Figure 1 Look out for these hazard symbols.

An exothermic reaction that has got out of control.

Combustion reactions often produce light and sound as well as heat.

Hand warmers and hot coffee

Sports people use heat packs to relax stressed muscles. One type contains a damp mixture of iron filings, salt and charcoal in a sealed bag. When the bag is opened and squeezed, oxygen gets in from the air and reacts with the iron, giving out energy. The other chemicals help to speed up the reaction.

iron + oxygen ⟶ iron oxide　*energy given out*

How can you get hot coffee from a cold can? Self-heating coffee cans have a hidden compartment full of calcium oxide. When a button is pressed, water is mixed with this and a strongly exothermic reaction occurs (see lesson C1 2.3). This quickly heats up the coffee.

calcium oxide + water ⟶ calcium hydroxide　*energy given out*

A student had three beakers, each containing 50 cm³ of copper sulfate solution at room temperature. She added an excess of powdered iron, magnesium and zinc to each in turn, measuring the temperature rise each time. Look at Table 1.

a What has happened?
b Which reaction gave out the most energy?
c Comment on any pattern you see, comparing it to other chemical reactions you know of.

Table 1 The student's results.

Metal added	Temperature rise/°C
Fe	9
Mg	28
Zn	15

Questions

1 If you dip an iron nail into copper sulfate solution it comes out coated in copper. **(a)** What kind of reaction is this? **(b)** How else might you be able to tell that a chemical reaction has occurred?

2 Do you think that iron rusting is an exothermic or endothermic reaction? Explain your answer.

3 Combustion reactions provide energy. What other forms of energy are sometimes given off as well?

4 If you mix acid and alkali solutions in a test tube you are holding, what clue will you get that a reaction has taken place?

5 Suggest some other uses for 'self-heating' cans.

6 The thermit reaction can be started with a simple fuse yet produces enough energy to make iron melt. What does this tell you about the reaction?

7 **(a)** Why do fuels need a 'spark' to get them burning?
   (b) What keeps the reaction going once it has started?
   (c) What problems could be caused if fuels did not need this 'spark'?

8 What are the main 'families' of exothermic reactions? Which can usefully be used as energy sources, and what are the potential dangers that go with their use?

Reactions that take in energy

Learning objectives

- explain why endothermic reactions need to take in energy from the surroundings
- explain the importance of endothermic reactions.

No ice is needed for this summer drink.

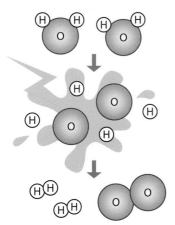

Figure 1 Electrolysis is an endothermic process

Sunlight energy powers this glucose factory.

The drink that cools itself

You can make a pleasant summer drink by adding a spoonful of baking soda to a glass of sweetened lemon juice. When you do this a chemical reaction occurs:

sodium + citric acid ⟶ sodium + carbon + water
hydrogen (lemon juice) carbonate dioxide
carbonate
(baking soda)

If you are holding the glass as you stir in the baking soda, you will feel it get cooler. A temperature change is an indication that a chemical reaction is happening. In this case, the temperature goes down, not up.

Supplying the energy to make things go

Reactions that take energy in from their surroundings are called **endothermic** reactions. It is fairly rare for endothermic reactions to 'go on their own'. You usually need to pump energy into them to make them work. Examples include materials that decompose when heated. We call these reactions thermal decomposition. Green copper carbonate ($CuCO_3$) will not break down on its own. Once heated, though, it quickly decomposes into black copper oxide (CuO) and carbon dioxide (CO_2). The energy transferred by heating you put in breaks the compound apart.

Thermal decomposition reactions are endothermic.

copper carbonate ⟨ energy ⟩ ⟶ copper oxide + carbon dioxide

The current way to purify metals

Electrolysis is an endothermic process. Electrical energy is used to split compounds into the ions that make them up. You can use electrolysis to extract aluminium from molten aluminium oxide, for example. It can also be used to make hydrogen and oxygen from water:

water ⟨ energy ⟩ ⟶ hydrogen + oxygen

The chemical energy store

One of the most important endothermic reactions of all is photosynthesis, in which plants use the energy from sunlight to build complex chemicals such as glucose from carbon dioxide and water:

carbon dioxide + wate ⟨ energy ⟩ ⟶ glucose + oxygen

This process stores up energy from sunlight as chemical energy that we can get back later in exothermic reactions such as combustion. Fossil fuels — coal, oil and gas — got their energy from sunlight in this way millions of years ago. When we burn fossil fuels we are getting that energy back; the problem is that we are using them at an increasing rate and they will soon be gone. However, if we grow plants today, for oils or foodstuffs, we can make new 'energy stores' whenever we need them.

Examiner feedback

Endothermic reactions take in energy from the surroundings whereas exothermic reactions release energy to the surroundings. Both types of reaction have practical and commercial applications.

Future energy storage?

Wind farms provide pollution-free energy. There's one big catch: the wind doesn't blow all the time. Then, at other times, it blows so hard you make more electricity than you need. Similarly, photovoltaic cells make electricity from sunlight by day but are not much use at night. What can be done? How can we store the extra energy?

What's the big problem with wind farms?

Unused electricity can be used to electrolyse water and make hydrogen. This can then be stored until needed, then used as a fuel to generate electricity. On a larger scale, this idea could be used to produce the hydrogen for the next generation of pollution-free cars, lorries and buses.

Science in action

A group of scientists in the United States are developing an energy storage system that can split water directly into hydrogen and oxygen, using sunlight and special catalysts. These gases can then be used to make electricity whenever it is needed, using a fuel cell.

Questions

1 Why is it unusual for endothermic reactions to 'go on their own'?
2 When you make toast, starch molecules break up into complex caramels and sugars. Is this reaction exothermic or endothermic? Explain your answer.
3 In the lemon juice/baking soda drink, what would you *see* happening that would suggest that a chemical reaction was occurring?
4 Write a balanced chemical equation for the decomposition of copper carbonate.
5 The hydrogen and oxygen ions are held together by strong bonds in water. What is the electrical energy doing in this reaction?
6 Balance the chemical equation for this reaction: $H_2O \longrightarrow H_2 + O_2$
7 How could the energy from wind farms or photovoltaic cells be stored?
8 Fossil fuels are used because of their exothermic combustion reactions. Explain how an endothermic reaction is the key to their formation, the 'stored' energy they contain and the way they are used.

Sports people use 'cold packs' to relieve injuries. These packs contain ammonium nitrate and a separate bag of water. When you squeeze the pack, the water mixes with the ammonium nitrate which starts to dissolve. The ammonium nitrate breaks up into ions, which is an endothermic process. The pack gets cold, drawing energy transferred by heating from the injured muscle.

warm

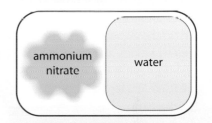

The water is kept separate in a thin plastic bag.

squeeze and burst inner bag

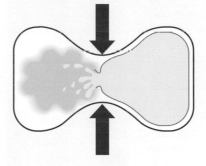

cold

The mixture cools as the ammonium nitrate dissolves.

Figure 2 Squeeze the cold pack to start the endothermic reaction.

Energy in reversible reactions

Practical

If you stand open beakers of concentrated hydrochloric acid and ammonia close to one another, white 'clouds' of ammonium chloride start to form in the air between them, as hydrogen chloride and ammonia gases diffuse out, meet and react. This can be used as a test for either gas.

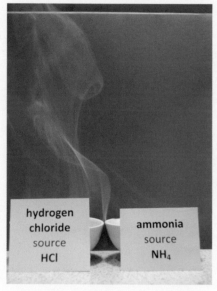

A test for ammonia or hydrogen chloride.

Breaking up and getting back together

If you heat ammonium chloride, the compound breaks up to form ammonia and hydrogen chloride gases. Like all thermal decomposition reactions, this reaction is endothermic. Energy has to be put into the reaction to break the chemical bonds.

ammonium chloride $\rightleftharpoons$ ammonia + hydrogen chloride

This reaction is easily reversible (see lesson C2 3.4). If the gases are allowed to mix when cool, they recombine to form ammonium chloride. The energy change is also reversed. The energy that had to be put in to break the chemical bonds is now given out as the chemical bonds are formed again. The 'back reaction' is exothermic.

$$\underset{\text{exothermic}}{\overset{\text{endothermic}}{NH_4Cl \quad \rightleftharpoons \quad}} NH_3 + HCl$$

In all reversible reactions, if the reaction in one direction is exothermic, then the reverse reaction will be endothermic. It is also important to note that the same amount of energy is transferred in each case.

The test for water

Blue copper sulfate has water molecules chemically bound up in its crystals. If you heat it, the water is driven out, forming white anhydrous copper sulfate.

If you add water to this white powder, it turns blue again. This is such a clear reaction that it is used as a test for water. The water molecules re-form the chemical bonds with the copper sulfate and new, blue crystals form. If you carry out this reaction in a test tube, you can feel the tube get hot. Energy is released as the bonds re-form.

$$\underset{\text{(blue)}}{\text{hydrated copper sulfate}} \underset{\text{exothermic}}{\overset{\text{endothermic}}{\rightleftharpoons}} \underset{\text{(white)}}{\text{anhydrous copper sulfate} + \text{water}}$$

Adding water to anhydrous copper sulfate causes an exothermic reaction. The colour change is a test for water.

Re-usable hand warmers

The iron-based hand warmers discussed in lesson C2 5.1 can only be used once, as the oxidation reaction is not easily reversible. However, some hand warmers use a reversible reaction and can be used over and over again.

Examiner feedback

Cobalt chloride has similar hydrated (blue) and anhydrous (pink) forms. You could explain the energy changes in the same way as for copper sulfate.

Sodium acetate melts to form a clear liquid if heated in a water bath. This is an endothermic reaction as the compound is breaking up into ions. What is unusual is that this clear liquid can be cooled back down to room temperature without recrystallising, forming a '**supercooled liquid'**. However, it only takes a 'seed' crystal to be dropped in for the crystallisation to start and rapidly spread. When this happens, bonds are re-forming so the reaction is exothermic. Energy is given out and the material heats up.

Re-usable hand warmers.

$$\text{solid sodium acetate} \underset{\text{exothermic}}{\overset{\text{endothermic}}{\rightleftharpoons}} \text{liquid sodium acetate}$$

Sodium acetate hand warmers are filled with the supercooled liquid. If a button on the pack is pressed, it 'seeds' the liquid and crystals start to grow. Energy is given out and the pack heats up. Unlike a one-time hand warmer, a reusable one can be 'reset' by putting it into boiling water for a few minutes, to re-melt the crystals. Once it has cooled down, the hand warmer is ready to be used again when needed.

Photosynthesis and respiration

In lesson C2 5.2 we saw how plants build carbon dioxide and water into glucose and oxygen in the endothermic process called photosynthesis. This also has an exothermic 'reverse' equivalent, called **respiration**. You are getting the energy you need for life by running the photosynthesis reaction in reverse.

$$\text{carbon dioxide} + \text{water} \;\; \text{energy} \;\; \underset{\text{exothermic respiration}}{\overset{\text{endothermic photosynthesis}}{\rightleftharpoons}} \text{glucose} + \text{oxygen}$$

Getting ready to reverse photosynthesis.

Questions

1   You pour a little water onto some white anhydrous copper sulfate in a cool tube. **(a)** Describe what you would see and feel. **(b)** Explain these effects.

2   It takes 2.5 kJ of energy to turn a quantity of hydrated copper sulfate into anhydrous copper sulfate. How much energy would be released if sufficient water was added to this white powder?

3   Explain why sodium acetate-based hand warmers are re-usable, whereas iron-based hand warmers are not.

4   Under certain conditions iron oxide can be reduced by hydrogen to form iron and water. Is this likely to be an endothermic or exothermic reaction? Explain your answer.

5   Calcium oxide (CaO) has an exothermic reaction with carbon dioxide (CO_2). Write this reaction as a balanced chemical equation. What reaction is this the reverse of?

6   What effect will raising the temperature have on an exothermic reaction and on an endothermic reaction? Explain your answers.

7   In what way does our use of vegetables as food 'reverse photosynthesis'?

8   If you heat ammonium chloride at the base of a long tube it seems to disappear. At the same time, a white powder starts to form at the cool end of the tube. Explain what is happening in terms of both the chemical reaction and the energy changes involved.

Science skills   If you put a 'used' sodium acetate hand warmer in hot water, the water will cool down as the crystals melt. If you let the hand warmer cool down, then activate it and put it in cold water, the water will warm up as the crystals re-form.

a   Design an experiment to show that the same amount of energy is involved on each occasion.
   i   What are your independent and dependent variables?
   ii   What other variables might affect your result?
   iii   How could you control them?

Assess yourself questions

1 A student put a beaker containing 200 cm³ of 1 M hydrochloric acid, and a watch glass with 1 g of marble chips, onto an electric balance. He then zeroed the display on the balance. After this he tipped the limestone into the acid and put the empty watch glass back on the balance next to the beaker. He recorded the readings from the scale every 20 seconds.

Table 1 The student's results.

Time/s	Balance reading/g
20	−0.11
40	−0.23
60	−0.33
80	−0.40
100	−0.43
120	−0.44
140	−0.44

(a) The student held a drop of limewater on the end of a glass rod over the beaker and it turned milky. What gas was given off? *(1 mark)*

(b) Complete the equation:
$$CaCO_3 + 2HCl \longrightarrow CaCl_2 + H_2O + \underline{\hspace{1cm}}$$ *(1 mark)*

(c) Explain why the reading on the balance scale dropped below zero. *(1 mark)*

(d) Plot a graph of the results. Put time along the *x*-axis and mass loss in grams up the *y*-axis. Draw a 'best fit' curve. *(3 marks)*

(e) What would the student have seen happening over the first minute or so? *(1 mark)*

(f) Suggest *two* ways in which the student would have known that the reaction had finished after 2 minutes. *(2 marks)*

2 Another student used the same reaction and apparatus as in question **1** to investigate how the reaction rate changed with temperature. In her experiment she timed how long it took for the mass loss to reach −0.20 g. She repeated the experiment at different temperatures, to see how this affected the reaction.

Table 2 Results.

Temperature/°C	0	12	25	37	50
Time to reach −0.20 g/s	150	72	37	19	10

(a) Suggest some safety precautions that would need to be taken for this experiment. How could the acid be heated safely, and how could it be cooled? *(3 marks)*

(b) Plot a graph of the results, with temperature along the *x*-axis. Draw a 'best fit' curve. *(3 marks)*

(c) Describe the pattern you see in words. How is 'the time it takes' changing as the temperature rises? *(1 mark)*

(d) What does this tell you about the way the speed of the reaction is changing as the temperature rises? *(1 mark)*

(e) Explain this effect in terms of what is happening at the particle level. *(2 marks)*

(f) Add a third row to the table and calculate the rate of the reaction at each temperature. (Rate = 0.20 g/time taken). *(2 marks)*

(g) Plot a new graph of rate against temperature and draw the 'best fit' line. *(3 marks)*

(h) From your 'best fit' line, how much do you need to raise the temperature by to double the rate of this reaction? Explain your reasoning. *(2 marks)*

(i) The student actually repeated each experiment three times at each temperature. The figures shown are the average of the three. Why did she do this? *(2 marks)*

3 **(a)** Sparklers are made from tiny iron filings glued onto an iron wire. Explain why the iron filings burn but the iron wire does not. *(2 marks)*

(b) A student wanted to find out which was more reactive, iron or nickel, so she put samples of each into tubes of 1 M hydrochloric acid, to see which bubbled the most. She had iron filings and nickel granules. Suggest why her results may not be reliable. *(2 marks)*

(c) Iron filings usually fizz faster in acid than iron nails. You might expect very fine iron dust to fizz even faster, but if it has settled out in a layer at the bottom that may not be the case.

  (i) Suggest a reason why. *(1 mark)*

  (ii) How could you easily get the iron dust to react faster again? Explain your answer. *(2 marks)*

(d) Iron acts as a catalyst in an important industrial reaction that makes ammonia.

  (i) Why does the iron have to be in very tiny pieces? *(1 mark)*

  (ii) Why does this iron have to be supported on a grid or mesh within the reacting gases? *(1 mark)*

4 A student was investigating the way hydrogen peroxide broke down when metal oxide powders were added. She had a flask containing hydrogen peroxide connected to a 100 cm³ gas syringe. She added a spoonful of oxide, quickly replaced the bung and took a reading every 20 seconds.

(a) What factors, apart from which metal oxide she chooses, must the student keep constant if this is to be a fair test? *(3 marks)*

(b) Copy and complete the equation for this reaction.
$$2H_2O_2 \longrightarrow 2H_2O + \underline{\hspace{2cm}}$$
(1 mark)

(c) The metal oxide acts as a catalyst for this reaction. Explain the term catalyst. *(2 marks)*

(d) Figure 1 shows the graph for manganese dioxide, nickel oxide and copper oxide.

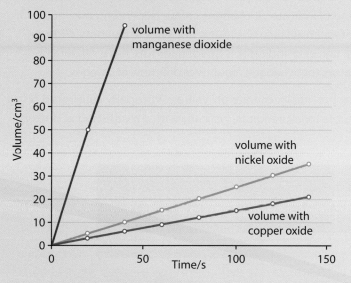

Figure 1 Graph of oxygen production.

Which is the most effective catalyst as shown by these results? *(1 mark)*

(e) Calculate the reaction rate for each of the catalysts. *(3 marks)*

5 For each of the following reactions, state whether it is exothermic or endothermic. If it is exothermic, state whether or not it needs a 'kick start' of energy.

(a) candle wax burning *(2 marks)*

(b) photosynthesis *(1 mark)*

(c) electrolysis of sodium chloride *(1 mark)*

(d) iron displacing copper from copper sulfate solution *(2 marks)*

(e) aluminium displacing iron from iron oxide powder *(2 marks)*

(f) baking soda dissolving in lemon juice *(1 mark)*

(g) caustic soda oven cleaner dissolving grease *(2 marks)*

(h) making sodium chloride from sodium hydroxide and hydrochloric acid *(2 marks)*

(i) gunpowder exploding *(2 marks)*

(j) paraffin burning in a jet engine *(2 marks)*

(k) making copper oxide by heating copper carbonate *(1 mark)*

(l) frying an egg *(1 mark)*

(m) limescale remover removing limescale *(2 marks)*

(n) a lead/acid battery discharging *(2 marks)*

(o) a lead/acid battery charging *(1 mark)*

6 The graphs in Figure 2 were produced by a datalogging set-up using a temperature probe.

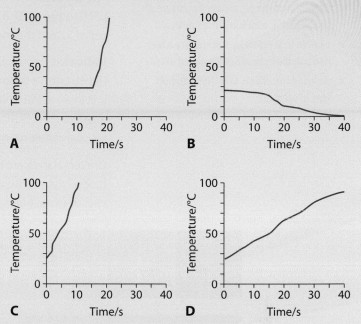

Figure 2 Temperature/time graphs.

Which reaction below could have produced which graph?

(a) solid sodium hydroxide dropped into hydrochloric acid *(1 mark)*

(b) a spoonful of baking soda dropped into a solution of citric acid *(1 mark)*

(c) iron filings dropped into copper sulfate solution *(1 mark)*

(d) a match head lit after 15 seconds *(1 mark)*

7 **(a)** Sodium metal is made by the electrolysis of molten sodium chloride. Is this reaction endothermic or exothermic? *(1 mark)*

(b) Write a balanced chemical equation for this reaction. *(1 mark)*

(c) If you know how much chlorine was produced by a given voltage and current for a given period of time, it is possible to calculate the amount of energy needed to break up a mole of sodium chloride. Explain how you could use this information to predict the energy change that occurs when sodium burns in chlorine. *(2 marks)*

8 Sherbert is a mixture of icing sugar, sodium hydrogen carbonate and citric acid. A student thought that the 'fizzy' feel when you put sherbert in your mouth might be caused by an endothermic reaction. Design an experiment that she could use to test this idea.
In this question you will be assessed on using good English, organising information clearly and using specialist terms where appropriate. *(6 marks)*

Acids and bases

The pH scale

The **pH scale** is a measure of the acidity or alkalinity of a solution. It runs from pH 0 to pH 14. The most **acidic** solutions have a pH of 0, neutral solutions have a pH of 7, and the most **alkaline** solutions have a pH of 14.

Indicator paper or an indicator solution can be used to give a rough measure of pH. Litmus is one kind of indicator. It changes colour depending on whether it is in an acidic solution or an alkaline one. But it does not reveal how acidic or alkaline the solution is; universal indicator is needed for this. It is a mixture of several different indicators that change colour over different pH ranges.

Universal indicator may provide inaccurate pH values. A properly **calibrated** pH meter gives more accurate measurements, typically with a **resolution** of one or two decimal places.

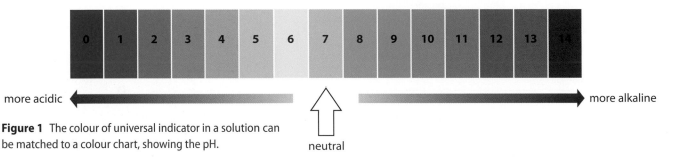

more acidic ← → more alkaline

neutral

Figure 1 The colour of universal indicator in a solution can be matched to a colour chart, showing the pH.

Bases and alkalis

Copper oxide and other metal oxides are **bases**. They react with acids and neutralise them. Sodium hydroxide and other metal hydroxides are also bases. A **soluble** metal hydroxide is called an **alkali**. Sodium hydroxide is soluble, so it is an alkali as well as a base.

Alkalis have certain properties in common. They:

- turn red litmus paper blue
- form solutions with a pH greater than 7
- release OH$^-$ ions in solution
- neutralise acids to make a salt and water.

Practical

The pH of a solution can be determined using universal indicator paper and a colour chart. It is unwise to dip the paper into the test solution because the dye can leach out and contaminate the test solution. Instead, a clean glass rod is dipped into the test solution. It is touched to the end of a piece of universal indicator paper. The aim is to leave a small spot rather than to soak the paper. The paper should be left for 30 seconds to allow the colour to develop, then the colour of the spot is matched to the nearest colour on the colour chart.

Special 'narrow range' indicator papers are also available. These determine pH values to within 0.5 of a pH unit or better. Typical pH ranges of these papers include 0–2.5, 3–6, 4–7, 7–10 or 10–13.

Alkalis release **hydroxide ions** (OH⁻) when they are dissolved in water. These ions make the solution alkaline. **Strong alkalis**, like sodium hydroxide, NaOH, and calcium hydroxide, Ca(OH)$_2$, produce solutions containing high concentrations of hydroxide ions. **Weak alkalis**, like ammonia solution, NH$_4$OH, produce solutions with lower concentrations of hydroxide ions.

Alkalis are corrosive and will damage skin and eyes, so they must be handled carefully. They feel soapy to the touch because they react with oils and fats to produce water-soluble products. This makes them useful ingredients in household cleaning products.

Acids

Acids release **hydrogen ions** (H⁺) when they are dissolved in water. These ions make the solution acidic.

Acids have certain properties in common. They:
- turn blue litmus paper red
- release H⁺ ions in solution
- form solutions with a pH less than 7
- react with bases to make a salt and water
- react with many metals to form a salt and hydrogen.

Hydrochloric acid (HCl) and sulfuric acid (H$_2$SO$_4$) are **strong acids**. They produce solutions containing high concentrations of hydrogen ions. They are corrosive and will damage skin and eyes, just like strong alkalis. **Weak acids**, such as citric acid, produce solutions with lower concentrations of hydrogen ions. Weak acids have a sharp taste. Think of the taste of lemons, which contain citric acid. Laboratory acids must never be tasted or swallowed.

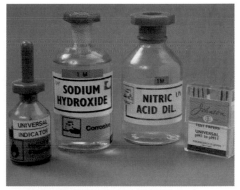

Universal indicator can be used to test if a solution is acidic, alkaline or neutral.

Questions

1. What is the pH scale?
2. Explain why copper oxide is a base but not an alkali.
3. Which ions do acids and alkalis produce in solution?
4. Compare the properties of acids with those of alkalis.
5. Describe the hazards presented by strong acids and strong alkalis.
6. Write balanced chemical equations for the production of H⁺ and Cl⁻ ions by hydrochloric acid, and for the production of H⁺ and SO$_4$$^{2-}$ ions by sulfuric acid.
7. About 1 in 56 000 ethanoic acid molecules release hydrogen ions in solution, whereas almost all nitric acid molecules do so. **(a)** Explain which of these acids is a weak acid. **(b)** Explain which acid will have the lower pH, if both are at the same concentration.
8. Answer these questions in terms of pH and ions. **(a)** The pH of some citric acid is 4 and the pH of some hydrochloric acid is 1. Which is the most strongly acidic, and why? **(b)** The pH of some ammonia solution is 10 and the pH of some sodium hydroxide solution is 14. Which is the most strongly alkaline, and why?

Science skills The pH scale was proposed in 1909 by a Danish chemist called Søren Sørenson. He knew that the acidity of a solution depended upon the concentration of hydrogen ions. He also knew that these concentrations varied enormously. They could be as low as 1×10^{-14} moles of H⁺ ions in each litre of solution to at least as high as 1 mole of H⁺ ions in each litre of solution. Sørenson's scale uses 'logarithms' so that an increase of one pH unit is due to an increase in the H⁺ ion concentration by 10 times. This idea made the scale much more manageable.

Neutralisation

Learning objectives

- use state symbols in balanced equations
- describe neutralisation reactions using an ionic equation
- describe how soluble salts can be made from acids by reacting them with alkalis.

State symbols

Consider these two equations:

sodium + water $\longrightarrow$ sodium hydroxide + hydrogen
$2Na$ + $2H_2O$ $\longrightarrow$ $2NaOH$ + H_2

The balanced chemical equation gives a lot more information than the word equation. It shows the formula of each substance, and the ratio in which they react or form. **State symbols** provide even more information. Here is the balanced chemical equation again, this time with its state symbols:

$$2Na(s) + 2H_2O(l) \longrightarrow 2NaOH(aq) + H_2(g)$$

State symbols tell you what state each substance is in: solid (s), liquid (l), gas (g) and dissolved in water — aqueous solution (aq).

The neutralisation reaction

Neutralisation reactions happen when acids and alkalis react together to produce a salt and water. For example:

sulfuric acid + sodium hydroxide $\longrightarrow$ sodium sulfate + water
$H_2SO_4(aq)$ + $2NaOH(aq)$ $\longrightarrow$ $Na_2SO_4(aq)$ + $2H_2O(l)$

In aqueous solution, acids produce hydrogen ions (H^+) and alkalis produce hydroxide ions (OH^-). In neutralisation, the hydrogen ions and hydroxide ions react together to produce water:

$$H^+(aq) + OH^-(aq) \longrightarrow H_2O(l)$$

This is called an **ionic equation** because it shows what happens to ions in a reaction. The same equation describes neutralisation, whatever acid or alkali is used.

 **Science skills** Neutralisation is an exothermic reaction. Table 1 shows the amount of energy transferred by heating to the surroundings for three different neutralisation reactions. Each one involves a different strong acid but the same strong alkali.

Table 1 Energy transferred by heating.

Neutralisation reaction	Energy transferred in kJ/mole
$HCl(aq) + NaOH(aq) \longrightarrow NaCl(aq) + H_2O(l)$	57.9
$HBr(aq) + NaOH(aq) \longrightarrow NaBr(aq) + H_2O(l)$	57.6
$HNO_3(aq) + NaOH(aq) \longrightarrow NaNO_3(aq) + H_2O(l)$	57.6

a The table provides evidence in support of the idea that neutralisation is the reaction between H^+ and OH^- ions to form water, rather than the reaction between ions to produce a particular salt. Explain why.

Making soluble salts

Acids and alkalis react together to make a soluble salt and water. When enough acid has been added to completely neutralise the alkali, the water can be evaporated from the salt solution to leave dry crystals of salt.

The **end-point** of the neutralisation reaction, when the acid and alkali have completely reacted, can be found using an indicator. Universal indicator's range of colours makes it difficult to tell when the end-point has been reached, so other indicators are used instead. For example, phenolphthalein is pink in alkaline solutions and colourless in acidic solutions. It is added to an alkaline solution in a conical flask. Acid is added drop-by-drop, from a teat pipette, and the flask swirled after each drop is added, until the pink colour has just disappeared at the end-point.

A white tile makes the colour change at the end-point easier to see.

Bromothymol blue is yellow in acidic solutions and blue in alkaline solutions.
Methyl orange is red in acidic solutions and yellow in alkaline solutions.
Phenolphthalein is pink in alkaline solutions and colourless in acidic solutions.

'Activated charcoal' – carbon powder that has been treated to give it a large surface area – is added to remove the phenolphthalein after it has done its job. The mixture is then filtered to remove the charcoal with the indicator bound to it. Water in the filtrate is evaporated to leave crystals of salt.

Examiner feedback

The pH of a reaction mixture can also be followed using a pH meter, or by removing drops of the reaction mixture to test its pH with an indicator.

Questions

1   What information do state symbols provide?

2   What is an ionic equation?

3   Give the names and formulae of the ions produced by acids and alkalis in solution.

4   Describe, with the help of an equation, what happens in neutralisation reactions.

5   Explain why the end-point might be determined using phenolphthalein, instead of using universal indicator solution.

6   Table 1 shows energy data. Explain why it would be useful to include data for reactions involving KOH(aq), rather than just reactions involving NaOH(aq).

7   **(a)** How can an indicator be removed from a salt solution? **(b)** Describe what will happen to the pH of an alkaline solution when excess acid is added to it gradually.

8   Describe how you could produce neutral potassium chloride solution using hydrochloric acid and potassium hydroxide solution.

Taking it further

A piece of glassware called a **burette** can be used to add precise volumes of acid to an alkali. A known volume of alkali is placed in a conical flask. An indicator is added to the alkali, and acid is added to the flask from the burette, until the end-point is reached. The **titre** is the volume of acid needed to exactly neutralise the alkali. Once the titre is known, the titration is repeated without the indicator. The water in the salt solution is then evaporated, as before, to leave crystals of salt.

More ways to make soluble salts

Learning objectives

- explain how soluble salts can be made by reacting acids with metals or with insoluble bases
- describe how salt solutions can be crystallised to produce a solid salt
- explain that the salt formed depends on the combination of acid and base used.

Examiner feedback

You do not need to recall the reactivity series of metals for the chemistry examinations, as the Data Sheet contains a copy of it.

Salts

Sodium chloride, common table salt, is the salt used to flavour and preserve food. It is soluble, i.e. it dissolves in water. Copper sulfate is another soluble salt. Copper sulfate solution is used as a **fungicide** spray for grapevines. The copper ions stop fungal spores germinating by affecting their enzymes, so the fungicide has to be applied before the fungus appears.

Copper sulfate solution is used to prevent fungal disease on grapevines.

Metal or metal oxide?

The reaction of an acid with a metal, or with a metal oxide, is a common way to make soluble salts like sodium chloride and copper sulfate. In general:

$$\text{metal} + \text{acid} \longrightarrow \text{salt} + \text{hydrogen}$$

$$\text{metal oxide} + \text{acid} \longrightarrow \text{salt} + \text{water}$$

Different acids produce different salts:

- Hydrochloric acid makes chlorides.
- Sulfuric acid makes sulfates.
- Nitric acid makes nitrates.

Sodium chloride might be made by reacting sodium with hydrochloric acid, and copper sulfate by reacting copper with sulfuric acid. However, sodium is a very reactive metal, so it would be dangerous to react it with acid. In contrast, copper is not reactive enough and would not form copper sulfate. To avoid problems like these, metal oxides are often used instead of metals.

Metal oxides are bases. They react with acids and neutralise them. The salt formed depends upon the combination of acid and base used. For example:

- copper oxide reacts with sulfuric acid to make copper sulfate and water
- nickel oxide reacts with sulfuric acid to make nickel sulfate and water.

most reactive

potassium
sodium
calcium
magnesium
aluminium
carbon
zinc
iron
tin
lead
hydrogen
copper
silver
gold
platinum

least reactive

Figure 1 Metals above hydrogen in the **reactivity series** usually react with acids, but those below hydrogen usually do not.

Practical

Many metal oxides are **insoluble**. When they react with acids, you know that the reaction is complete when there is some metal oxide left over.

The diagrams show how copper sulfate is made by reacting copper oxide with sulfuric acid. Here is the equation for the reaction:

$$\text{copper oxide } + \text{ sulfuric acid } \longrightarrow \text{ copper sulfate } + \text{ water}$$
$$CuO(s) \quad + \quad H_2SO_4(aq) \quad \longrightarrow \quad CuSO_4(aq) \quad + \quad H_2O(l)$$

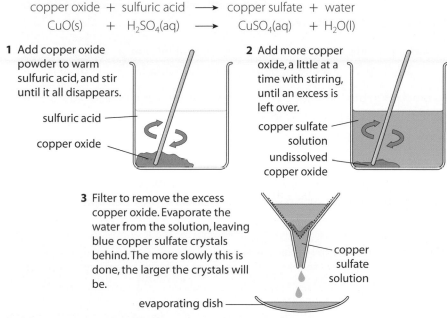

1 Add copper oxide powder to warm sulfuric acid, and stir until it all disappears.

sulfuric acid

copper oxide

2 Add more copper oxide, a little at a time with stirring, until an excess is left over.

copper sulfate solution

undissolved copper oxide

3 Filter to remove the excess copper oxide. Evaporate the water from the solution, leaving blue copper sulfate crystals behind. The more slowly this is done, the larger the crystals will be.

copper sulfate solution

evaporating dish

Figure 2 Making copper sulfate from copper oxide and sulfuric acid.

Questions

1 Name two soluble salts and give one use of each.

2 Name the salts formed when nitric acid reacts with: **(a)** magnesium **(b)** copper oxide.

3 Explain why you should make potassium chloride using potassium hydroxide, not potassium.

4 Explain why you could make silver nitrate using silver oxide, but not using silver.

5 Describe how you could make copper sulfate using copper oxide and sulfuric acid.

6 Calcium chloride can be made from hydrochloric acid by reaction either with calcium or with calcium oxide. **(a)** Write a word equation for each reaction. **(b)** Describe what you would expect to observe in each reaction. **(c)** Give a reason, other than the reactivity of calcium, why using calcium oxide may be safer.

7 Zinc nitrate ($Zn(NO_3)_2$) can be made by reacting nitric acid (HNO_3) with zinc, or with insoluble zinc oxide (ZnO). Write balanced chemical equations for the two reactions involved.

8 Magnesium oxide, MgO, is only sparingly soluble in water. Magnesium sulfate, $MgSO_4$, is also known as Epsom salts. It is used in bath water to ease aches and pains. Describe how you could prepare dry magnesium sulfate from magnesium oxide. Include a balanced chemical equation in your answer.

Science skills

It is important that scientists can communicate their ideas and discoveries to each other easily and clearly. The French chemist Antoine Lavoisier invented a system, published in 1787, to name chemicals. It led directly to the systems used today. Before Lavoisier's work, it was very difficult to work out what a substance was, or what it contained, from its name.

For example, copper sulfate had several names including 'Roman vitriol', 'blue stone' and 'super-vitriolated copper'. Sulfuric acid had even more names. 'Acid of sulphur' was probably easy enough to understand, but other names were not. Chemistry would be much more difficult to understand if you still had to remember names like vitriolic acid, oil of vitriol and spirit of vitriol.

Taking it further

Copper sulfate is white when anhydrous, but blue when hydrated. This is because a copper ion forms a 'complex ion' in solution. Each copper ion becomes chemically bonded to water molecules by 'coordinate bonds', a type of covalent bond.

Route to A*

Transition metals, such as copper and iron, often form more than one ion. The difference is shown using Roman numbers in brackets: Cu^+ is copper(I) and Cu^{2+} is copper(II). The numbers are not used where a metal commonly forms just one ion.

Making insoluble salts

Learning objectives

- explain how insoluble salts can be made by precipitation reactions
- identify the substances needed to make a particular insoluble salt
- describe how precipitation reactions can be used to remove unwanted ions from solutions.

A yellow precipitate of lead iodide forms when sodium iodide solution is mixed with lead nitrate solution.

Examiner feedback

To make an insoluble salt XY, a combination of X nitrate solution and sodium Y solution will always work.

Precipitation reactions

Insoluble salts can be made by reacting together solutions of two soluble salts. Lead nitrate and sodium iodide are soluble salts. They both dissolve in water to form clear, colourless solutions, but clouds of yellow solid appear when they are mixed together. The yellow solid is a **precipitate** of insoluble lead iodide. It is formed in a **precipitation reaction**:

$$\text{sodium iodide} + \text{lead nitrate} \longrightarrow \text{sodium nitrate} + \text{lead iodide}$$
$$2NaI(aq) + Pb(NO_3)_2(aq) \longrightarrow 2NaNO_3(aq) + PbI_2(s)$$

The brackets in the formula $Pb(NO_3)_2$ show that each unit of lead nitrate contains two nitrate ions. So a unit of lead nitrate contains one lead atom, two nitrogen atoms and six oxygen atoms.

The ions from each soluble salt have 'swapped partners', and when the lead ions and iodide ions meet they form insoluble lead iodide. Any insoluble salt can be made using such precipitation reactions, provided appropriate solutions of soluble salts are available.

Table 1 The main soluble and insoluble salts.

Soluble	Insoluble
all sodium, potassium and ammonium salts	
all nitrates	
most chlorides, bromides and iodides	lead chloride, bromide and iodide silver chloride, bromide and iodide
most sulfates	barium, calcium and lead sulfates
sodium, potassium and ammonium carbonates	most carbonates
sodium, potassium and ammonium hydroxides	most hydroxides

Preparing an insoluble salt

Most chlorides are soluble, but silver chloride is not. Two soluble salts are needed to make it. One must contain silver ions and the other chloride ions. From the table, you can see that all nitrates are soluble, so silver nitrate will be soluble. In addition all sodium, potassium and ammonium salts are soluble, so sodium chloride will be soluble. This is what happens when they are mixed:

$$\text{silver nitrate} + \text{sodium chloride} \longrightarrow \text{silver chloride} + \text{sodium nitrate}$$
$$AgNO_3(aq) + NaCl(aq) \longrightarrow AgCl(s) + NaNO_3(aq)$$

A white precipitate of silver chloride (AgCl) forms. This can be separated by filtration, washed with water and dried in a warm oven.

Practical

Silver chloride crystals can be prepared by mixing silver nitrate solution with sodium chloride solution. The silver chloride precipitate is filtered and washed with distilled water to remove any unreacted substances.

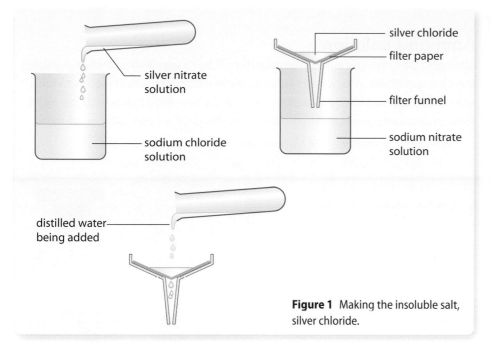

Figure 1 Making the insoluble salt, silver chloride.

Science in action

Silver chloride, silver bromide and silver iodide are light sensitive. The energy supplied by visible light or X-rays is enough to decompose them, forming black crystals of silver. This property makes them useful for traditional photographic film. The increasing use of digital photography, including for X-ray imaging in hospitals, has led to a large decrease in the use of silver for photography.

Science in action

It is important to use the correct amounts of the chemicals when treating water. In 1988 there was an accident at the water works supplying drinking water to Camelford, a town in Cornwall. A tanker driver who was unfamiliar with the site delivered 20 tonnes of aluminium sulfate into the wrong tank by mistake. The dissolved aluminium ions reduced the pH of the water, releasing toxic metals from the pipes. Around 20 000 homes were affected for several weeks afterwards. Health problems included green hair from copper compounds, mouth ulcers, vomiting and rashes. Studies later showed that some people had suffered brain damage due to the aluminium, causing symptoms such as short-term memory loss.

Water treatment

Precipitation reactions can be used to clean up waste **effluent** from factories. This water may contain dissolved transition metal ions such as copper, chromium, cadmium or mercury ions. Transition metal carbonates are insoluble but sodium carbonate is soluble. The effluent can be treated by adding sodium carbonate. Insoluble metal carbonates form precipitates that can be removed by filtering. For example, to treat water contaminated with cadmium:

cadmium ions $+$ carbonate ions $\longrightarrow$ cadmium carbonate

$$Cd^{2+}(aq) \quad + \quad CO_3^{2-}(aq) \quad \longrightarrow \quad CdCO_3(s)$$

Water from a reservoir must be treated so that it is suitable for drinking. Large pieces of solid such as leaves are removed by sieving. But smaller solids remain suspended in the water, making it cloudy. They stay apart because they are negatively charged and are too small to settle out. Aluminium sulfate is added to clump these particles together. Its positively charged aluminium ions attract the tiny particles, forming a precipitate that settles out faster.

Questions

1 Name two soluble salts and two insoluble salts.

2 Explain what a precipitate is.

3 **(a)** Write a word equation for the reaction between ammonium sulfate and lead nitrate. **(b)** Name the precipitate formed in the reaction.

4 Describe how you would make dry lead chromate from lead nitrate solution and potassium chromate solution.

5 Describe three practical uses of precipitation reactions.

6 Explain, with the aid of a suitable example, what happens in a precipitation reaction involving two soluble salts.

7 Mercury sulfate ($HgSO_4$) is insoluble. Explain how waste water from a factory might be treated with sodium sulfate solution to remove soluble Hg^{2+} ions. Include a balanced chemical equation in your answer.

8 Name two soluble salts that could react together to make silver iodide. Write a word equation and a balanced chemical equation for the precipitation reaction.

Making fertilisers and salts

Ammonia solution

Ammonia (NH_3) is a gas at room temperature. It has a very sharp smell and irritates the eyes and throat. Ammonia is very soluble in water. It readily dissolves to form ammonia solution. When it dissolves, some of the ammonia reacts with the water to form ammonium ions and hydroxide ions. This produces an alkaline solution:

ammonia + water $\longrightarrow$ ammonium ions + hydroxide ions

$NH_3(g)$ + $H_2O(l)$ $\longrightarrow$ $NH_4^+(aq)$ + $OH^-(aq)$

Ammonia solution is alkaline. It must be handled with care because it can damage skin and eyes.

Ammonia solution iturns universal indicator solution blue-green.

Ammonium salts

It is often convenient to show ammonia solution as $NH_4OH(aq)$, the compound that would be formed if ammonium ions and hydroxide ions reacted together. Remember, though, that these ions are separate in solution and it is not possible to make solid ammonium hydroxide. Ammonia solution can neutralise acids, just as the other alkalis can. For example:

ammonia solution + sulfuric acid $\longrightarrow$ ammonium sulfate + water

$2NH_4OH(aq)$ + $H_2SO_4(aq)$ $\longrightarrow$ $(NH_4)_2SO_4(aq)$ + $2H_2O(l)$

The volume of hydrochloric acid needed to neutralise the ammonia solution can be determined using a titration (see lesson C3 4.3). It is usual to add a slight excess of ammonia solution to ensure that the reaction is complete. The ammonia is lost as the water is evaporated from the solution.

Ammonium salts as fertilisers

As plants grow, they absorb minerals and water through their root hair cells from the soil. Over time, the soil can become deficient in minerals. These must be replaced to allow the plants to grow well. **Fertilisers** are substances added to soil to replace minerals used up by plants. Ammonium salts are very important artificial fertilisers.

Ammonium salts are water soluble, so they are readily absorbed by roots. The ammonium ion also supplies the plant with nitrogen in a usable form. Plants cannot absorb or use nitrogen gas, but nitrogen is essential for making proteins. The growth of plants in nitrogen-deficient soil is stunted.

'Nitrogenous' fertilisers contain ammonium salts such as ammonium nitrate, ammonium sulfate and ammonium phosphate. Of these three, ammonium nitrate NH_4NO_3 contains the largest percentage of nitrogen, making it the best fertiliser for providing nitrogen.

The crop on the left was grown using nitrogen fertiliser. The crop on the right was grown without it. What differences can you see?

Which method for which salt?

Insoluble salts are made using precipitation reactions involving two solutions. There are three ways to make soluble salts. These involve reactions between an acid and:

- a soluble base (alkali)
- an insoluble base
- a metal.

The flowchart summarises how an appropriate method could be chosen. Remember to correctly identify the acid needed:

- hydrochloric acid for chlorides
- sulfuric acid for sulfates
- nitric acid for nitrates.

Examiner feedback

You must be able to choose an appropriate method to make a named salt, given relevant information. You do not need to recall the flowchart below. This is provided to help you practise choosing appropriate methods when you prepare for the examination.

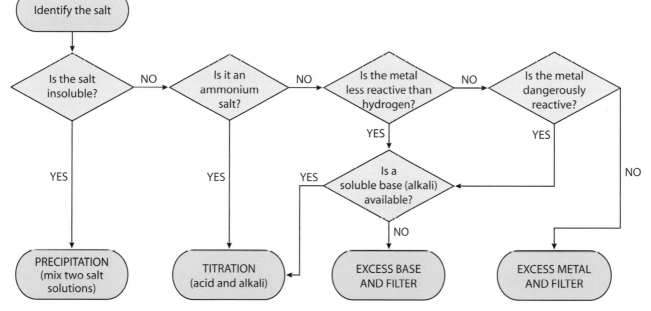

Figure 1 Choosing how to make a particular salt.

Questions

1 Explain, with the help of an equation, why ammonia solution is alkaline.

2 Explain why ammonium salts are useful as fertilisers.

3 Describe how you could prepare solid zinc sulfate using a metal.

4 Explain which method you would use to make:
   (a) potassium chloride **(b)** copper chloride.

5 Ammonia solution itself may be injected into the soil, where it acts as a fertiliser. Describe a benefit and a drawback of doing this.

6 Soluble barium salts such as barium chloride are toxic. However, barium sulfate is useful in medical imaging because it is insoluble and it absorbs X-rays.

It allows doctors to examine the structure of their patients' intestines. Explain how you would make dry barium sulfate.

7 Ammonium nitrate NH_4NO_3 and ammonium sulfate $(NH_4)_2SO_4$ are both used as nitrogenous fertilisers.
   (a) Calculate the percentage of nitrogen by mass in each substance. **(b)** Use your answer to part **(a)** to explain which salt is likely to be most useful as a nitrogenous fertiliser.

8 Ammonium phosphate, made using phosphoric acid, H_3PO_4, is useful as a fertiliser. Explain how you could prepare dry ammonium phosphate. Include a balanced chemical equation in your answer.

Electrolysis

Splitting water

Alessandro Volta invented the electric battery in 1800. Scientists were immediately curious to investigate the effect of electricity on different substances. An English chemist called William Nicholson discovered that electricity broke down acidified water into its elements, hydrogen and oxygen. This process is called electrolysis. It also works with other substances.

What is electrolysis?

Electrolysis is the breaking down of an ionic substance into simpler substances using an electric current. The ions in the ionic substance must be free to move for it to work. In an ionic solid, strong electrostatic forces keep the ions in place. However, when the substance is in aqueous solution or is molten, the ions are able to move.

The molten substance or its solution is called the **electrolyte**. Positive and negative **electrodes** carry the electric current into the electrolyte. Negatively charged ions move to the positive electrode and positively charged ions move to the negative electrode. **Oxidation** and **reduction** are often described in terms of gaining or losing oxygen, but they can also involve electrons:

- oxidation is loss of electrons, and it happens to the negatively charged ions at the positive electrode
- reduction is gain of electrons, and it happens to the positively charged ions at the negative electrode.

Examiner feedback

It helps to remember OIL RIG: oxidation *is* *l*oss of electrons, and *r*eduction *is* *g*ain of electrons. For example, chloride ions lose electrons and are oxidised to chlorine. Zinc ions gain electrons and are reduced to zinc atoms.

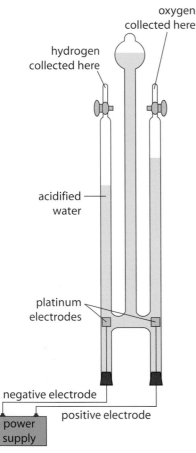

Figure 1 The Hoffman voltameter is often used to demonstrate the electrolysis of water, acidified to increase its **conductivity**.

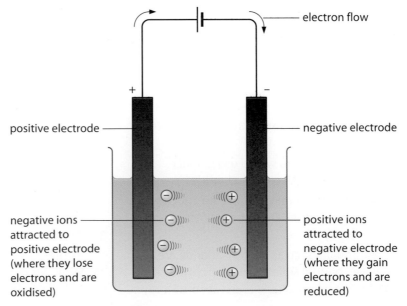

Figure 2 An overview of electrolysis.

Molten ionic substances

Zinc chloride is often used to demonstrate the electrolysis of a molten ionic substance, as it melts at a fairly low temperature (283 °C). Zinc ions move to the negative electrode, where they gain electrons and are **discharged** as zinc metal. Chloride ions move to the positive electrode, where they lose electrons and are discharged as molecules of chlorine gas (Cl_2).

Examiner feedback

Electrolysis is an endothermic process. The energy source is electrical energy rather than thermal energy.

Zinc metal is extracted from zinc compounds by electrolysis. Zinc is coated onto steelwork to help stop it rusting.

Half equations

The reactions at the positive and negative electrodes (**electrode reactions**) can be represented by **half equations**:

At the negative electrode: $Zn^{2+} + 2e^- \longrightarrow Zn$

At the positive electrode: $2Cl^- - 2e^- \longrightarrow Cl_2$

Electrons are shown as e^- in half equations. Notice that zinc ions gain electrons and are reduced, while chloride ions lose electrons and are oxidised. The number of electrons in a balanced half equation must be enough to make the total charge on each side zero.

Taking it further

Half equations can be recombined to give the full balanced equation, if you make sure both half equations have the same number of electrons. If you add both left-hand sides together in the example here, and then both right-hand sides, you get:

$Zn^{2+} + 2e^- + 2Cl^- - 2e^- \longrightarrow Zn + Cl_2$

The electrons cancel out to give:

$Zn^{2+} + 2Cl^- \longrightarrow Zn + Cl_2$ or

$ZnCl_2 \longrightarrow Zn + Cl_2$

Questions

1   In chemistry, what is electrolysis?

2   Explain why electrolysis works if an ionic substance is molten or dissolved in water, but not if it is solid.

3   Explain which products form at each electrode during the electrolysis of molten sodium chloride.

4   Molten lead bromide is sometimes used to demonstrate electrolysis. **(a)** Correctly balance these half equations:

$Pb^{2+} + e^- \longrightarrow Pb \qquad Br^- - e^- \longrightarrow Br_2$

(b) Explain what you would see at each electrode.

5   During the electrolysis of water, hydrogen forms at the negative electrode. Balance this half equation:
$H^+ + e^- \longrightarrow H_2$

6   Balance this half equation for the reaction at the positive electrode during the electrolysis of water:
$OH^- - e^- \longrightarrow H_2O + O_2$

7   Potassium was first isolated by Sir Humphry Davy in 1807. He passed electricity through molten potassium hydroxide, KOH. **(a)** Write a balanced half equation for the reaction at the negative electrode. **(b)** Explain whether oxidation or reduction happens at this electrode.

8   Magnesium is produced commercially by the electrolysis of magnesium chloride, ($MgCl_2$), which is abundant and melts at 714 °C. **(a)** Write a balanced half equation for the reaction at the negative electrode. **(b)** The reaction at the positive electrode can be shown as $2Cl^- - 2e^- \longrightarrow Cl_2$. Explain why it may be shown instead as $2Cl^- \longrightarrow Cl_2 + 2e^-$. **(c)** Suggest and explain two reasons why the production of magnesium is expensive.

Electroplating

A thin layer of metal

Electroplating involves using electrolysis to create a thin layer of metal over the surface of an object. This may be done to improve an object's appearance or to make it more resistant to corrosion or abrasion. It may also be done to reduce costs: for example, inexpensive jewellery is sometimes made from a fairly cheap metal such as nickel, and then plated with gold or silver.

The object to be electroplated is used as the negative electrode. The positive electrode is a piece of the electroplating metal. The electrolyte is a solution containing ions of the electroplating metal. During electrolysis, the metal ions are attracted to the negatively charged object. They gain electrons and are discharged as a thin layer of metal atoms on its surface.

High-end electrical and computer connectors are gold-plated. Gold is corrosion-resistant and a very good conductor of electricity.

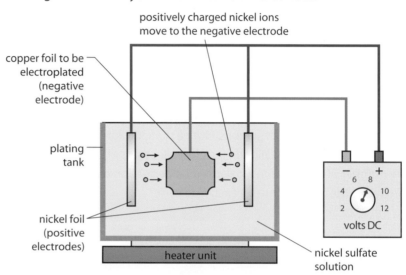

Figure 1 Electroplating copper with nickel. The negative electrode is a piece of copper foil and the positive electrode is a piece of nickel. The electrolyte is a solution of nickel sulfate.

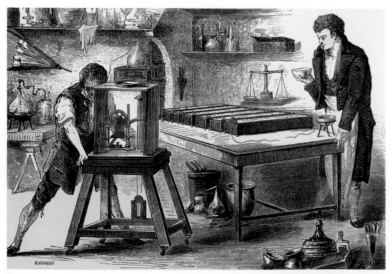

In 1807 Humphry Davy (on the right) became the first person to isolate potassium, using electrolysis. He produced only hydrogen at the negative electrode until he changed the electrolyte from potassium hydroxide solution to molten potassium hydroxide.

Electrolysis of solutions

Electrolysis of a solution of metal ions does not always produce pure metal at the negative electrode. In the electrolysis of copper chloride solution, copper forms at the negative electrode and chlorine at the positive electrode. However, if potassium chloride is used instead, hydrogen forms at the negative electrode rather than potassium. Why is this?

The reason concerns the reactivity series (see lesson C2 6.3). A solution of a metal chloride contains two different positive ions: the metal ions themselves, and hydrogen ions H^+ from the water. Copper is less reactive than hydrogen. When the positive ions reach the negative electrode, copper ions are discharged in preference to hydrogen ions. On the other hand, hydrogen is less reactive than potassium. Hydrogen ions are discharged at the negative electrode in preference to potassium ions.

Which non-metal?

Non-metals are produced at the positive electrode during electrolysis. In the electrolysis of copper chloride solution, chlorine is produced at the positive electrode. However, when copper sulfate or copper nitrate solution are used, oxygen is produced instead.

This happens because of differences in how easily different negative ions are discharged in solution. Copper chloride solution contains chloride ions Cl^-, plus hydroxide ions OH^- from the water. Chloride ions are more easily discharged than hydroxide ions. The other halide ions, bromide Br^- and iodide I^-, are also more easily discharged. On the other hand, hydroxide ions are more easily discharged than sulfate ions SO_4^{2-} or nitrate ions NO_3^-. So oxygen is produced during the electrolysis of copper sulfate solution or copper nitrate solution.

Questions

1. Explain why hydrogen is produced when sodium chloride solution is electrolysed.

2. Explain why oxygen is produced when sodium nitrate solution is electrolysed.

3. Name the ions present in the following solutions: **(a)** lead chloride **(b)** silver nitrate **(c)** potassium hydroxide.

4. Name the products formed at each electrode during the electrolysis of each solution in question **3**.

5. Describe how a steel knife could be electroplated with silver using silver nitrate solution.

6. Explain why the electrolysis of dilute hydrochloric acid $HCl(aq)$ produces hydrogen and chlorine, but the electrolysis of dilute sulfuric acid $H_2SO_4(aq)$ produces hydrogen and oxygen.

7. Copper nitrate dissolves in water to produce a blue solution of ions. It can be electrolysed using graphite electrodes. **(a)** Identify the ions in the solution. **(b)** Write a balanced half equation for the reaction that takes place at the negative electrode. **(c)** Suggest why the blue colour fades.

8. Food cans are made from steel electroplated with tin. The tin forms a protective layer, preventing food reacting with the steel. **(a)** Suggest, with reasons, suitable substances for each electrode and for the electrolyte. **(b)** Describe what you would see during the electroplating process. **(c)** Write a balanced half equation for the discharge of tin ions, Sn^{2+}.

A*

Aluminium manufacture

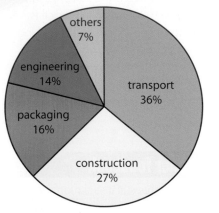

Figure 1 Aluminium is strong, lightweight and resists corrosion. It has many uses.

Science in action

Aluminium is a reactive metal but, unlike iron and steel, it is resistant to corrosion. It is protected from corrosion by a natural microscopically thin layer of aluminium oxide. This stops air and water reaching the metal below and reacting with it. The ability of aluminium to resist corrosion, combined with its strength, low density and ability to conduct electricity and heat well, makes it a very useful metal. For example, aluminium is used to make aircraft and trains, window frames, cooking foil and drinks cans, television aerials and overhead electricity cables.

Examiner feedback

Make sure you understand why cryolite is used in the manufacture of aluminium, including its economic and environmental benefits.

Making aluminium by electrolysis

The American chemist Charles Hall and the French chemist Paul Héroult were both born in 1863 and both died in 1914. In 1886, working on opposite sides of the Atlantic, the two men independently invented a way to produce aluminium by electrolysis. Aluminium is still manufactured today using the Hall–Héroult process. Around 39 million tonnes of aluminium are manufactured in the world each year.

Aluminium smelting uses huge amounts of electricity. This smelting plant was built next to a hydroelectric power plant to provide cheap, plentiful electricity.

Before Hall and Héroult

Aluminium is the most abundant metal in the Earth's crust. It is strong, has a low density and is resistant to corrosion. However, it is high in the reactivity series and is naturally found combined with other elements. For example, **bauxite** is the most abundant aluminium **ore**. The aluminium is present as aluminium hydroxide, which can be processed to make aluminium oxide (Al_2O_3).

Aluminium is too reactive to be easily extracted from aluminium oxide so, for many years, the existence of aluminium was suspected but could not be confirmed. Pure aluminium was first isolated in 1825 by heating aluminium chloride with potassium, an even more reactive metal. Potassium has to be extracted using electrolysis, so the whole process was very expensive. Aluminium was more expensive at that time than silver. The work of Hall and Héroult led to a big reduction in the cost of aluminium manufacture and it became economic to produce aluminium on a large scale.

The Hall–Héroult process

Aluminium oxide is insoluble in water, so it must be molten to allow an electric current to pass through it. Unfortunately, aluminium oxide has a very high melting point, 2054 °C. Heating the oxide to such a high temperature would be very costly.

Hall and Héroult discovered that aluminium oxide dissolves in molten **cryolite** at 1012 °C, a much lower temperature than the melting point of aluminium oxide. Using a molten mixture of aluminium oxide and cryolite for the electrolyte involves less heating. Since the electrolyte is heated electrically, this reduces the electricity costs of manufacturing aluminium. Lower electricity use also means reduced emissions of waste gases to the environment, if the electricity is generated using fossil fuels.

Electrolysis takes place in a steel container, known as a cell. This is lined with carbon, which acts as the negative electrode. The positive electrodes are also made from carbon. They dip into the electrolyte, the molten mixture of aluminium oxide and cryolite. An electric current of around 200 000 A is passed through the electrolyte. This is equivalent to the current drawn by around 15 000 domestic kettles at once.

Balanced half equations for the aluminium cell

Aluminium ions are positively charged and move to the negative electrode. Here they gain electrons and are reduced to molten aluminium:

$$Al^{3+} + 3e^- \longrightarrow Al$$

Oxide ions are negatively charged and move to the positive electrode. Here they lose electrons and are oxidised to oxygen:

$$2O^{2-} \longrightarrow O_2 + 4e^-$$

The oxygen reacts with the hot carbon electrodes to produce carbon dioxide. This reaction gradually wears away the electrodes, so they must be replaced regularly.

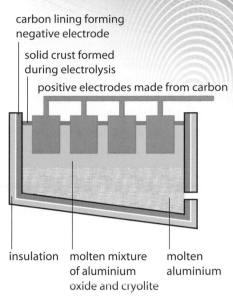

carbon lining forming
negative electrode

solid crust formed
during electrolysis

positive electrodes made from carbon

insulation   molten mixture   molten
             of aluminium   aluminium
             oxide and cryolite

Figure 2 A cross-section through an aluminium cell.

Questions

1   Explain why aluminium is manufactured using electrolysis.

2   Describe what happens at each electrode during aluminium manufacture.

3   Suggest why aluminium costs around three times as much per tonne as steel, even though it is the most abundant metal in the Earth's crust.

4   Explain why cryolite is used in the manufacture of aluminium.

5   Write a balanced half equation for the reaction at each electrode, and an overall balanced chemical equation for the production of aluminium from aluminium oxide.

6   Explain why aluminium smelters are often sited near sources of hydroelectricity.

7   **(a)** Explain why, in aluminium manufacture, reduction happens at the negative electrode and oxidation happens at the positive electrode.
    (b) Describe an environmental problem associated with a product of aluminium manufacture by electrolysis.

8   Titanium is another abundant metal that is difficult to extract from its ore. A new method of manufacture involves electrolysing a mixture of titanium dioxide and molten calcium chloride. **(a)** Balance this half equation and explain at which electrode it takes place: $TiO_2 + e^- \longrightarrow Ti + O^{2-}$ **(b)** Explain why oxide ions move through molten calcium chloride. **(c)** Explain, with the help of two equations, why carbon dioxide is produced at the other electrode, which is made from carbon.

Science skills   Table 1 compares some of the properties of copper and aluminium.

Table 1 Properties of copper and aluminium

Property	Copper	Aluminium
density in g/cm³	8.9	2.7
cost in £/tonne	4400	1320
electrical conductivity	1.6	1.0

a Use the information to explain why aluminium is used for overhead electricity cables, rather than copper.

The chlor-alkali industry

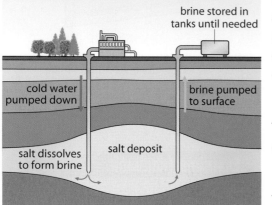

Figure 1 The concentrated sodium chloride solution produced by solution mining is called brine.

Mining solutions

Deep beneath the fields of Cheshire the buried remains of an ancient dried-out sea are being mined. Water is forced through boreholes in underground deposits of rock salt. The water dissolves the salt, forming a saturated salt solution. This is pumped to the surface where it becomes the raw material for the chlor-alkali industry.

Electrolysing sodium chloride solution

Sodium chloride solution contains a mixture of ions: positively charged sodium and hydrogen ions (Na^+, H^+) and negatively charged chloride and hydroxide ions (Cl^-, OH^-). During electrolysis, the positively charged ions move to the negative electrode. Since hydrogen is less reactive than sodium, the hydrogen ions are discharged and hydrogen gas is produced:

$$2H^+ + 2e^- \longrightarrow H_2$$

The negatively charged ions move to the positive electrode. Chloride ions are discharged rather than the hydroxide ions, and chlorine gas is produced:

$$2Cl^- \longrightarrow Cl_2 + 2e^-$$

The remaining sodium ions and hydroxide ions form sodium hydroxide solution. This is the alkali in the 'chlor-alkali' name. The following equation summarises the overall process:

$$\text{sodium chloride} + \text{water} \longrightarrow \text{hydrogen} + \text{chlorine} + \text{sodium hydroxide}$$
$$2NaCl(aq) + 2H_2O(l) \longrightarrow H_2(g) + Cl_2(g) + 2NaOH(aq)$$

Examiner feedback

Remember that sodium metal is too reactive to be produced from sodium chloride solution, but it would be produced if molten sodium chloride were electrolysed instead.

Practical

The electrolysis of sodium chloride solution is easily demonstrated in the laboratory. The presence of hydrogen is confirmed using a lighted splint, which ignites the gas with a 'pop'. The presence of chlorine is confirmed using damp starch iodide paper (or by using damp litmus paper that is bleached), which turns black. The alkaline solution turns universal indicator solution purple.

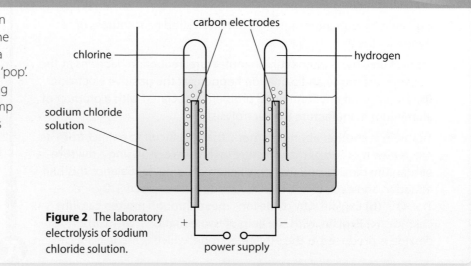

Figure 2 The laboratory electrolysis of sodium chloride solution.

The industrial process

The three products of the chlor-alkali process should not be allowed to mix; otherwise unwanted **by-products** such as sodium chlorate are made. There are different solutions to the problem. One design is called the membrane cell. It contains a polymer sheet, the membrane, which allows positively charged ions to pass through, but not negatively charged ions. As a result, the chlorine and sodium hydroxide cannot react with each other.

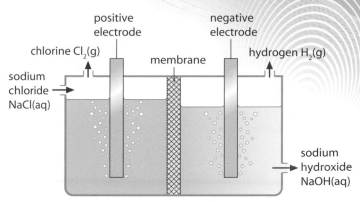

Figure 3 Electrolysis of sodium chloride solution in a membrane cell.

Using the three products

Hydrogen is used in the manufacture of ammonia. It is also increasingly important as a fuel. Chlorine is used to kill bacteria in drinking water. It is used in the manufacture of chloroethene, the monomer for poly(chloroethene) or PVC. Chlorine and sodium hydroxide make sodium chlorate, used as bleach. Sodium hydroxide is used in the manufacture of paper and soap.

Chlorine is used to kill bacteria in swimming pool water.

Questions

1 Write a word equation for the overall process of electrolysing sodium chloride solution.

2 **(a)** Describe two uses for each product from the electrolysis of sodium chloride solution. **(b)** Explain why chlorine should not be allowed to mix with the other two products.

3 Write a balanced half equation for the reaction at each electrode during the electrolysis of sodium chloride solution, and an overall balanced chemical equation.

4 Explain why sodium hydroxide is formed by the electrolysis of sodium chloride solution, but not by the electrolysis of molten sodium chloride.

5 The three products of the chlor-alkali process are formed in a fixed ratio, with equal amounts of the two gases and double the amount of the alkali. Suggest, giving your reasons, two problems that this might cause the chlor-alkali industry.

6 Hydrogen produces only water vapour when it burns. Suggest another reason why it may be viewed as an 'environmentally friendly' fuel.

7 Worldwide, only 20% of electricity is generated using renewable resources. The rest is generated using fossil fuels or nuclear fuels. The chlor-alkali process uses around one per cent of the world's electricity. To what extent might this information affect the view of hydrogen as an 'environmentally friendly' fuel?

8 An older design of cell, the mercury cell, uses a layer of flowing mercury as one of its electrodes. Sodium is produced and dissolves in the mercury, forming a mixture called an amalgam. **(a)** Write a balanced half equation for the production of sodium from sodium ions. **(b)** Explain which electrode, positive or negative, the mercury is. **(c)** The amalgam is mixed with water, letting the sodium react with water to form sodium hydroxide and hydrogen. Write a balanced chemical equation for this reaction. **(d)** Write a balanced chemical equation for the production of sodium chlorate, $NaClO$, from sodium hydroxide and chlorine.

ISA practice: the best metal for hand-warmers

Many metals in the reactivity series react with 1 mole per dm^3 copper sulfate solution to transfer measurable amounts of energy into the solution. This raises the temperature of the solution.

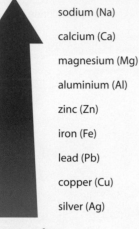

most reactive

sodium (Na)

calcium (Ca)

magnesium (Mg)

aluminium (Al)

zinc (Zn)

iron (Fe)

lead (Pb)

copper (Cu)

silver (Ag)

least reactive

Figure 1 The metal reactivity series.

Section 1

1   Write a hypothesis about the position of a metal in the reactivity series and how it may affect the amount of energy transferred to copper sulfate solution. Use information from your knowledge of displacement reactions to explain why you made this hypothesis.

(3 marks)

2   Describe how you could carry out an investigation into this factor.

You should include:

- the equipment that you would use
- how you would use the equipment
- the measurements that you would make
- a risk assessment
- how you would make it a fair test.

You may include a labelled diagram to help you to explain the method.

In this question you will be assessed on using good English, organising information clearly and using specialist terms where appropriate.   *(9 marks)*

3   Design a table that you could use to record all the data you would obtain during the planned investigation.

(2 marks)

Total for Section 1: 14 marks

Section 2

Two students, Study Group 1, carried out an investigation into the metals. They investigated the first two metals in the first lesson, and two days later tested the other three. Their results are shown in Figure 1.

> Magnesium raised temperature of solution by 11°C
>
> Iron raised temperature of solution by 2°C
>
> Zinc raised temperature of solution by 3°C
>
> Copper did not raise temperature of solution at all
>
> Aluminium raised temperature of solution by 7°C

Figure 1 Study Group 1's results.

4   **(a)** Plot a graph of these results.   *(4 marks)*

   (b) What conclusion can you draw from the investigation about a link between the temperature change and the reactivity series? You should use any pattern that you can see in the results to support your conclusion.   *(3 marks)*

   (c) Do the results support the hypothesis you wrote in answer to question 1? Explain your answer. You should quote some figures from the data in your explanation.   *(3 marks)*

Opposite are the results from three more study groups.

Table 1 shows the results of Study Group 2, two other students who investigated the hypothesis. They used five metals and recorded the temperature change of their solutions after 10 minutes.

Table 1 Study Group 2's results.

Metal	Temperature change of solution 10 minutes after adding metal/°C
magnesium	6
iron	0
zinc	1
copper	0
aluminium	3

Study Group 3 was a third group of students. Their results are given in Table 2. They used 50 cm³ of solution and 5 g of metal.

Table 2 The results obtained by Study Group 3.

Metal	Temperature change of the solution/°C			
	Test 1	Test 2	Test 3	Mean of tests
copper	0	1	0	0
silver	1	0	1	2
zinc	6	5	6	6
magnesium	22	24	21	22
calcium	5	4	4	5

Study Group 4 were researchers for the hand-warmer manufacturers looked on the internet and found out how much each of the metals would cost to buy. This information is shown in Table 3.

Table 3 Cost of each of the metals used in the investigations.

Metal	Cost of 25 g of the metal/£
magnesium	0.04
iron	0.02
zinc	0.04
copper	0.15
aluminium	0.04
calcium	2.58
silver	0.43

Study Group 4 also measured the energy released by each metal when 10 g of metal is displaced, and when 1 mole of the metal (A_r) is displaced. Table 4 shows the results.

Table 4 Energy released by each metal.

Metal	Energy released/kJ	
	10 g displaced	1 mole of metal (A_r) displaced
magnesium	461	1105.4
iron	131.5	736.4
zinc	122.7	797.5
copper	n/a	n/a
aluminium	430.6	1162.6
calcium	n/a	n/a
silver	n/a	n/a

5 (a) Draw a sketch graph of the results from Study Group 2. *(3 marks)*

   (b) Look at the results from Study Groups 2 and 3. Does the data support the conclusion you reached about the investigation in question 5(a)? Give reasons for your answer. *(3 marks)*

   (c) The data contain only a limited amount of information. What other information or data would you need in order to be more certain whether the hypothesis is correct or not?
   Explain the reason for your answer. *(3 marks)*

   (d) Look at Study Group 4's results. Compare them with the data from Study Group 1. Explain how far the data supports or does not support your answer to question 5(b). You should use examples from Study Group 4's results and from Study Group 1. *(3 marks)*

6 (a) Compare the results of Study Group 1 with Study Group 2. Do you think that the results for Study Group 1 are *reproducible*?
   Explain the reason for your answer. *(3 marks)*

   (b) Explain how Study Group 1 could use results from other groups in the class to obtain a more *accurate* answer. *(3 marks)*

7 Applying the results of the investigation to a context.

   Suggest how ideas from the original investigation and the other studies could be used by the manufacturers to decide which metal they should use to obtain a temperature rise of 20 °C using only 25 g of metal. *(3 marks)*

   Total for Section 2: 31 marks
   Total for the ISA: 45 marks

Assess yourself questions

1. Explain what the symbols (s), (l), (g) and (aq) are, and what they mean. *(5 marks)*

2. Figure 1 shows how copper sulfate crystals can be made.

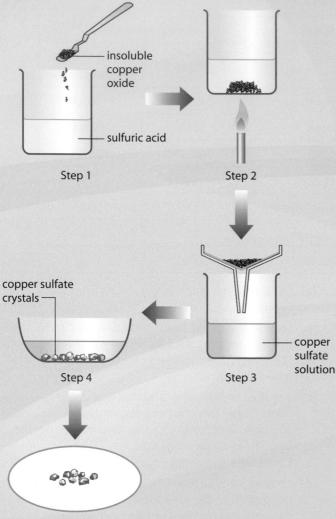

insoluble copper oxide

sulfuric acid

Step 1

Step 2

copper sulfate crystals

Step 4

Step 3

copper sulfate solution

Step 5

Figure 1 Making copper sulfate crystals.

(a) Explain why the mixture of sulfuric acid and copper oxide is warmed. *(1 mark)*

(b) Write a word equation for the reaction. *(2 marks)*

(c) Why is the copper sulfate solution filtered? *(1 mark)*

(d) Describe how you could make copper sulfate crystals form quickly from the solution. *(1 mark)*

3. The information below describes how silver chloride, an insoluble salt, can be made.

Mix silver nitrate solution with an excess of sodium chloride solution. Filter the precipitate of silver chloride from the mixture. Wash it several times with de-ionised water. Dry the silver chloride and store it in a dark bottle.

(a) What is a precipitate? *(1 mark)*

(b) Why was an *excess* of sodium chloride used in the preparation? *(1 mark)*

(c) How do you know, from the information in the paragraph above, that silver chloride is insoluble rather than soluble in water? *(1 mark)*

(d) Describe one method that could be used to dry the silver chloride precipitate. *(1 mark)*

(e) Write a word equation for the reaction between silver nitrate and sodium chloride. *(2 marks)*

4. Table 1 shows which substances are likely to be soluble and which are likely to be insoluble.

Table 1 Soluble and insoluble.

Soluble	Insoluble
all sodium salts	
all nitrates	
most other chlorides	lead chloride and silver chloride
most other sulfates	lead sulfate and barium sulfate
sodium hydroxide	most other hydroxides

Use the information in the table to help you answer the following questions.

(a) Name *two* insoluble lead salts. *(2 marks)*

(b) Which substance, barium nitrate or barium hydroxide, would be suitable to mix with sodium sulfate solution to make barium sulfate? Give reasons for your answer. *(2 marks)*

(c) Name *two* soluble substances that could be mixed together to make lead chloride. *(2 marks)*

5. Soluble salts can be made from acids by reacting them with metals, with insoluble bases, or with alkalis. Explain, giving reasons, the most suitable method to prepare the following salts from hydrochloric acid.

(a) potassium chloride *(3 marks)*

(b) copper chloride *(3 marks)*

6. Different acids produce different salts.

(a) Name the salt produced by each of the following pairs of reacting substances.

   (i) sodium hydroxide and hydrochloric acid *(1 mark)*

   (ii) potassium hydroxide and sulfuric acid *(1 mark)*

   (iii) zinc oxide and sulfuric acid *(1 mark)*

(b) Name the acid needed to produce sodium nitrate. *(1 mark)*

7 Ammonium salts are important as fertilisers.

    (a) What is a fertiliser? *(1 mark)*

    (b) Name the salt produced by reacting ammonia solution with the following acids.

        (i) sulfuric acid *(1 mark)*

        (ii) nitric acid *(1 mark)*

    (c) Ammonium chloride can be made from ammonia solution.

        (i) Name the acid needed. *(1 mark)*

        (ii) Describe what would happen to the pH of the ammonia solution if excess hydrochloric acid was added to it. *(1 mark)*

8 The presence of certain ions makes solutions acidic or alkaline.

    (a) Give the name or formula of the following ions.

        (i) The ion that makes solutions acidic. *(1 mark)*

        (ii) The ion that makes solutions alkaline. *(1 mark)*

    (b) Write a balanced chemical equation, involving these ions, to represent neutralisation. Include the correct state symbols. *(3 marks)*

9 Figure 2 shows the electrolysis of molten lead iodide.

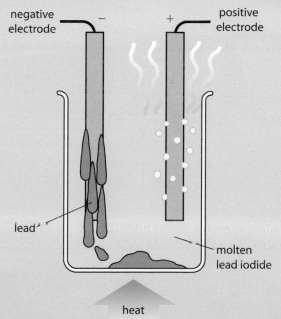

Figure 2 Molten lead iodide can be electrolysed using carbon electrodes.

    (a) What is meant by the word *electrolysis*? *(3 marks)*

    (b) Explain why the positively charged ions move to the negative electrode. *(1 mark)*

    (c) Name the product formed at the positive electrode. *(1 mark)*

    (d) Explain why an electric current will pass through molten lead iodide, but not through solid lead iodide. *(2 marks)*

10 Aluminium is manufactured by the electrolysis of a molten mixture of aluminium oxide and cryolite, as seen in Figure 3.

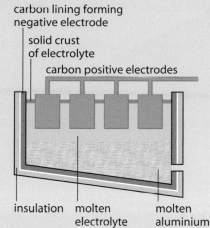

Figure 3 Aluminium is extracted from aluminium oxide by electrolysis.

    (a) Why is molten cryolite used? *(2 marks)*

    (b) Aluminium forms at the negative electrode.

        (i) What does this tell you about aluminium ions? *(1 mark)*

        (ii) Explain, in terms of electrons, whether the aluminium ions are oxidised or reduced. *(2 marks)*

    (c) Explain why carbon dioxide is produced at the positive electrode. *(2 marks)*

11 When Humphry Davy first attempted to produce potassium by electrolysis in the 1800s, he used potassium hydroxide solution. He was unsuccessful until he used molten potassium hydroxide.

    (a) Explain why an electric current can pass through potassium hydroxide solution. *(2 marks)*

    (b) Explain why hydrogen, rather than potassium, forms at the negative electrode during the electrolysis of potassium hydroxide solution. *(2 marks)*

    (c) Name the gas that would form at the positive electrode during both of Humphry Davy's experiments. *(1 mark)*

    (d) Name the gas that forms at the positive electrode during the electrolysis of:

        (i) copper chloride solution *(1 mark)*

        (ii) copper sulfate solution *(1 mark)*

12 The electrolysis of sodium chloride solution is an important industrial process.

    (a) Name the gas formed at the negative electrode. *(1 mark)*

    (b) Complete and balance the half equation below.

$$Cl^- \longrightarrow Cl_2 \qquad \textit{(2 marks)}$$

    (c) Name the alkali produced during the electrolysis of sodium chloride solution, and give one important use for it. *(2 marks)*

GradeStudio Route to A*

Here are three students' answers to the following question:

The electrolysis of sodium chloride solution is very useful industrially.

(a) Explain the meaning of *electrolysis*. *(2 marks)*

(b) (i) Name Gas A.

   (ii) Name the alkali formed.

   (iii) Complete the half equation for the reaction at the positive electrode, and explain whether it is an oxidation or a reduction reaction:

   $$Cl^- \longrightarrow Cl_2 \quad (4 \text{ marks})$$

In this question you will be assessed on using good English, organising information clearly and using specialist terms where appropriate.

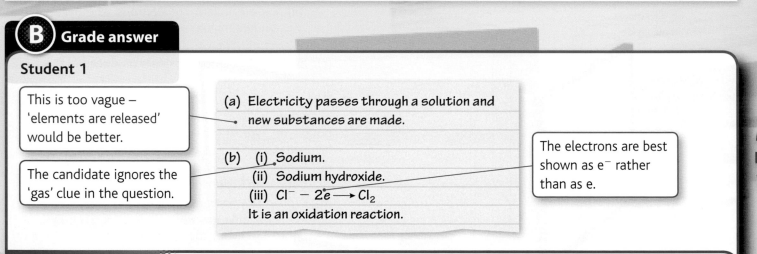

Figure 1 A membrane cell for the electrolysis of salt solution.

Read the answers together with the examiner comments. Then check what you have learnt and try putting it into practice in any further questions you answer.

B Grade answer

Student 1

> This is too vague – 'elements are released' would be better.

> The candidate ignores the 'gas' clue in the question.

> The electrons are best shown as e⁻ rather than as e.

(a) Electricity passes through a solution and new substances are made.

(b) (i) Sodium.

   (ii) Sodium hydroxide.

   (iii) Cl⁻ − 2e ⟶ Cl₂

   It is an oxidation reaction.

Examiner comment

The candidate has not answered the question carefully enough. In part (a) he knows that electricity is involved and gains one mark. However, the remainder of his answer is too vague to gain the second mark. He fails to mention that the electric current passes through a molten or aqueous ionic compound.

In part (b) (i), the candidate forgets that sodium is too reactive to form in aqueous solution and that hydrogen will form instead. He is unable to correctly balance the half equation in part (b) (iii) but correctly states that it is an oxidation reaction. However, he forgets to explain why it is.

MOVING UP THE GRADES

Read the whole question carefully.

• Look at the number of marks available for each section so you know how many different points to make.

• Use both your knowledge *and* the information given to you in the question.

• Make sure you use scientific terminology correctly.

• Take care not to write a correct and an incorrect answer to the same question: give one answer and move on.

A Grade answer

Student 2

(a) This is when an electric current passes through an ionic compound in solution.

This is one of the ions in solution not the alkali itself.

(b) (i) Hydrogen.
   (ii) Hydroxide.
   (iii) $2Cl^- \longrightarrow Cl_2 - 2e^-$

The candidate shows the correct number of electrons but on the wrong side.

It is an oxidation reaction because it loses electrons.

Take care with the use of the word 'it', which means two different things here.

Examiner comment

The candidate has answered part (a) well. She correctly states that an electric current must pass through, and she links this to an ionic compound in solution. She might have added that the ionic compound could also be molten, or that decomposition takes place. Nevertheless, she gets full marks here. She correctly answers part (b) (i) but then starts to go wrong. Hydroxide ions are produced by alkalis in solution, but they are not alkalis.

The candidate should have identified the alkali itself, sodium hydroxide. In part (b) (ii) an almost correct answer was spoiled by a silly mistake, as the electrons are shown on the wrong side of the equation. She does correctly explain that the reaction is oxidation. However, the answer would have been clearer if the second 'it' had been 'chloride ions' instead.

A* Grade answer

Student 3

(a) Electrolysis is when an electric current is passed through a molten or aqueous ionic compound. This causes the ionic compound to decompose.

This is a more precise term than 'break down' or 'split up'.

(b) (i) Hydrogen.
   (ii) Sodium hydroxide, NaOH.
   (iii) $2Cl^- \longrightarrow Cl_2 + 2e^-$

The candidate begins her answer with what she knows.

Oxidation is loss of electrons, so it is an oxidation reaction because the chloride ions lose electrons.

The candidate gave the correct formula but was only asked to name the alkali.

Examiner comment

The candidate has answered in great detail, taking care to use scientific terms correctly. Her answer to part (a) covers three marking parts but of course she can only gain a maximum of two marks. This is alright when the answer is short, like this one, but she could run out of time if she wrote too much in longer answers. She makes good use of words such as molten, aqueous, ionic and decompose.

The candidate gains full marks in part (b), too. She correctly identifies the gas in part (b) (i) and the alkali in part (b) (ii). However, she was only asked to name the alkali but also gave its

formula. Luckily this formula was correct, otherwise she would have lost the mark because the examiner would have to choose between a correct and an incorrect answer. In part (b) (iii) the candidate gives a more sophisticated balanced half equation in which the electrons are shown on the right-hand side. The Specification shows $2Cl^- - 2e^- \longrightarrow Cl_2$ as one of its examples, and this would have been acceptable too. The candidate gives a clear and complete explanation of why the reaction is an oxidation reaction.

Examination-style questions

1 Hydrocarbons are molecular substances containing carbon and hydrogen only.

(a) Butane is a hydrocarbon with the molecular formula C_4H_{10}. The atoms in a butane molecule are held together by covalent bonds. A diagram of a butane molecule is shown below.

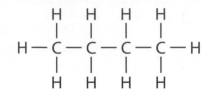

Butane has a low boiling point and is a gas at room temperature. Explain why butane has a low boiling point.

(2 marks)

(b) Propane is a hydrocarbon with the molecular formula C_3H_8. Calculate the percentage by mass of carbon in this compound.

(1 mark)

(c) Cyclohexane is another hydrocarbon.

(i) Analysis showed that it contains 85.7% carbon by mass. Calculate the empirical formula of cyclohexane.

(4 marks)

(ii) The mass spectrum of cyclohexane is shown below. Use the data from the molecular ion to find the relative formula mass of cyclohexane.

(1 mark)

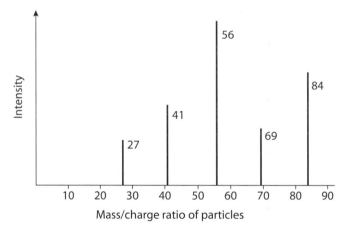

2 Tungsten (W) is a metal. It has many uses, including use in the production of X-rays, as the filament in filament light bulbs and in some electrical contacts. It has a very high melting point and conducts electricity.

(a) Describe the structure and bonding in the metal tungsten, and use it to explain why tungsten conducts electricity and has a high melting point.

(6 marks)

Tungsten metal is often extracted from the compound tungsten oxide (WO_3) produced from the ore wolframite. In the final stage of the extraction process, WO_3 is reacted with hydrogen.

$$WO_3 + 3H_2 \longrightarrow W + 3H_2O$$

(b) Calculate the mass of hydrogen needed to react with 1 kg of tungsten oxide (WO_3).
(Relative atomic masses: H = 1, O = 16, W = 184)

(3 marks)

(c) The maximum theoretical yield for the extraction of tungsten from 200 g of tungsten oxide is 159 g. In an extraction process from 200 g of tungsten oxide, only 140 g of tungsten was formed.

(i) Calculate the percentage yield for this reaction.

(1 mark)

(ii) Give one reason why the yield is less than 100%.

(1 mark)

3 (a) Potassium bromide is an ionic compound made from the reaction between potassium, a group 1 metal, and bromine, a group 7 non-metal. It is used by vets to treat epilepsy in dogs. Potassium bromide has a high melting point and conducts electricity when molten or dissolved in water.

(i) Which of the following is the correct formula of potassium bromide?

KBr_2, K_2Br, KBr, K_2Br_2 *(1 mark)*

(ii) Explain why potassium bromide has a high melting point. *(2 marks)*

(b) Diamond and graphite are forms of the element carbon. They both have a giant covalent structure. Describe and explain one similarity in, and one difference between, the physical properties of diamond and graphite by a consideration of their structure and bonding.

In this question you will be assessed on using good English, organising information clearly and using specialist terms where appropriate. *(6 marks)*

(c) Poly(ethene) is a thermosoftening polymer used to make plastic bags and bottles. Poly(ethene) softens and melts on heating. Melamine is a thermosetting polymer that is used in the manufacture of kitchen worktops. Melamine does not soften or melt on heating. Explain why poly(ethene) softens on heating, but melamine does not, by considering their structure and bonding. *(2 marks)*

4 A student investigated how the rate of reaction between zinc granules and sulfuric acid varied with temperature. The reaction can be represented by the equation:

$Zn(s) + H_2SO_4(aq) \longrightarrow ZnSO_4(aq) + H_2(g)$

She used a gas syringe and a stopclock to time how long it took to produce 20 cm³ of gas at different temperatures.

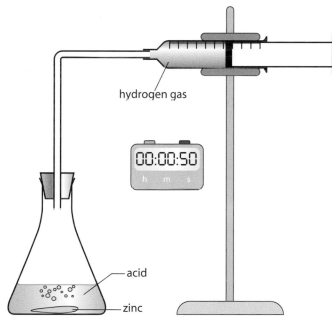

hydrogen gas

00:00:50

acid

zinc

(a) Suggest three variables that the student needed to keep constant in this reaction. *(3 marks)*

Here are the results of her experiment:

Temperature (°C)	0	25	37	50	63	76
Time taken (s)	799	197	101	49	25	12

(i) Plot a graph of her results, adding a line of best fit. *(3 marks)*

(ii) Use your line to predict the time to produce 20 cm³ of gas at 12°C in this experiment. *(1 mark)*

(c) Use your graph to describe, in detail, how the time taken to produce 20 cm³ of gas in this reaction changes with temperature. *(2 marks)*

5 Read the advertisement below and then answer the question.

> # Whizzo re-usable hand-warmers
>
> Fed up with wasting money on those 'use once – throw them away' handwarmers? Our liquid sodium acetate-filled hand-warmer can be used over and over again. When you need it, press the button and watch the crystals grow as the pouch steadily warms up. Then, to 'refill' it with energy to produce heat for next time, simply pop it into a bowl of hot water. The crystals will take in energy as they melt, storing it up for future use. The sodium acetate will stay liquid, even when it cools down. That is, until you press that button again! Use as often as you like and save money!

Use information in the advertisement and your knowledge of the processes involved to answer this question.

Explain, as fully as you can, how energy is stored in this device, how it can be released by starting the process of crystallisation, and how and why it may be re-used over and over again.

In this question you will be assessed on using good English, organising information clearly and using specialist terms where appropriate.   *(6 marks)*

6 Potassium phosphate, K_3PO_4, is used as a fertiliser. It can be made by reacting phosphoric acid, H_3PO_4, with an alkali.

(a) Name an alkali that could react with phosphoric acid to make potassium phosphate and water.

(1 mark)

(b) Write a balanced symbol equation for the reaction between this alkali and phosphoric acid.

(2 marks)

(c) Identify the ions that make solutions acidic or alkaline, and write a balanced equation for the reaction between them.   *(3 marks)*

7   Read the information in the box and then answer the question.

Sodium chloride solution contains sodium ions, Na^+, hydrogen ions, H^+, chloride ions, Cl^-, and hydroxide ions, OH^-. The electrolysis of sodium chloride solution produces three important reagents for the chemical industry. The diagram below shows the apparatus used to electrolyse sodium chloride solution.

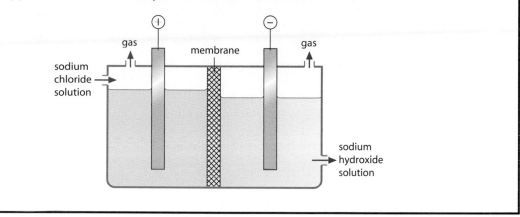

(a) Use information in the box and your knowledge of this process to answer this question.

Explain, as fully as you can, how chlorine, hydrogen and sodium hydroxide solution are formed in this process.

In this question you will be assessed on using good English, organising information clearly and using specialist terms where appropriate.   *(6 marks)*

(b) Explain why sodium chloride solution can undergo electrolysis, but solid sodium chloride cannot.

(2 marks)

(c) Describe an industrial use of sodium hydroxide.   *(1 mark)*

C3 Looking for patterns

When you first turned this page, the picture probably appeared to show a tree trunk. However, if you look carefully you will see a gecko on the tree trunk. Once you know, it probably looks obvious.

An even worse problem faced the early chemists as they searched for patterns in the mass of information they were uncovering about the different elements. They did not know how many elements there were, so they did not know how much information was still missing. And they had no idea why elements were different, as the structure of the atom was not discovered until the 20th century. It is only thanks to their pioneering work that you are now in the fortunate position of knowing what to look for.

In this section you will learn about how the periodic table was discovered and what it can do to help you understand the patterns that lie behind the complex world of chemical reactions.

You will then use your understanding of these underlying principles of chemistry to look at two key resource areas: water supply and energy from chemical reactions (including fuels).

Test yourself

1   What does the group number of the periodic table tell you about the number of electrons in the highest energy level?

2   What are the obvious differences between group 1 metals (such as sodium or potassium) and transition metals (such as iron or copper)?

3   Most fuels contain the elements carbon and hydrogen. What are the two simple chemicals that are formed when these fuels burn in air?

4   What is the difference between an exothermic and an endothermic reaction? Give an example of each type.

Objectives

By the end of this unit you should be able to:

- evaluate the modern periodic table as a tool for the deeper understanding of chemical reactions
- explain the trends found within and across the periodic table groups
- explain how our water supply is purified to make it fit to drink
- evaluate different methods of water treatment
- evaluate the social, economic and environmental consequences of our use of fuels
- explain what is happening in both exothermic and endothermic reactions.

The early periodic table

Newlands' law of octaves

In the early 1800s, about 30 elements were known. Several chemists had spotted that sets of elements, such as lithium, sodium and potassium, had similar properties. However, no one had found an overall pattern in the behaviour of the elements.

By 1864, just over 50 elements were known. The British chemist, John Newlands, spotted a pattern when he arranged these elements in order of atomic mass. He noticed that the properties of the elements seemed to repeat every eighth element. He called this the 'law of octaves', comparing it to musical scales.

H Li Be B C N O F Na Mg Al Si P S Cl K Ca

Figure 1 Newlands' law of octaves worked for the lighter elements he knew: arranged in order on a piano keyboard, lithium, sodium and potassium appear an octave apart, for example.

Unfortunately this simple idea broke down after calcium. For example, in the first four rows of Newlands' table, very unreactive copper appears in the same column as the highly reactive lithium, sodium and potassium. Also, the metals cobalt and nickel had to share the same position as their atomic masses were so similar – yet ended up in the column containing the non-metal gases hydrogen, fluorine and chlorine. Because of these inconsistencies, Newlands' ideas were not accepted at the time.

Table 1 The first four rows of Newlands' table of elements.

H	Li	Ga	B	C	N	O
F	Na	Mg	Al	Si	P	S
Cl	K	Ca	Cr	Ti	Mn	Fe
Co, Ni	Cu	Zn	Y	In	As	Se

Mendeleev's achievement

In 1869, Dmitri Mendeleev, a Russian chemist, devised a table that is the basis of the **periodic table** as we know it. The first part is shown below. Although he did not know about Newlands' arrangement, his basic idea was the same. However, he realised that some elements had not yet been discovered and he left gaps for them (shown by * in the table). This decision enabled him to organise the elements in a repeating pattern. Elements with similar properties occur in **groups** at regular intervals – **periods**.

Figure 2 There are many stories that Mendeleev cut his hair only once a year. When the Russian Tsar ruled that all Russians should shave, Mendeleev refused.

	Group i	Group ii	Group iii	Group iv	Group v	Group vi	Group vii	Group viii
Period 1	H							
Period 2	Li	Be	B	C	N	O	F	
Period 3	Na	Mg	Al	Si	P	S	Cl	
Period 4	K / Cu	Ca / Zn	* / *	Ti / *	V / As	Cr / Se	Mn / Br	Fe Co Ni
Period 5	Rb / Ag	Sr / Cd	Y / In	Zi / Sn	Nb / Sb	Mo / Te	* / I	Ru Rh Pd

Mendeleev went further than earlier scientists and used the patterns in his table to predict the properties of the elements he thought had yet to be discovered. Three of these elements were discovered in the next few years, and his predictions were found to be accurate. Table 3 shows his predictions about the properties of the element between silicon and tin, which he called 'eka-silicon', Es. It also shows the actual properties of the element, called germanium, when it was discovered in 1886.

Table 3 Mendeleev's predictions and the actual properties of germanium.

Property	'Eka-silicon' (Es)	Germanium (Ge)
appearance	grey metal	grey-white metal
melting point	over 800 °C	947 °C
atomic mass	72	72.3
density	5.5 g/cm^3	5.47 g/cm^3
formula of oxide	EsO_2	GeO_2
formula of chloride	$EsCl_4$	$GeCl_4$

Changing the order

In devising his table, Mendeleev did not follow the atomic mass order strictly. For example, he swapped iodine (I) and tellurium (Te) round because iodine fitted better with fluorine, chlorine and bromine than tellurium did. Following the discovery of protons, neutrons and electrons in the early 20th century, it became clear that the elements should be placed in order of atomic number, not atomic mass. Mendeleev thought that the atomic masses of some elements had been measured inaccurately, but he had actually placed the elements in order of atomic number, even though he did not know it.

Mendeleev's table was accepted because many of his remarkable predictions about undiscovered elements proved to be correct. His table has been modified as more elements, including the noble gases, have been discovered, but the modern **periodic table** of the elements is based on Mendeleev's.

Science skills Scientists do not automatically accept new ideas. Newlands was ignored because in general scientists could see no point in his idea. Even Mendeleev's table remained a mere curiosity until its useful predictive power became clear.

Questions

1 Roughly what proportion of the elements we now know was available to John Newlands? Was he aware of this?

2 **(a)** What property did Newlands use to arrange the elements? **(b)** What did he discover when he arranged them in this way? **(c)** Why were his ideas ignored at the time?

3 **(a)** Which as yet undiscovered group of elements was missing from Newlands' model? **(b)** Would his model have worked any better had he known about them?

4 **(a)** What did Mendeleev do differently from Newlands? **(b)** What convinced people that Mendeleev's table was useful?

5 **(a)** In what order did Mendeleev put the elements? **(b)** Why did he not put the elements in strict atomic mass order?

6 Compare the properties of germanium and eka-silicon. How accurate were Mendeleev's predictions?

7 The modern periodic table is ordered by atomic number (proton number). Why did Mendeleev not use this method?

8 Evaluate the contributions of Newlands and Mendeleev to the development of the modern periodic table.

A*

The modern periodic table

The structure of the modern periodic table

The elements in the modern periodic table are ordered by proton number rather than by increasing atomic mass. In this order, the elements naturally fall into groups with similar properties that repeat over periods. Beyond the first 20 elements, the simple pattern of vertical groups and horizontal periods is separated by a block of metals with similar properties – these are called the **transition metals**. The additional blocks that fit in after elements 57 and 89 have been left out at this level, including element 92, uranium.

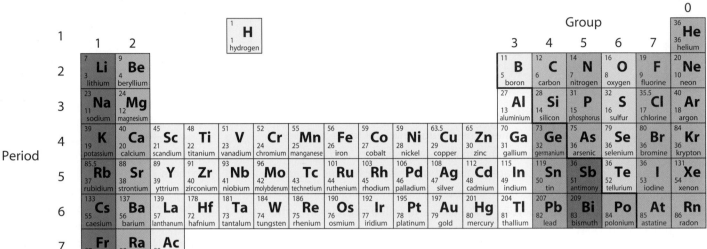

Figure 1 The modern periodic table.

Electronic structure and the periodic table

The atomic number describes the number of protons in the atom, and therefore also the number of electrons and their arrangement around the nucleus. The pattern of the periodic table is precisely linked with the way in which the electrons are arranged around the atoms of each element – the **electronic structure** – see lesson C1 1.3.

The period number tells you how many energy levels containing electrons the element has. The group number tells you how many electrons an element has in its outer energy level. For example, all the elements in period 2 have two energy levels. Lithium is in group 1, so it only has one electron in its outer energy level; neon is in the eighth group (group 0) so it has eight electrons in its outer energy level, which is therefore full.

The power of the periodic table lies in the way it can be used to predict the chemical properties of elements, which are governed by their electronic structure.

So why is it that, in period 4, groups 2 and 3 are split apart by a block of 10 extra elements? The answer is that the third energy level has another 10 electron places to fill, but these are filled *after* the first two places of level 4. This pattern is repeated for periods 5–7.

Metals

More than three-quarters of elements are metals. Metals are found in the middle and on the left of the full periodic table.

Groups 1 and 2 contain very reactive but quite soft metals with low melting points, such as sodium (group 1) and calcium (group 2).

The transition metal block contains the 'everyday' metals such as iron and copper. They are not as reactive as those in groups 1 and 2, but they are harder and stronger and have higher melting points.

Non-metals

Less than a quarter of the elements are non-metals. They are found to the right of the periodic table.

Groups 6 and 7 contain reactive non-metals such as oxygen (group 6) and chlorine (group 7). Groups 6 and 7 contain gases at the top of the group but solids further down.

Group 0 contains the noble gases, a family of highly unreactive non-metals.

Groups 3, 4 and 5

The properties of the elements in groups 3–5 are not so simply defined. The elements at the top are non-metals, but as you move down the table the elements become more metallic. The boundary between metals and non-metals zigzags down to the right. Elements on the boundary show intermediate properties between metals and non-metals; for example, silicon is a semiconductor.

Table 1 The number of electrons in the outer energy level of period 2 elements.

Element	Group	Electrons in outer shell
lithium	1	1
beryllium	2	2
boron	3	3
carbon	4	4
nitrogen	5	5
oxygen	6	6
fluorine	7	7
neon	0	0 (8 in full level below)

Examiner feedback

You will not be asked about the electronic structures of elements beyond calcium in your exam, so just remember that for groups 1–7 the group number is equal to the number of electrons in the highest occupied energy level.

Questions

1. Are the following elements metals or non-metals?
   (a) Lithium (atomic number 3). **(b)** Nitrogen (6).
   (c) Calcium (20). **(d)** Titanium (22). **(e)** Palladium (46).
   (f) Iodine (53).

2. Describe the likely chemical properties of the following elements. **(a)** Xenon (54).
   (b) Strontium (38). **(c)** Cobalt (27). **(d)** Fluorine (9).
   (e) Rubidium (37).

3. Explain why helium is put in with group 0, even though it only has two electrons in its shell.

4. How many electrons do the following elements have in their outer energy level? Ba, Te, Pb, B, Fr.

5. How many energy levels are in use for the following elements? Ra, Mn, W, Ne, Mo.

6. Why have argon (Ar) and potassium (K) been given their positions in the periodic table? What property does not fit this pattern?

7. Explain how the structure of the periodic table relates to electron configuration.

8. To which group (or block) are the following elements likely to belong? Explain your answers. **(a)** X, a silver-grey solid that conducts heat and electricity. It is very hard and will not melt in a Bunsen flame. **(b)** Y, a silvery solid that tarnishes in air. It fizzes in water, giving off hydrogen gas. The atoms lose two electrons to form double-positive ions. **(c)** Z, a brown liquid. Its atoms join together in pairs to form covalent molecules. Its atoms can also gain an electron to form a single-negative ion.

Group 1 – the alkali metals

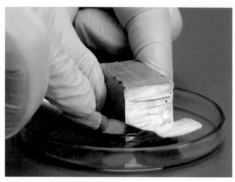

Sodium is easily cut with a knife.

Physical properties

The elements in group 1 of the periodic table are lithium (Li), sodium (Na), potassium (K), rubidium (Rb), caesium (Cs) and francium (Fr). They are known as the **alkali metals**. Francium is radioactive and very rare.

The elements in group 1 all have similar physical properties. They are all soft metals that can be cut with a knife and that conduct heat and electricity. The freshly cut metal is shiny and silvery, but it soon tarnishes in air.

For metals, they have a low density. The first three elements in group 1, lithium, sodium and potassium, all float on water. For metals, they also have relatively low melting and boiling points. As you go down the group, the melting and boiling points get lower, as shown in Table 1.

Table 1 Properties of the alkali metals.

Element	Density / g/cm³	Melting point/°C	Boiling point/°C
lithium (Li)	0.53	181	1342
sodium (Na)	0.97	98	883
potassium (K)	0.86	63	760
rubidium (Rb)	1.53	39	686
caesium (Cs)	1.88	29	669

Typical chemical reactions

The alkali metals also have similar chemical properties. They are all very reactive, and react with water and with the oxygen in air (which is why they tarnish). To stop these reactions, the alkali metals are all stored in oil.

The alkali metals all have one electron in their outer energy level (electron shell). When they react, they lose this electron and form 1^+ ions. When they react with non-metals, they form ionic compounds. These compounds are white and dissolve in water to form colourless solutions. For example, sodium reacts with chlorine to form sodium chloride, which is a white ionic compound containing Na^+ ions. Sodium chloride dissolves in water to form a colourless solution.

$$2Na + Cl_2 \rightarrow 2Na^+Cl^-$$
sodium + chlorine → sodium chloride

When the alkali metals react with water, they form a metal hydroxide and hydrogen. The metal hydroxides dissolve in water, making an alkaline solution.

alkali metal + water → metal hydroxide + hydrogen

For example:

$$2Na + 2H_2O \rightarrow 2NaOH + H_2$$
sodium + water → sodium hydroxide + hydrogen

Trends in chemical properties

These reactions get more and more vigorous as you go down Group 1, as the descriptions of their reactions with water show:

- lithium fizzes steadily
- sodium gets so hot from the reaction that it melts, forming a silver ball
- potassium gets so hot that the hydrogen gas catches fire
- the caesium reaction is so violent that an explosion occurs

This is because alkali metals react by losing their lone outer electron. As you go down the group, the atoms get bigger as more and more energy levels are added. This means that the lone outer electron is further away from the positive nucleus that is holding it in place, increasingly screened by the inner electron shells. This reduces the attraction from the nucleus to the outer electron and makes the electron easier to lose. The bigger the atom, the more reactive the metal.

Sodium reacting with water.

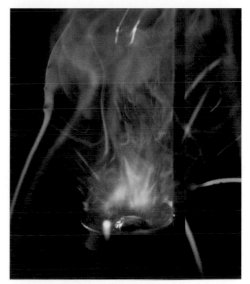

Potassium reacting with water.

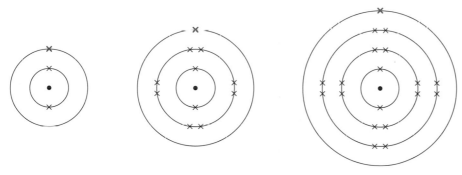

lithium 2,1   sodium 2,8,1   potassium 2,8,8,1

Figure 1 The bigger the atom, the 'looser' the outer electron.

A similar pattern can be seen in group 2, with reactions getting more vigorous from beryllium and magnesium, down through calcium to strontium and barium.

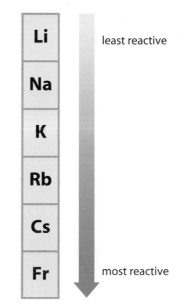

Figure 2 Trend in reactivity of the alkali metals.

Questions

1. Draw bar charts of the melting and boiling points of the alkali metals in order and comment on the trends shown.

2. Which element in group 1 has an anomalous density?

3. List some similarities and differences between the physical properties of group 1 metals and 'everyday' metals such as copper or iron.

4. Explain why the alkali metals are always stored under oil.

5. What is the charge on an alkali metal ion? Explain why this is.

6. Write word and balanced equations for the reaction of potassium with chlorine.

7. Write word and balanced equations for the reaction of caesium in water. Why is this reaction not demonstrated in school laboratories?

8. Describe and explain the trend of reactivity in group 1.

C3 1.4 The transition metals

Learning objectives

- describe the physical and chemical properties of the transition metals and compare them with the alkali metals
- describe some uses for transition metals
- explain why the transition metals have special properties.

Route to A* $\bigstar$

You will need a clear understanding of the range of transition metal 'special properties'.

Transition metals are hard and strong.

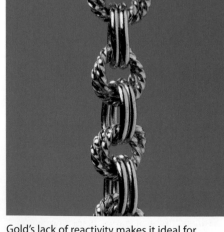

Gold's lack of reactivity makes it ideal for jewellery, as it does not tarnish.

The 'everyday' metals

The transition metals are a block of elements between group 2 and group 3 in the periodic table. Many common metals, including iron, copper, nickel, zinc, silver, gold and platinum, are transition metals. They include most of the metals used for construction, machinery, vehicles, wiring and a host of other everyday uses.

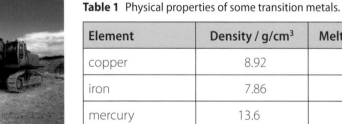

Group

												3
1	2											
		Sc	Ti	V	Cr	Mn	Fe	Co	Ni	Cu	Zn	
		Y	Zr	Nb	Mo	Tc	Ru	Rh	Pd	Ag	Cd	

Figure 1 The transition metals in the periodic table.

The transition metals all have similar physical properties. They are hard and strong with high melting and boiling points (excepting mercury), they have high densities, and they conduct heat and electricity.

Table 1 Physical properties of some transition metals.

Element	Density / g/cm³	Melting point/°C	Boiling point/°C
copper	8.92	1083	2567
iron	7.86	1535	2759
mercury	13.6	−39	357
gold	18.88	1064	3080

They also have similar chemical properties. They react slowly, if at all, with water and oxygen. For example, iron reacts with oxygen and water very slowly when it rusts. Other transition metals, such as gold, do not react at all with water or oxygen.

Compared with group 1 metals, the transition metals:

- have higher melting points (except mercury, which is a liquid at room temperature)
- have higher densities
- are stronger and harder
- are much less reactive, for example with oxygen and water.

Special properties of transition metals

Transition metals have some special properties that other metals do not have. They are useful as **catalysts** – they speed up certain chemical reactions without being used up themselves (see lesson C2 4.5). Some important examples are shown in Table 2.

Table 2 Reactions catalysed by transition metals.

Process	Equation	Catalyst
making margarine	vegetable oils + hydrogen → margarine	nickel
making ammonia	hydrogen + nitrogen → ammonia	iron
catalytic converters	carbon monoxide + nitrogen monoxide → carbon dioxide + nitrogen	platinum

Compounds containing transition metals are usually coloured. Some statues and roofs are made of copper, which reacts with substances in the air to produce an attractive green compound containing copper. Many gemstones are coloured by transition metal compounds. Coloured transition metal compounds are also used in some paint pigments and ceramic glazes.

Transition metals can form ions with different charges. For example, copper can form Cu^+ and Cu^{2+} ions, and iron can form Fe^{2+} and Fe^{3+} ions.

The Statue of Liberty is green because of a compound containing copper.

Rubies are red because of a chromium compound.

Taking it further

You do not need to know about the electronic structure of transition metals for your exam. However, learning about the electronic structure will help you to understand why transition metals have special properties.

Calcium has two electrons in the fourth energy level. However, in the next 10 elements after calcium, electrons slot into the third energy level. The third energy level is not full until it has 18 electrons. This pattern is repeated for the next two periods, forming the transition metals block, which fits in between groups 2 and 3 of the periodic table.

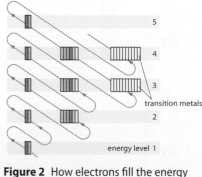

Figure 2 How electrons fill the energy levels.

Questions

1. Give two physical properties of transition metals that are **(a)** similar to those of the alkali metals and **(b)** different from those of the alkali metals.

2. What simple property gives a first indication as to whether a compound contains an alkali or a transition metal?

3. In what ways are the properties of mercury **(a)** typical and **(b)** unusual for a transition metal?

4. **(a)** Describe how transition metals react with water and oxygen. **(b)** How does this compare with the alkali metals?

5. What is odd about the way some transition metals form ions?

6. Transition metals have useful physical properties, but what chemical property makes them useful in industry?

7. Aluminium (Al) and gallium (Ga) both have three electrons in their outer energy level. What is different about their electron configuration?

8. Potassium (K)/calcium (Ca) and cobalt (Co)/nickel (Ni) are pairs of metals that occur side-by-side in group 4 of the periodic table. Compare the chemical properties of the metals in each pair and explain why the pattern is different for the two pairs.

Group 7 – the halogens

The halogens have coloured, diatomic vapours.

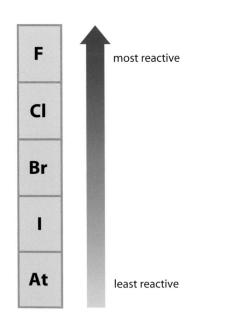

Figure 1 Trend in reactivity of the halogens.

Coloured non-metals

The elements in group 7 of the periodic table are fluorine (F), chlorine (Cl), bromine (Br), iodine (I) and astatine (At). They are known as the **halogens**. Astatine is radioactive and very rare.

Group 7 elements all have similar physical properties. They are all non-metals with coloured vapours. They do not conduct heat or electricity. They are all made of molecules that contain two atoms (**diatomic molecules**). For example, chlorine molecules have the formula Cl_2. They all have low melting and boiling points, because the forces between the molecules are weak. The trend in melting and boiling points can be seen in Table 1.

Table 1 Some physical and chemical trends in the halogens.

Halogen	Appearance at room temperature	Melting point/°C	Boiling point/°C	Reaction with hydrogen
Fluorine	pale yellow	−220	−188	explodes when mixed
Chlorine	green gas	−101	−35	mixture explodes in sunlight
Bromine	red-brown liquid	−7	59	mixture reacts if heated
Iodine	purple-black	114	184	partial reaction only

Halogen chemical properties

The halogens all have similar chemical properties. This is because all halogen atoms have seven electrons in their outer energy level (shell).

The halogens all react with metals to form ionic compounds containing halide ions. Electrons are transferred from the metal atoms to the halogen atoms, forming halide ions with a 1− electric charge. For example, chlorine reacts with the metal sodium to form the ionic compound sodium chloride, which contains chloride ions (Cl^-).

The halogens react with non-metals to form molecular compounds containing covalent bonds. Electrons are shared between the atoms. For example, chlorine reacts with the non-metal hydrogen to form the molecular compound hydrogen chloride.

$$H_2 + Cl_2 \rightarrow 2HCl$$

Reactivity in group 7

The halogens become less reactive as you go down the group. For example, flourine reacts explosively with hydrogen, whereas iodine only reacts partially, as seen in Table 1. Once again, the reason is the electronic structure. When the halogens react, the halogen atoms gain one electron. As Figure 2 shows, the atoms get bigger down the group, so the electron gained is further from the nucleus. This means that the attraction of the nucleus for the electron being gained is weaker and so the electron is harder to gain. This explains why the halogens become less reactive as you go down the group.

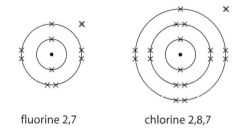

fluorine 2,7   chlorine 2,8,7

Figure 2 As the atoms get bigger, the electron gained (in red) is further from the nucleus.

A more reactive halogen will displace a less reactive halogen from a compound in a **displacement reaction**. For example, chlorine will displace bromine from a solution of potassium bromide because chlorine is more reactive than bromine:

$$Cl_2 + 2KBr \rightarrow 2KCl + Br_2$$

chlorine + potassium bromide → potassium chloride + bromine

(green)   (colourless)   (colourless)   (orange-brown)

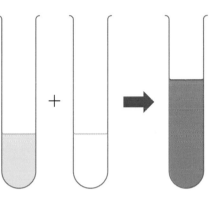

chlorine   bromide ions   bromine + chloride ions

Figure 3 Displacement of bromide ions by chlorine is shown by a colour change from green to orange-brown.

Questions

1. Draw bar charts of the melting and boiling points of the halogens in order and comment on the trends shown.

2. Explain why the halogens have low melting points.

3. What type of compound is formed from the following reactions? Explain your answer in each case. **(a)** Chlorine reacts with magnesium. **(b)** Fluorine reacts with nitrogen. **(c)** Bromine reacts with calcium.

4. Explain why chlorine is more reactive than bromine.

5. Write a word equation for the following reactions, or write 'no reaction' if nothing happens. Explain your answer in each case, describing anything you would expect to see. **(a)** Iodine + sodium chloride. **(b)** Fluorine + potassium chloride. **(c)** Chlorine + sodium bromide.

6. Describe the reaction between bromine and hydrogen. Write a word equation and a balanced symbol equation for this reaction.

7. Explain why the halogens form diatomic molecules, but the noble gases exist as single atoms.

8. Describe and explain the trend of reactivity in group 7.

Route to A*

A clear understanding of the reasons for the trends in reactivity of the halogens will be needed when answering higher level questions.

Taking it further

Displacement reactions like this are used in the production of elemental bromine and iodine from solutions of bromide or iodide salts.

Assess yourself questions

1 Table 1 gives the melting and boiling points of the first 38 elements of the periodic table.

Table 1 The first 38 elements.

Atomic (proton) number	Element	Melting point/°C	Boiling point/°C
1	H	−259	−253
2	He	−272	−269
3	Li	181	1342
4	Be	1278	2969
5	B	2300	3658
6	C	3652	4827
7	N	−210	−196
8	O	−218	−183
9	F	−220	−188
10	Ne	−248	−246
11	Na	98	883
12	Mg	649	1090
13	Al	661	2467
14	Si	1410	2355
15	P	44	280
16	S	386	445
17	Cl	−101	−35
18	Ar	−189	−186
19	K	63	774
20	Ca	839	1484
21	Sc	1814	2831
22	Ti	1660	3287
23	V	1887	3377
24	Cr	1857±20	2672
25	Mn	1244	1962
26	Fe	1535	2750
27	Co	1495	2870
28	Ni	1453	2732
29	Cu	1084	2567
30	Zn	420	907
31	Ga	30	2403
32	Ge	938	2830
33	As	sublimes	616
34	Se	217	685
35	Br	−7	59
36	Kr	−156	−152
37	Rb	39	688
38	Sr	769	1384

(a) By hand or using a computer, plot out graphs of melting point and boiling point against atomic number. *(4 marks)*

(b) How many of these are:

   (i) gases at room temperature? *(1 mark)*

   (ii) liquids at room temperature? *(1 mark)*

(c) (i) What do you notice about the melting and boiling points of elements 2, 10, 18 and 36? *(1 mark)*

   (ii) What group do these elements belong to? *(1 mark)*

   (iii) Why are their melting and boiling points like this? *(1 mark)*

(d) (i) What do you notice about the melting and boiling points of elements 6, 14 and 32? *(1 mark)*

   (ii) What group do these elements belong to? *(1 mark)*

   (iii) Why are their melting and boiling points like this? *(1 mark)*

(e) (i) Plot a separate graph of the melting points of the group 1 metals and comment on the trend. *(2 marks)*

   (ii) How do the chemical properties of these metals vary along this line? *(1 mark)*

   (iii) The next group 1 metal is caesium (Cs), atomic number 55). Predict its melting point and chemical properties. *(2 marks)*

(f) (i) Plot a separate graph of the melting points of the halogens and comment on the trend. *(2 marks)*

   (ii) How do the chemical properties of the halogens elements vary along this line? *(1 mark)*

   (iii) The next halogen is iodine (I), atomic number 53). Predict its chemical properties. *(1 mark)*

   (iv) Do you think iodine will be a solid, liquid or gas at room temperature? Explain your answer. *(2 marks)*

(g) Predict the chemical and physical properties of element number 37. *(1 mark)*

(h) Supersonic aircraft suffer great frictional heating. Suggest one advantage of using titanium rather than aluminium in their construction. *(1 mark)*

2 (a) How many electrons are there in the outer shell of the following elements?

   Lithium, chlorine, silicon, calcium, argon. *(5 marks)*

(b) How many electron shells do each of the following elements have?

   Sodium, hydrogen, potassium. *(3 marks)*

3 Sodium reacts with water:

sodium + water → sodium hydroxide + hydrogen

(a) What would happen if you put litmus or other indicator paper in the water after this reaction? Why? *(1 mark)*

(b) What is the name given to group 1? Why do you think it is called this? *(2 marks)*

(c) Write a balanced equation for this reaction. *(3 marks)*

(d) Lithium fizzes steadily in water. What gas is given off? *(1 mark)*

(e) What is the name of the alkali that forms during this reaction? *(1 mark)*

4 Element Q is a colourless gas that boils at −108 °C. It is very unreactive.

(a) To which group does it belong? *(1 mark)*

(b) How many electrons must it have in its outer shell? *(1 mark)*

(c) Is the gas made of individual atoms or molecules? *(1 mark)*

5 Chlorine reacts explosively with hydrogen in sunlight to form hydrogen chloride. This gas is very soluble in water.

(a) What would happen if you put litmus or other indicator paper in the solution? *(1 mark)*

(b) What name is given to this acid? *(1 mark)*

(c) What compound would form from bromine and hydrogen? *(1 mark)*

(d) Would you expect this reaction to be faster or slower than that with chlorine? Explain your answer. *(3 marks)*

6 (a) What name is given to the block of metals that wedge in between calcium (20) and gallium (31)? *1 mark)*

(b) Normally the properties of the elements change dramatically as you move along the same period. What is unusual about this block? *(1 mark)*

(c) Iron is a hard, magnetic metal with a high melting point that forms coloured compounds. Nickel is a hard, magnetic metal with a high melting point that forms coloured compounds. Predict the properties of cobalt (Co). *(1 mark)*

7 The periodic table pattern matches the way electrons fill the electron shells.

(a) Use this idea to explain why there are eight elements in period 2, but just two elements in period 1. *(2 marks)*

(b) What do all group 2 elements have in common? *(1 mark)*

(c) What do all elements in period 3 have in common? *(1 mark)*

8 Explain carefully why group 1 metals become more reactive down the group in terms of their electron configuration.

In this question you will be assessed on using good English, organising information clearly and using specialist terms where appropriate. *(6 marks)*

9 Explain carefully why group 7 elements (the halogens) become less reactive down the group in terms of their electron configuration.

In this question you will be assessed on using good English, organising information clearly and using specialist terms where appropriate. *(6 marks)*

10 Figure 1 shows how chlorine combines by covalent bonding with carbon, and by ionic bonding with sodium. (The diagrams show only the outer electron shells.)

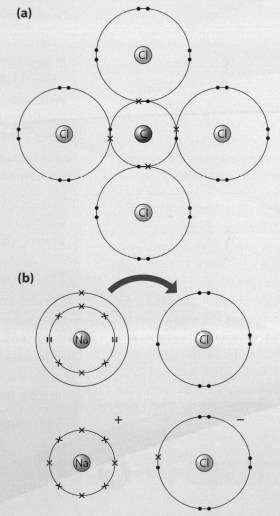

Figure 1 (a) Covalent bonds and **(b)** ionic bonds.

Draw similar diagrams and write balanced equations for the reactions between:

(a) fluorine and carbon *(2 marks)*

(b) fluorine and sodium. *(2 marks)*

Hard water

What is hard water?

Rainwater contains hardly any dissolved salts. However, when it seeps through the ground it dissolves compounds from the rocks. In many parts of the country these compounds contain calcium or magnesium ions, both of which lead to **hard water**.

Hard water has some advantages. Hard drinking water provides some of the calcium we need for our bones and teeth and has been found to reduce heart illnesses. Many people also think that it tastes better than soft water, and brewers prefer it for making beer.

Unfortunately the dissolved compounds in hard water react with soap to form scum. This forms an unsightly deposit in the bath or sink, but also, and more importantly, it wastes soap, which costs money. **Soft water**, without dissolved salts, gives a fantastic lather with just a small amount of soap.

☐ Soft to moderately soft: 0–100 mg/dm³ as calcium carbonate equivalent

☐ Slightly hard to moderately hard: 100–200 mg/dm³ as calcium carbonate equivalent

☐ Hard to very hard: above 200 mg/dm³ as calcium carbonate equivalent

Figure 1 Hard water areas in England and Wales.

To overcome this problem, scientists developed a range of soapless detergents that do not form scum with hard water. These are used in shampoos and washing-up liquid.

The chemistry of scum

The calcium and magnesium ions that cause hard water are not removed when water is purified at the water works, so they are found in tap water. One of the compounds in soap is called sodium stearate, which is soluble in water. However, when soap meets hard water, the sodium stearate reacts with the calcium or magnesium ions in the water to form insoluble calcium stearate or magnesium stearate. This is what is in the scum:

calcium ions from water + stearate ions from soap → calcium stearate (scum)
(soluble)   (soluble)   (insoluble)

You can show whether these ions are present in hard water by adding small amounts of salt solution to soap solution. If the water is hard, a scum forms when the mixture is shaken. If the water is soft, lather forms at once.

Practical

To find out just how hard or soft a water sample is, you can measure how much soap solution is needed to make a permanent lather.

The same volume of water is used in each experiment. Soap solution is added 0.1 cm^3 at a time and the water shaken vigorously after each addition to see whether a permanent lather forms. Table 1 shows some sample results.

Table 1 Water hardness in different areas.

Water source	Volume of soap solution needed to produce a permanent lather/cm^3
London	7.2
Exeter	0.5
Manchester	3.7
Distilled water (control)	0.2

Comparing the hardness of water samples.

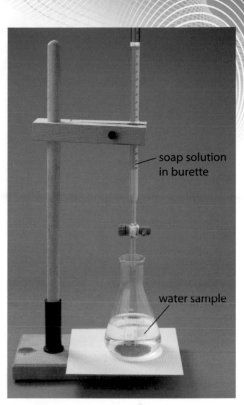

soap solution in burette

water sample

The effect of soap on hard and soft water.

Examiner feedback

In an exam, you may well get data of the content of different water supplies to interpret and discuss.

Questions

1   From Table 1, which area has the hardest water? How can you tell?
2   Which ions are responsible for hard water and how do they get there?
3   How could you check if you have hard water at home?
4   Before shampoo was invented, people in hard water areas often used to collect rainwater to wash their hair with. Why was that?
5   What is the chemical name of the compound that builds up in kettles in hard water areas?
6   Write a word equation for the reaction of magnesium ions and soap.
7   Strontium is in the same group as calcium and magnesium in the periodic table. Would you expect its compounds to produce hard water? Explain your answer.
8   Describe an experiment you could use to investigate the hardness of an unknown water sample, compared with some known samples. What are the independent, dependent and control variables?

Taking it further

Getting a good lather might be great in the bath, but would be a problem in a washing machine or dishwasher. Scientists have had to develop non-foaming detergents for these uses.

Water softening

All rain is acid rain

Even in the most unpolluted areas, rain is slightly acidic. This is because carbon dioxide gas dissolves in water, forming a weak solution of carbonic acid:

$$CO_2(g) + H_2O(l) \rightarrow H_2CO_3(aq)$$
carbon dioxide + water → carbonic acid

Dissolving the rocks

When rain falls onto rocks, this weak acidity sets off the slow chemical reactions that cause chemical weathering. Some of the minerals in igneous rocks are slowly turned to clay, for example, and sodium, calcium and magnesium ions are released. However, the weak carbonic acid has a more direct effect on rocks such as limestone that are made of calcium carbonate. These rocks are slowly dissolved away as soluble calcium hydrogencarbonate forms:

$$CaCO_3(s) + H_2CO_3(aq) \rightarrow Ca(HCO_3)_2(aq)$$
calcium carbonate + carbonic acid → calcium hydrogencarbonate

Stalactites form when soluble calcium hydrogencarbonate is re-deposited as calcium carbonate (limestone).

Limescale problems

Hard water gives rise not only to stalactites and stalagmites in caves, but also unsightly limescale around taps and toilet bowls. As hard water evaporates (or is heated: see below), some carbon dioxide is lost with the water, and calcium carbonate reforms. This process is enhanced if the water is heated, so kettles, hot water tanks, central heating boilers and pipes all suffer from limescale in hard-water areas. In a kettle, this is a minor problem as the limescale can be easily removed by dissolving it in a weak acid.

Temporary and permanent hardness

If you boil hard water containing hydrogencarbonate ions (HCO_3^-) you can completely reverse the carbonic acid + limestone reaction, driving off the carbon dioxide and precipitating all of the calcium ions as calcium carbonate. This makes the water soft again, so this type of hardness is called **temporary hardness**. Sometimes hard water contains sulfate ions (SO_4^{2-}) as well as calcium (Ca^{2+}) or magnesium (Mg^{2+}) ions, however, and in these cases boiling has no effect. This is known as **permanent hardness**.

The limescale build-up on this heating element from a washing machine will make it very inefficient.

Water softening

In hard-water areas, prevention can be better than cure. If you don't like greasy scum forming in your bath, just add bath salts. Baths salts are made of sodium carbonate. This softens the water immediately by precipitating the calcium (and magnesium) ions in a double decomposition reaction – one in which the ions 'swap partners':

$$Na_2CO_3(aq) \quad + \quad Ca^{2+}(aq) \quad \rightarrow \quad CaCO_3(s) \quad + \quad 2Na^+(aq)$$

sodium carbonate + calcium ions → calcium carbonate + sodium ions

The sodium ions left in solution do not react with soap or form limescale deposits. Unfortunately this method of softening water will not work for water pipes or central heating systems, which can be totally ruined by limescale.

Bath salts soften water by removing the calcium ions.

Instead, an expensive water softener containing an **ion-exchange column** may have to be used. This contains sodium ions (provided by common salt) or hydrogen ions, which can swap places with calcium or magnesium ions in the hard water. It can run continuously for a long time, removing the magnesium and calcium ions from the water. Dishwashers have a system like this built in, which is why they must have salt added to them regularly.

However, the water that comes out of a water softener now contains sodium ions, which are not good for our health. You should avoid drinking softened water, since too much sodium can lead to high blood pressure and heart problems.

Route to A*

A*

If mixed ions can form an insoluble combination, they will.

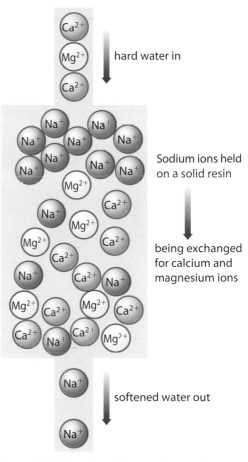

Figure 1 Exchanging calcium and magnesium ions for sodium ions.

Questions

1. In what way is all rain 'acid rain', even when it is unpolluted?

2. What chemical is responsible for most temporary water hardness?

3. How do stalactites form?

4. Some rocks contain magnesium carbonate ($MgCO_3$). What effect will rainwater have on these rocks? Write the balanced chemical equation for this reaction.

5. Hard water that has been boiled will usually form a much richer lather with soap than unboiled water. Explain why.

6. Citric acid is sometimes used as a kettle descaler. Write a word equation for this reaction. Why not use a strong acid?

7. Write word and balanced chemical equations for the reaction between sodium carbonate and hard water containing magnesium sulfate ($MgSO_4$).

8. Some houses have ion-exchange columns fitted to their water supply to soften it. How does this work, and why must they have a separate tap delivering unsoftened water?

Drinkable water

A water reservoir in Wales.

Table 1 Where all the water goes: domestic water use.

Used for:	dm³/day
drinking	2
cooking	3
washing up	15
laundry	15
washing/bathing	50
flushing the toilet	50
other	10
Total	**145**

Finding a source

Drinking water has to be free of poisons and disease-causing microorganisms. The first step towards clean, safe drinking water is to find and test a suitable source. In Britain the source is usually rainwater, which is collected from groundwater or rivers and is often stored in reservoirs. In drier countries the water often comes from deep underground. Wells were once dug by hand, but today boreholes are drilled down to the water.

How long must it have taken to dig out this step-well in Rajasthan, India?

Making it fit to drink

Whatever the source, the water will need careful treatment before it is fit to drink. At East Ham in London, for example, the water is obtained from the chalk rocks underground. Chalk is a natural filter, so the water is already of good quality. The water is pumped through boreholes to the water treatment plant. There it is processed through various filters called **filter beds** to remove smaller and smaller solids. (If the source had been a river it would have been passed through a sieve first, to remove any large objects.)

The result is 'clear' water – but this is still not fit to drink, as it may contain harmful bacteria and other microorganisms. These are killed by adding a controlled amount of chlorine, which leaves water that is safe to drink.

More filters

The water that comes out of our taps is perfectly safe, but it is not absolutely pure as it still contains some harmless dissolved salts such as calcium hydrogencarbonate (see lesson C3 2.1). Some people, encouraged by much advertising, prefer to drink bottled mineral water. The reasons for this vary: some think it is 'purer' or 'safer', while others say they prefer the taste. For whatever reason, many offices have water coolers, supplied with bottled water, and the market for bottled water has grown dramatically.

An alternative option is to buy filtering systems for the home. These vary from simple filter jugs to larger systems that are plumbed into the kitchen. These filters use **activated carbon** and/or microscopic silver.

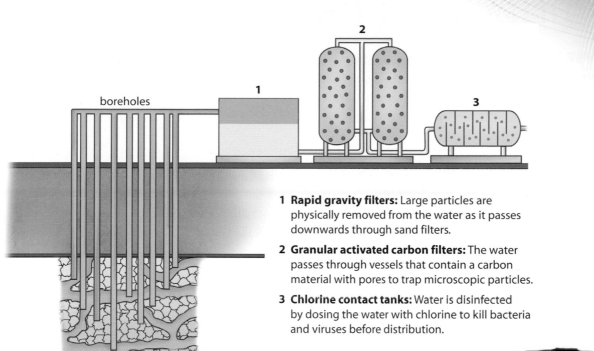

1 Rapid gravity filters: Large particles are physically removed from the water as it passes downwards through sand filters.

2 Granular activated carbon filters: The water passes through vessels that contain a carbon material with pores to trap microscopic particles.

3 Chlorine contact tanks: Water is disinfected by dosing the water with chlorine to kill bacteria and viruses before distribution.

Figure 1 Modern water treatment plant.

Activated carbon is charcoal that has been broken up into granules to increase the surface area and then treated to give the pieces a slight positive charge. Some contaminants in the water stick to the carbon, which has to be replaced on a regular basis.

Silver is used in the form of porous ceramic filters impregnated with microscopic particles of silver. The ceramic and silver filter purifies the water effectively, as the silver kills any microorganisms that might otherwise grow in the filter.

Ion-exchange resins can also be used for water purification, softening the water by removing calcium and magnesium ions, as described in lesson C3 2.2.

A commercial filter jug; the filter cartridges have to be replaced regularly.

Questions

1 Estimate (to one significant figure) the amount of water used every day in homes across the United Kingdom (population 60 million). What proportion of this needs to be of 'drinking water' quality?

2 What problems could be caused for the British water system if global warming affects our climate?

3 How is drinking water sterilised?

4 Explain why the level of chlorine used to treat the water needs to be carefully controlled.

5 Is tap water 'pure'? Explain your answer.

6 Why is the carbon in activated charcoal filters broken up into tiny pieces?

7 Microscopic silver is a very effective sterilising agent. Suggest a reason why all of our water supply is not treated in this way.

8 How do 'home filtration' systems work? Explain your own views about whether they are needed or not.

Route to A*

Understanding the relationship between particle size and total surface area is crucial to a wide range of scientific topics.

Purify or enhance?

Science in action

It has been seriously suggested that one way to overcome water shortages is to tow icebergs from the Arctic.

a What problems could you envisage with such a plan?

There are over 1.3 billion cubic kilometres of water on Earth. Only about 1 per cent is usable fresh water.

Water, water everywhere...

We live on a world covered by water, but most of it is far too salty to drink. Sea water cannot be made fit to drink by filtration. Just 3 per cent of the water on Earth is fresh water, and two-thirds of that is locked up in glaciers and ice caps.

Distillation

One way of purifying sea water is to evaporate the water and allow it to re-condense elsewhere, leaving the salt and impurities behind. This process is called **distillation**, and is carried out in a **still**. In a solar still, the energy to evaporate the water is provided by the Sun. Solar stills are used along desert coastlines to produce drinking water from sea water.

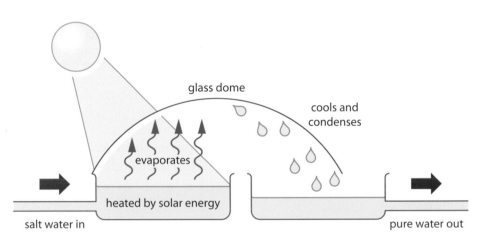

Figure 1 A simple solar still.

It is also possible to provide the energy for distillation by burning fuels such as oil. Flash distillation plants heat water under pressure, then allow it to evaporate fast. The vapour is condensed and stored. Oil-rich Saudi Arabia makes much of its drinking water in this way. It is, however, a very expensive way to purify water; it uses up our dwindling stocks of fossil fuels and pours carbon dioxide into the atmosphere. Flash distillation also produces concentrated salty water as a waste product, which has to be disposed of carefully. If it were just poured back into the sea it could seriously damage the ecology of the region.

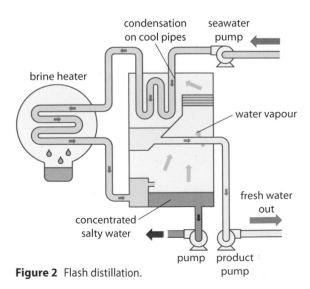

Figure 2 Flash distillation.

Pure – but is it good for you?

Hard water, with its dissolved calcium and magnesium salts, appears to be better for your health than soft water. It is easy to see that the calcium in the water might be good for your bones and teeth, but various medical studies have also shown that hard water:

- reduces the risk of heart problems
- reduces the risk of some intestinal cancers
- reduces the risk of lead poisoning (from old lead water pipes)
- increases overall lifespan.

So would drinking distilled water all the time damage health? There are few studies, as not many people in the world drink nothing but distilled water, so the evidence is inconclusive and more research is needed.

What *is* clear is that drinking water softened using ion-exchange resins has the potential to cause health problems. Ion-exchange resins replace calcium ions with sodium ions. Sodium ions are an essential component of your diet, usually taken in the form of salt. The problem arises if you take in too much sodium, as it is known to cause heart problems. Drinking nothing but softened water could give you up to a third of your recommended daily amount (rda) of sodium before you eat anything at all.

Adding fluoride to water

In the 1950s, it was realised that people living in areas with high levels of natural fluoride ions in their drinking water had much healthier teeth than the rest of the population. Ever since then, debate has raged over whether to add fluoride to drinking water. Some people were concerned that fluoride ions could cause bone weakening and tooth mottling. They also raised the ethical issue of whether we should be 'forced' to take fluoride in our drinking water whether we wanted to or not.

Others questioned the link between fluoride ions and better dental health. In 1994, the World Health Organisation took the view that there were no associated health risks from appropriate water **fluoridation**, but the debate continues. At present, only a few areas of Britain have fluoride ions added to the public water supply.

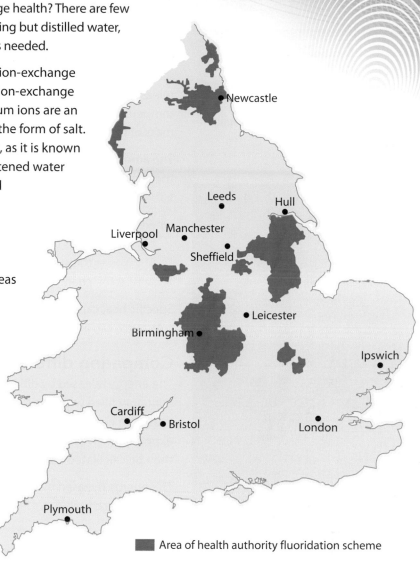

■ Area of health authority fluoridation scheme

Figure 3 Fluoride is added to the public water supply in just a few areas of Britain.

Questions

1   The Earth is not short of water. Why is most of it unsuitable for drinking?

2   Design a solar still made from everyday materials.

3   What health concerns have been raised about drinking distilled water?

4   Why can drinking too much softened water be bad for your health?

5   What are the main health arguments for and against the fluoridation of drinking water?

6   Discuss the ethical issues associated with the fluoridation of the public water supply.

7   Discuss the major economic and environmental issues associated with the flash distillation of water.

Taking it further

You may be aware of the dangers society faces as we run out of fossil fuels. However, with growing global populations and increased demand, some people think that running out of suitable water supplies could be an even bigger problem – and one that might start to bite even sooner.

Calorimetry

Learning objectives

- describe how the energy released in chemical reactions can be measured using calorimetry
- calculate the energy released in chemical reactions.

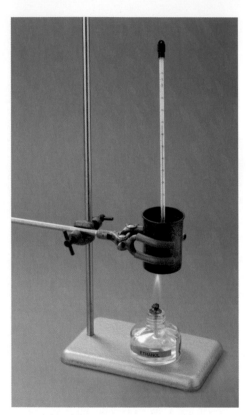

Calorimetry using a copper can and spirit burner.

Calorimetry

Calorimetry is a method for measuring the energy released by chemical reactions. A simple type of calorimetry can be used to compare the energy released when different fuels burn. This can be done by measuring the temperature change in a known mass of water when it is heated by a certain mass of fuel.

The amount of energy absorbed by water depends upon the mass of water and the change in temperature. It can be calculated from this equation:

$$Q = mc\Delta T$$

where
Q = energy in joules, J
m = mass of water in grams, g
c = the **specific heat capacity** of water
ΔT = change in temperature in °C.

Specific heat capacity is a measure of how much energy a certain amount of a substance can absorb. Water has a very high specific heat capacity of 4.2 J/g.°C.

Comparing different fuels

The energy released by different fuels can be measured using a glass or metal container filled with a known mass of water. Fuel burned in a 'spirit burner' is used to warm the water. Weighing the burner at the start and end of the experiment gives the mass of fuel used. The energy absorbed by the water can then be calculated using $Q = mc\Delta T$.

The energy released by different fuels can be compared if Q is worked out per mole for each fuel.

Example 1

Table 1 Heating 100 g of water using methanol, CH_3OH.

Mass of burner/g		Temperature/°C		Mass of fuel used/g	Temperature change/°C
at start	at end	at start	at end		
240.9	240.1	22	42	0.8	20

The results in Table 1 can be used to calculate the energy absorbed by the water when burning 0.8 g of methanol.

Using $Q = mc\Delta T$:
$$Q = 100 \times 4.2 \times 20$$
$$= 8400 \text{ J} = 8.4 \text{ kJ}$$

Assuming that the water absorbed all the energy, the energy released per gram of methanol is $8.4 \div 0.8 = 10.5$ kJ/g

This can also be expressed in kJ/mole of methanol.
$$M_r \text{ of methanol} = 32$$
$$\text{Energy released per gram} = 10.5 \text{ kJ}$$
$$\text{So energy per mole} = 10.5 \times 32 = 336 \text{ kJ/mole}$$

The accepted value for the energy released by burning methanol is 726 kJ/mole. The experimental value in the worked example is less than half this. This is because not all the energy from the fuel went into heating the water. Some heated up the **calorimeter** (the copper can) and the thermometer, and some was released into the surroundings.

a Describe two ways in which energy losses could be reduced.
b Despite the energy losses, this method can still be used to *compare* fuels. Describe two factors that must be kept constant for the comparison to be fair.

Energy from reactions in solution

Calorimetry can also be used to measure the energy produced or absorbed by reactions in solution. The reagents are mixed in an insulated container, such as a polystyrene cup fitted with a lid, and the temperature change is recorded. This method is suitable for neutralisation reactions and for reactions of solids with water.

Example 2

When 50 cm³ of 0.2 moles/dm³ copper(II) sulfate was mixed with an excess of zinc powder, the temperature of the mixture increased by 25 °C. Calculate the energy change for the reaction in kJ/mole of $CuSO_4$. Assume that 1 cm³ of solution contains 1 g of water, and c is 4.2 J/g.°C.

Mass of water = 50 g

$Q = mc\Delta T = 50 \times 4.2 \times 25 = 5250\ J = 5.25\ kJ$

Amount of $CuSO_4$ = concentration $\times$ volume

$= 0.20 \times (50 \div 1000) = 0.01\ moles$

So energy released = $5.25 \div 0.01 = 525\ kJ/mole$

Calorimetry for reactions in solution.

In the following questions, assume that $c = 4.2$ J/g.°C.
$A_r(H) = 1, A_r(C) = 12, A_r(N) = 14, A_r(O) = 16$

1 Calculate the energy needed, in kJ, to heat 100 cm³ of water by 30 °C.

2 100 g of chocolate contains 500 kcal of energy. If 1 kcal = 4.2 kJ, how much energy in kJ does 1 g of chocolate contain?

3 When 100 cm³ of water is heated by 1 g of burning propanol, its temperature rises by 35 °C. Calculate the energy absorbed per g of fuel.

4 When 50 cm³ of water is heated by 0.5 g of burning methanol, CH_3OH, its temperature rises by 25 °C. Calculate the energy released in kJ per mole of fuel.

5 In a neutralisation reaction, 0.1 moles of hydrochloric acid reacts with an excess of sodium hydroxide solution. The total mass of liquid is 100 g and its temperature rises by 9 °C. Calculate the energy released in kJ/mole of acid.

6 When 1.6 g of ammonium nitrate, NH_4NO_3, dissolves in 25 g of water, the temperature falls by 4 °C. Calculate the energy change in kJ/mole of NH_4NO_3.

7 A manufacturer of portable stoves decides to investigate two alcohols as fuels: ethanol, C_2H_5OH, and propanol, C_3H_7OH. They used the same stove to heat 0.5 kg of water from 16 °C to 70 °C using each fuel in turn. The results are shown in Table 2.

Table 2 Alcohols as fuels

Fuel	Mass of stove/g	
	at start	at end
ethanol	1161.4	1151.7
propanol	1162.1	1153.4

(a) For each fuel, calculate the energy released in kJ/g, and in kJ/mole of fuel.

(b) From the results, which fuel might be better for a portable stove and why?

Energy level diagrams

The US Army uses the exothermic reaction between powdered magnesium and water to heat food in its 'flameless ration heaters'.

Examiner feedback

Don't forget which way round energy level diagrams are drawn. The reactants have more energy than the products in exothermic reactions.

Self-cooling ice packs make use of endothermic reactions.

Breaking and making bonds

During a chemical reaction, bonds in the reactants are broken, and bonds in the products are formed. Energy is involved in these processes.

- Energy must be supplied to break bonds.
- Energy is released when bonds are formed.

These changes are shown on **energy level diagrams**.

Exothermic reactions

In an **exothermic reaction**, energy is transferred to the surroundings, often heating them. Combustion and neutralisation are typical examples of exothermic reactions. Figure 1 shows a simple energy level diagram for such a reaction.

There is more chemical energy in the reactants than in the products, so the energy level of the reactants is higher on the diagram. Energy is transferred to the surroundings as the reaction proceeds. This is shown by the downwards pointing arrow. The change in energy is shown as ΔH. It is calculated by subtracting the energy level of the products from the energy level of the reactants. For exothermic reactions, ΔH is negative.

Endothermic reactions

In an **endothermic reaction**, energy is absorbed from the surroundings. Thermal decomposition reactions are exothermic. Figure 3 shows a simple energy level diagram for such a reaction.

There is less chemical energy in the reactants than in the products, so the energy level of the reactants is lower on the diagram. Energy is absorbed from the surroundings as the reaction proceeds. This is shown by the upward-pointing arrow. For endothermic reactions, ΔH is positive.

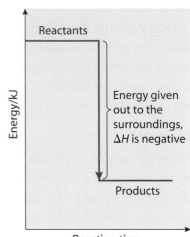

Figure 1 A simple energy level diagram for an exothermic reaction.

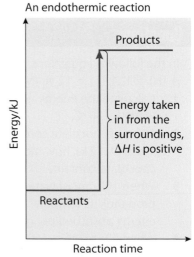

Figure 2 A simple energy level diagram for an endothermic reaction.

Activation energy and catalysts

Many exothermic reactions need to absorb energy before they can start. For example, fuels need a hot spark or a flame before they will ignite. The minimum amount of energy that colliding reactant particles must have for a reaction to occur is called the **activation energy**.

You learned in lesson C2 4.5 that a suitable **catalyst** can speed up the rate of some reactions. Catalysts work by providing a different pathway for the reaction, which has a lower activation energy. The activation energy of a reaction with and without a catalyst can be shown on an energy level diagram (see Figure 3).

If a reaction is reversible, the change in the amount of energy for the forward and backward reactions is the same. The only difference is the sign. For example, if the forward reaction is exothermic and ΔH is negative, the backward reaction is endothermic and ΔH is positive.

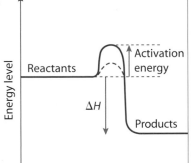

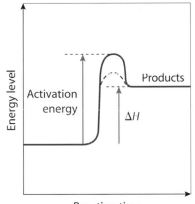

Figure 3 Activation energies in an exothermic reaction (top) and in an endothermic reaction (bottom). The dashed red line represents the activation energy in the presence of a catalyst.

Taking it further

There are two main types of catalyst. Heterogeneous catalysts are in a different state to the reactants. They are usually solid while the reactants are gaseous or in aqueous solution. Heterogeneous catalysts provide a large surface area with many active sites that bring reactants closer together. Homogeneous catalysts are in the same state as the reactants, usually in aqueous solution. They reduce the activation energy via reversible redox reactions.

Science in action

The term *activation energy* was first introduced by the Swedish scientist Svante Arrhenius in 1889. Arrhenius was interested in mathematical relationships between experimental data. He produced evidence in support of the 'Arrhenius rate equation', which links the rate of a reaction to its activation energy and other factors. His early work on electrolysis and ions was not well received by other scientists, but it was gradually accepted and he was awarded the 1903 Nobel Prize in Chemistry.

Questions

1   What is an energy level diagram?

2   How is the overall energy change in a reaction shown on an energy level diagram?

3   Draw a simple energy level diagram for the combustion of ethanol:

    ethanol + oxygen → carbon dioxide + water

4   Draw a simple energy level diagram for the thermal decomposition of calcium carbonate:

    calcium carbonate → calcium oxide + carbon dioxide

5   What is the activation energy for a reaction, and how is it shown on energy level diagrams?

6   Hydrochloric acid and sodium hydroxide solution immediately react together when mixed:

    $HCl(aq) + NaOH(aq) \rightarrow NaCl(aq) + H_2O(l)$

The reaction releases 57.9 kJ/mole of energy to the surroundings.

Draw an energy level diagram to represent this reaction that includes all the information given.

7   The electrolysis of water absorbs 256 kJ/mole of energy from the surroundings:

    $H_2O(l) \rightarrow H_2(g) + \frac{1}{2}O_2(g)$

Draw an energy level diagram to represent this reaction that includes all the information given.

8   Methane molecules contain 1740 kJ/mole of chemical energy, oxygen molecules 498 kJ/mole, carbon dioxide 1610 kJ/mole and water molecules 928 kJ/mole. The activation energy is 130 kJ/mole. **(a)** Write a balanced equation for the complete combustion of methane, CH_4. **(b)** Calculate the total energy contained in the reactants, and in the products. **(c)** Use your answers to part (b) to calculate the energy change in kJ/mole. **(d)** Draw an energy level diagram to show all the information.

Bond energy calculations

Table 1 A selection of bond energies.

Bond	Bond energy / kJ/mole
C–C	347
C–H	413
C=C	612
C–O	358
C=O	805
C–Br	290
Br–Br	193
H–Br	366
H–H	436
O–H	464
O=O	498
N–H	391
N≡N	945

Bond energies

Energy is needed to break bonds. It is released when bonds form. The **bond energy** is the amount of energy needed to break a particular chemical bond. Different bonds have different bond energies. The amount of energy needed to break a particular bond is the same as the amount of energy released when the bond forms.

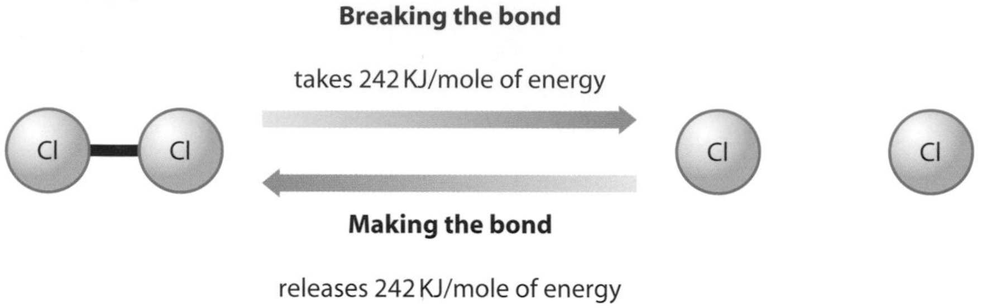

Breaking the bond

takes 242 KJ/mole of energy

Making the bond

releases 242 KJ/mole of energy

Figure 1 The same amount of energy is involved whether a bond is broken or formed.

Bond energies are measured in kJ/mole of *bonds*, not molecules.

Science skills Bond energies given in tables are usually mean values, calculated from the bond energies of the same type of bond in a range of different compounds.

a The mean bond energy for the C–H bond in methane, CH_4, is 435 kJ/mole. What does this tell you about the strength of the C–H bond in methane compared with the same bond in other compounds?

b Suggest why different tables of bond energies may contain slightly different values for the same type of bond.

Breaking and making bonds

The overall energy change in a reaction can be calculated using bond energies.

Step 1 Calculate the total amount of energy needed to break all the bonds in the reactants (the 'energy in').

Step 2 Calculate the total amount of energy released in making all the bonds in the products (the 'energy out').

Step 3 Overall energy change = energy in − energy out

Example 1

Methane burns completely in oxygen to form carbon dioxide and water. Use bond energies from Table 1 to calculate the overall energy change in the reaction.

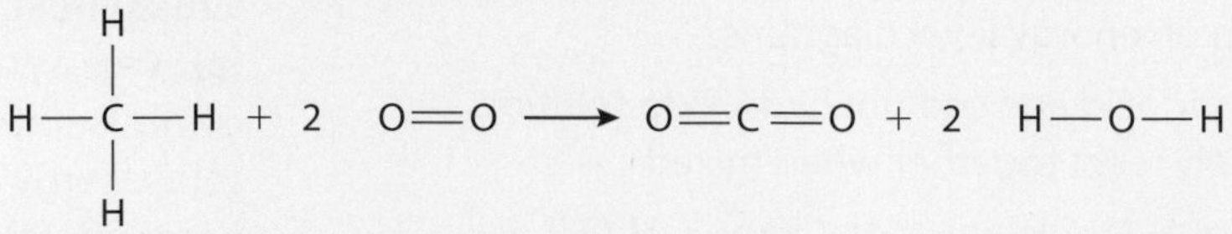

$$H-\underset{\underset{H}{|}}{\overset{\overset{H}{|}}{C}}-H \; + \; 2 \; O=O \longrightarrow O=C=O \; + \; 2 \; H-O-H$$

Figure 2 The bonds in the reactants and products.

Natural gas is nearly all methane. On this drilling rig, a gas flare burns off excess methane when the pressure gets too high.

Step 1 Energy in (bonds broken)

$4 \times (C–H) = 4 \times 413 = 1652$ kJ/mole

$2 \times (O=O) = 2 \times 498 = 996$ kJ/mole

Total $= 1652 + 996 = 2648$ kJ/mole

Step 2 Energy out (bonds made)

$2 \times (C=O) = 2 \times 805 = 1610$ kJ/mole

$4 \times (O–H) = 4 \times 464 = 1856$ kJ/mole

Total $= 1610 + 1856 = 3466$ kJ/mole

Step 3 Overall energy change $= 2648 - 3466 = -818$ kJ/mole

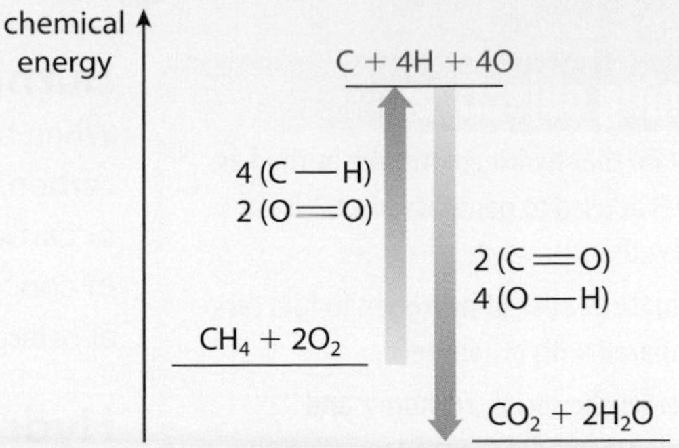

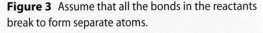

Figure 3 Assume that all the bonds in the reactants break to form separate atoms.

Exothermic and endothermic reactions

In the worked example, the overall energy change ΔH is negative. This shows that the reaction is exothermic. However, a negative value does not necessarily mean that the reaction will happen at room temperature, or that it will happen quickly. If the reaction has a high activation energy; it may not happen without an input of energy or the use of a suitable catalyst.

Combining the ideas from energy level diagrams and bond energy calculations:

- in an exothermic reaction, the energy needed to break existing bonds is *less* than the energy released from forming new bonds

- in an endothermic reaction, the energy needed to break existing bonds is *more* than the energy released from forming new bonds.

Questions

Use the bond energies in the table to help you answer these questions.

1 **(a)** How much energy is needed to break 1 mole of C=C bonds? **(b)** How much energy is released when 1 mole of C=C bonds forms?

2 Study the bond energies in Table 1, **(a)** Explain which bond is the weakest. **(b)** Compare the bond energies of the single bonds, double bonds and triple bond. What do you notice?

3 Hydrogen and bromine react to produce hydrogen bromide:

H—H + Br—Br → 2 × [H—Br]

Calculate the overall energy change for the reaction.

4 Nitrogen and hydrogen react to produce ammonia:

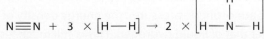

Calculate the overall energy change for the reaction.

5 Ethene and bromine react to produce dibromoethane:

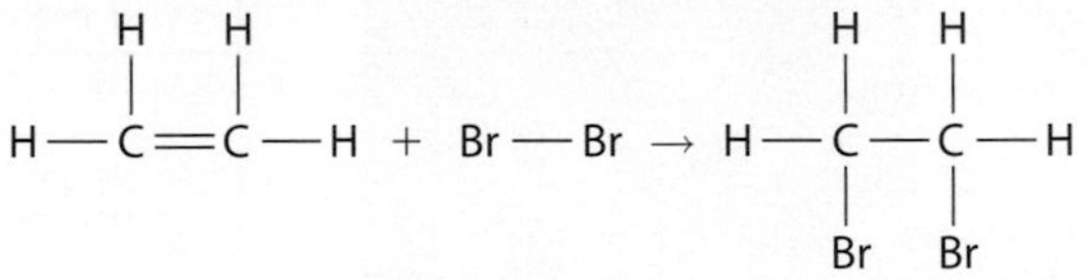

$$\begin{array}{ccc} H & H & \\ | & | & \\ H-C{=}C-H + Br-Br \rightarrow \end{array} \begin{array}{cc} H & H \\ | & | \\ H-C-C-H \\ | & | \\ Br & Br \end{array}$$

Calculate the overall energy change for the reaction, and state whether it is exothermic or endothermic.

6 Ethene, $CH_2{=}CH_2$, reacts with hydrogen to form ethane, CH_3CH_3. Calculate the overall energy change in the reaction.

7 Ethane, CH_3CH_3, reacts with oxygen to form carbon dioxide and water. Calculate the overall energy change in the reaction.

8 Ethanol, CH_3CH_2OH, reacts with oxygen to form carbon dioxide and water. **(a)** Write a balanced equation for the complete combustion of ethanol. **(b)** Draw a diagram to show all the bonds in the reactants and products. **(c)** Calculate the overall energy change for the reaction. **(d)** Explain how you know that the reaction is exothermic.

Hydrogen power

Burning hydrogen

When hydrogen burns in air, the only product is water. As it does not contain carbon, hydrogen burns with an invisible flame. There are no emissions of smoke or carbon dioxide. There is interest in hydrogen as an alternative fuel because of concern that burning of **fossil fuels** produces **greenhouse gases** and is accelerating **climate change**.

Hydrogen as a fuel

Rockets such as the European Space Agency's Ariane V use liquid hydrogen to lift them into space. Ariane V uses 28 tonnes of liquid hydrogen at $-253\,°C$. The rocket also carries 162 tonnes of liquid oxygen at $-183\,°C$. This allows the hydrogen to burn, even in space.

A drawback of hydrogen as a fuel is that it must be liquefied or pressurised so that it occupies a smaller volume. Fuels may be compared by their **energy densities**. In terms of energy density per gram, hydrogen releases far more energy than other fuels. However, it is a gas at room temperature, so its energy density per cm^3 is far lower than that of petrol and other liquid fuels. Hydrogen-fuelled vehicles would need huge tanks if they stored hydrogen as a gas at atmospheric pressure.

The main engines of the Ariane V rocket use hydrogen fuel.

Science skills

Table 1 The energy densities of some common fuels.

Fuel	Energy density / kJ/g	Energy density / kJ/cm³
hydrogen (gas)	143	0.012
hydrogen (liquid)	143	10
natural gas	54	0.037
petrol	46	34
biodiesel	42	33
ethanol	30	24

a Draw a suitable graph to compare the energy densities in kJ/g.

b Use the information in the table to explain the advantages of petrol and diesel over the other fuels.

c Suggest why cars may be adapted to run on liquefied petroleum gas (LPG), rather than natural gas.

Hydrogen refuelling stations provide hydrogen under pressure.

Hydrogen-powered vehicles

Hydrogen is used as a fuel for some cars and buses, but its use is still experimental. If a road vehicle uses liquid hydrogen, it needs an expensive insulated tank to keep the fuel cold. So vehicles may use pressurised hydrogen instead. At 350 bar (350 times atmospheric pressure), the energy density of hydrogen is about one-quarter that of liquid hydrogen. This is still much less than the energy density of petrol, so scientists are researching other ways to store hydrogen. These include binding it reversibly to certain metals, or even storing it in heat-treated chicken feathers.

Fuel cells

Water decomposes to form hydrogen and oxygen when electricity is passed through it. The reverse process works, too. **Fuel cells** produce electricity by the reaction of hydrogen with oxygen. Like burning hydrogen, water is the only waste product. Unlike burning, hydrogen and oxygen do not react together directly in a fuel cell. They are kept apart by a partially permeable membrane that only lets hydrogen ions through.

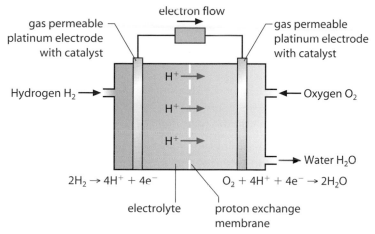

Figure 1 How a hydrogen fuel cell works.

Fuel cells are very efficient, and they produce electricity as long as they are provided with fuel and oxygen. They run electric motors in hydrogen-powered vehicles, removing the need to burn the fuel in an engine.

Making hydrogen

At the moment hydrogen is not a truly 'clean' fuel because of the way it is made. Most of the world's hydrogen is made from fossil fuels, either directly, or indirectly using electricity. So using hydrogen as a fuel still uses **non-renewable resources** and produces carbon dioxide emissions.

Scientists are working to develop ways of producing hydrogen that do not use fossil fuels. One idea being investigated is to modify photosynthesis so that **algae** produce hydrogen.

A hydrogen-powered car.

Questions

1 **(a)** Write a word equation for the combustion of hydrogen. **(b)** Explain why carbon dioxide may be produced by the use of hydrogen.

2 Explain what a fuel cell is.

3 'Petrol is a highly flammable fuel that produces explosive vapours. It must be transported, stored and handled with care.' To what extent would this statement also apply to hydrogen?

4 Explain why hydrogen-powered cars may need larger fuel tanks than petrol-driven cars.

5 Hydrogen and oxygen can be produced from water, using electricity. Fuel cells can produce electricity and water from hydrogen and oxygen. To what extent would you agree that using hydrogen is just a way to store and transport electricity?

6 **(a)** Write a balanced equation for the combustion of hydrogen. **(b)** From the equation, what is the ratio of moles of hydrogen to moles of oxygen needed to exactly use up both gases? **(c)** Suggest why rockets use a hydrogen-to-oxygen ratio of 2.75 : 1.

7 What is your opinion of hydrogen as an 'environmentally friendly' fuel? Include at least three benefits and drawbacks in your answer.

ISA practice: calorimetry of alcohols

Alcohols are liquid fuels. They release heat energy when they burn. The amount of heat energy released can be measured using a method called calorimetry.

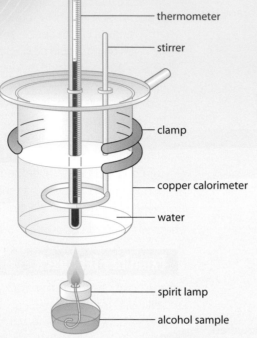

Figure 1 Simple calorimetry.

Table 1 shows four suitable alcohols with their relative formula masses, M_r.

Table 1 Four alcohols and their relative formula masses.

Alcohol	methanol	ethanol	propanol	butanol
M_r	32	46	60	74

A fuel supplier believes that the amount of energy released on combustion is related to the relative formula mass of the alcohol. Your task is to investigate if the relative formula mass, M_r, of an alcohol affects the amount of energy it releases when it burns.

Section 1

1   Write a hypothesis about the amount of energy released and the relative formula mass of an alcohol. Use information from your knowledge of the combustion of alcohols to explain why you made this hypothesis.
    (3 marks)

2   Describe how you are going to do your investigation.
    You should include:
    - the equipment that you are going to use
    - how you will use the equipment
    - the measurements that you are going to make
    - a risk assessment
    - how you will make it a fair test.

You may include a labelled diagram to help you to explain your method.

In this question you will be assessed on using good English, organising information clearly and using specialist terms where appropriate.
    (9 marks)

3   Design a table that you could use to record all the data you would obtain during the planned investigation.
    (2 marks)

Total for Section 1: 14 marks

Section 2

Study group 1 were a group of students who carried out an investigation into the hypothesis. They used methanol, ethanol, propanol and butanol. They burned each alcohol for 5 minutes and recorded the temperature change in each case. Figure 2 shows their results.

> methanol goes up 6°C
> ethanol goes up 8°C
> propanol goes up 11°C
> butanol goes up 13°C

Figure 2 Study Group 1's results.

4   **(a)** Plot a graph of these results.   *(4 marks)*

    (b) What conclusion can you draw from the investigation about a link between the temperature change and the reactivity series? You should use any pattern that you can see in the results to support your conclusion.   *(3 marks)*

    (c) Do the results support the hypothesis you put forward in answer to question 1? Explain your answer. You should quote some figures from the data in your explanation.   *(3 marks)*

Study Group 2 was a second group of students that carried out a similar investigation. They too burned each alcohol for 5 minutes, but then calculated the temperature rise that would be achieved if they burned the formula mass of each alcohol in grams. Figure 3 shows their results.

> 32g of methanol: temperature goes up 6.80°C
> 46g of ethanol: temperature goes up 4.10°C
> 60g of propanol: temperature goes up 3.20°C
> 74g of butanol: temperature goes up 2.85°C

Figure 3 Study Group 2's results.

Study Group 3 was a team of research scientists, who carried out the investigation using a bomb calorimeter. They burned 0.1 moles of each alcohol, calculated the energy transferred in kilojoules, and then repeated each trial three times.

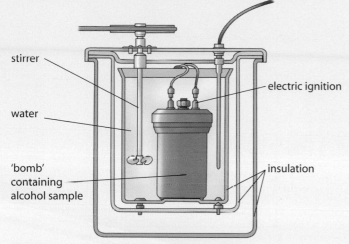

Figure 4 A bomb calorimeter.

The results are given in Table 2.

Table 2 Study Group 3's results.

Alcohol	Energy transferred/kJ			
	Trial 1	Trial 2	Trial 3	Mean
methanol	74.35	70.62	71.96	72.31
ethanol	131.25	137.23	138.69	135.72
propanol	205.36	199.25	148.25	184.29
butanol	270.42	263.23	268.52	267.39

Study Group 3 then plotted their results on a graph, Figure 5.

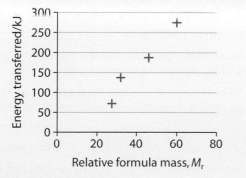

Figure 5 Graph of Study Group 3's results.

Study Group 4 was a second team of scientists, who burned 10 g of six different alcohols in a bomb calorimeter. They repeated each measurement three times and recorded the mean energy transferred for each alcohol. They then produced a graph to show the amount of energy released.

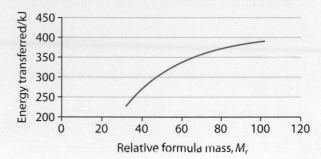

Figure 6 Graph of Study Group 4's results.

5   (a)   Draw a sketch graph of the results from Study Group 2. *(3 marks)*

    (b)   Look at the results from Study Groups 2 and 3. Does the data support the conclusion you reached about the investigation in question 5(b)? Give reasons for your answer. *(3 marks)*

    (c)   The data contain only a limited amount of information. What other information or data would you need in order to be more certain whether or not the hypothesis is correct?

          Explain the reason for your answer. *(3 marks)*

    (d)   Look at Study Group 4's results. Compare them with the data from Study Group 1. Explain how far the data support or do not support your conclusion you came to in answer to question 5(b). You should use examples from Study Group 4's results and from Study Group 1. *(3 marks)*

6   (a)   Compare the results of Study Group 1 with Study Group 2. Do you think that the results for Study Group 1 are *reproducible*?

          Explain the reason for your answer. *(3 marks)*

    (b)   Explain how Study Group 1 could use results from other groups in the class to obtain a more *accurate* answer. *(3 marks)*

7   Suggest how ideas from the original investigation and the other studies could be used by the fuel supplier to decide which alcohol to use so that the maximum energy can be obtained from the minimum mass of fuel transported. *(3 marks)*

Total for Section 2: 31 marks

Total for the ISA: 45 marks

Assess yourself questions

1 Drinking water is soft in some parts of the UK but hard in other parts of the UK. Choose words once only from this list to complete the sentences below.

   detergent   lather   soap   scum   calcium

   (a) Soft water readily forms .... with soap, but hard water reacts with soap to form .... *(2 marks)*

   (b) A soapless .... does not form scum. *(1 mark)*

2 Hard water contains dissolved compounds of certain metal ions.

   (a) Name two metal ions that cause hardness in water. *(2 marks)*

   (b) Explain how these compounds may get into the water. *(2 marks)*

3 Four different water samples were tested to investigate hardness. In the first test, soap flakes were added, one by one, and the mixture shaken each time. The number of soap flakes needed to produce lather was recorded. In the second test, the water was boiled and the results recorded.

Table 1 The results of this investigation.

Sample	Number of soap flakes needed	Observation on boiling
A	1	no change seen
B	7	white precipitate seen
C	7	no change seen

   (a) Explain why the same volume of each sample of water should be used in the investigation. *(1 mark)*

   (b) Which water sample was the softest? Explain your answer. *(2 marks)*

   (c) What is a precipitate? *(1 mark)*

   (d) Which sample was temporary hard water? Explain your answer. *(2 marks)*

   (e) Describe the difference, if any, in the number of soap flakes needed to obtain lather in sample C after boiling and cooling it. Give reasons for your answer. *(2 marks)*

4 Temporary hard water contains hydrogen carbonate ions, HCO_3^-. These decompose on heating to produce carbonate ions, CO_3^{2-}.

   (a) Correctly balance the equation below:

   $HCO_3^- \rightarrow CO_3^{2-} + CO_2 + H_2O$ *(1 mark)*

   (b) Name the insoluble compound formed when carbonate ions react with calcium ions. *(1 mark)*

   (c) Use your answers to parts (a) and (b) to explain why temporary hard water is softened by boiling. *(2 marks)*

5 One of the drawbacks of temporary hard water is that it can produce limescale. This reduces the efficiency of heating systems. Figure 1 shows the effect of limescale on a certain type of domestic heating boiler.

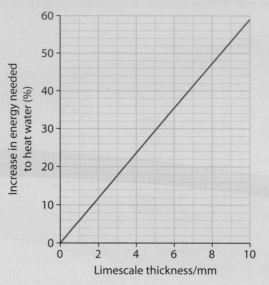

Figure 1 The effect of limescale on the amount of energy needed to heat water in the boiler.

   (a) Describe the effect of limescale on the amount of energy needed to heat water in the boiler. *(2 marks)*

   (b) How much more energy is needed if there is a 2-mm-thick layer of limescale in the boiler? *(1 mark)*

   (c) What thickness of limescale increases the amount of energy needed by 50%? *(1 mark)*

   (d) Apart from its effect on heating systems, describe one other reason why the use of hard water can increase costs. *(1 mark)*

   (e) Describe one benefit of hard water. *(1 mark)*

6 Permanent hard water can be softened by adding sodium carbonate, or by passing the water through an ion exchange column.

   (a) Explain why permanent hard water is softened by adding sodium carbonate to it. *(2 marks)*

   (b) Explain why permanent hard water is softened by passing it through an ion exchange column. *(2 marks)*

7 Water must be treated to make it safe to drink.

   (a) Explain why chlorine is added to drinking water. *(1 mark)*

   (b) Explain why fluoride compounds are added to drinking water. *(1 mark)*

   (c) Some people pass their tap water through water filters containing carbon or ion exchange resins.

   (i) Name another element, apart from carbon, that is found in such water filters. *(1 mark)*

(ii) Explain why people use water filters at home.
(1 mark)

(d) Pure water can be produced by distillation. Give one reason why this process is expensive. *(1 mark)*

8 Simple calorimetry can be used to measure the amounts of energy released when substances burn. In one such experiment, 0.92 g of ethanol was used to heat 200 g of water in a copper container. The temperature of the water increased from 22 °C to 38 °C.

(a) Use the expression $Q = mc\Delta T$ to calculate the energy, in kJ, absorbed by the water. Assume that $c = 4.2$ J/g.°C. *(4 marks)*

(b) Calculate the number of moles of ethanol in 0.92 g of ethanol? *(2 marks)*

(c) Use your answers to parts (a) and (b) to calculate the heat energy released by the burning ethanol in kJ/mole. *(1 mark)*

(d) A book containing scientific data gives the energy released by burning ethanol as 1360 kJ/mole. Apart from incorrect calculations, suggest a reason why your answer to part (c) differs from the value in the data book. *(1 mark)*

9 50 cm³ of 1 mole/dm³ hydrochloric acid was added to 50 cm³ of 2 moles/dm³ sodium hydroxide. The temperature of the mixture rose by 7 °C.

(a) Use the expression $Q = mc\Delta T$ to calculate the energy, in kJ, released in the reaction. Assume that 1 cm³ of the reaction mixture has a mass of 1 g, and $c = 4.2$ J/g.°C. *(4 marks)*

(b) Calculate the number of moles of hydrochloric acid in 50 cm³ of 1 mole/dm³ hydrochloric acid. *(2 marks)*

(c) Use your answers to parts (a) and (b) to calculate the energy released by the reaction in kJ per mole of hydrochloric acid. *(1 mark)*

10 During a chemical reaction, bonds in the reactants are broken and bonds are formed in the products.

(a) When bonds are broken, is energy released or supplied? *(1 mark)*

(b) In terms of energy, and bonds broken and formed, what is the difference between an exothermic reaction and an endothermic reaction? *(4 marks)*

11 Figure 2 shows an energy level diagram for a reaction.

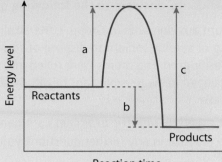

Figure 2 An energy level diagram.

(a) Which arrow (a, b or c) shows the activation energy of the reaction? *(1 mark)*

(b) Does the diagram represent an exothermic reaction or an endothermic reaction? Explain your answer. *(3 marks)*

(c) Describe how the energy level diagram would change if a catalyst were present in the reaction. *(2 marks)*

12 Hydrogen can be burned as a fuel in engines. The equation for the reaction can be shown using structural formulae:

2 (H–H) + O=O → 2 (H–O–H)

Table 2 Some bond energies.

Bond	Bond energy / kJ/mole
H–H	436
O–H	464
O=O	498

Use the bond energies in Table 2 to calculate the energy change when hydrogen burns. *(3 marks)*

13 Carbon monoxide, CO, is a toxic pollutant produced in vehicle engines. Catalytic converters fitted to the exhaust system can oxidise carbon monoxide to carbon dioxide:

2 (C≡O) + O=O → 2 (O=C=O)

The table shows some bond energies.

Table 3 More bond energies.

Bond	Bond energy / kJ/mole
C≡O	1077
O=O	498
C=O	805

(a) Calculate the energy change for the reaction. *(3 marks)*

(b) Explain whether the reaction is exothermic or endothermic. *(1 mark)*

Here are three students' answers to the following question:

Lithium and sodium are elements in Group 1, the alkali metals. After World War II ended, the US Army had 9000 kg of surplus sodium to dispose of. It decided to dump the drums of sodium into Lake Lenore in Washington. The drums were rolled into the frozen lake. They fell through the ice and, once water seeped in, huge explosions began. Steamy corrosive clouds rose several hundred metres over the lake.

Read the answers together with the examiner comments. Then check what you have learnt and try putting it into practice in any further questions you answer.

Figure 1 The warning label on the sodium barrels.

In this question you will be assessed on using good English, organising information clearly and using specialist terms where appropriate.

(6 marks)

MOVING UP THE GRADES

- Do not change chemical formulae to balance symbol equations.
- Underline the key points in the question to help you identify and address all the ideas required. Tick them off as you check your answer.
- Make sure you use scientific terminology correctly.
- Use the word 'it' sparingly to avoid confusion between different ideas.
- Take care to use comparative words when writing an answer to a comparison question.

B Grade answer

Student 1

> It would be better to name these.

> The candidate wrongly changes the symbol for sodium to balance the equation.

Sodium reacts really quickly with water makes an alkalie and gas.

$$Na_2 + 2H_2O \longrightarrow 2NaOH + H_2$$

Sodium atoms are big so there is less attraction from the nucleus, making it difficult to lose the outer electron.

> Poor grammar and spelling can sometimes affect grades.

> The candidate has forgotten that it is a comparison question.

Examiner comment

The candidate's answer is very short with little attention to detail, and they include few relevant specialist terms. Their misspelling, while unfortunate, would not affect their marks, as it is close enough when spoken aloud and could not be confused with another specialist word. The candidate implies, but does not state, that sodium is very reactive. To balance the equation, they should have written 2Na instead of Na_2.

As the question requires a comparison of sodium with lithium, the candidate should use words like 'bigger' rather than 'big'. They make a good point about there being less attraction from the nucleus, but forget to state clearly that the attraction is for the outer electron. The wording implies that the outer electron is unable to leave the atom, which would make sodium unreactive.

A Grade answer

Student 2

The candidate uses their knowledge about sodium to explain something not given to them in the question.

Sodium is a very reactive metal that reacts violently with water. The reaction makes sodium hydroxide and hydrogen.

$$Na + H_2O \longrightarrow NaOH + \frac{1}{2}H_2$$

It is acceptable to use fractions to correctly balance equations, if you prefer.

It has a larger atom so the outer electron is not attracted to the nucleus as much and it is more easily lost.

The word 'it' refers to two different things in this sentence, sodium and the outer electron.

Examiner comment

The candidate has used appropriate specialist terms and, apart from the brief last sentence, the answer is structured well. The candidate correctly names the products, but they could have related them to the observations. The equation is correctly balanced, in this case using a fraction rather than putting numbers in front of the other formulae. Some students find this easier to do.

The candidate's explanation of the difference in reactivity of sodium and lithium suffers because of the word 'it'. As the question was about sodium, 'it' is taken to mean 'sodium' by the examiner. The word cannot also mean 'outer electron' in the same sentence.

A* Grade answer

Student 3

This starts the answer well. It provides useful information to explain why the reaction starts after the barrels enter the water.

Sodium is a very reactive metal. It is usually stored under oil to keep water and air away from it. Once water reaches the sodium, a reaction happens that makes sodium hydroxide and hydrogen gas. $2Na + 2H_2O \longrightarrow 2NaOH + H_2$

The candidate provides information to suggest why the clouds were corrosive.

Sodium hydroxide is a strong alkali, so that is probably why the clouds were corrosive. Hydrogen is very flammable, which is probably why the explosions happened.

The candidate provides information to suggest why explosions are observed.

A sodium atom is larger than a lithium atom. It has more shells, so its outer electron is further from the nucleus. This means there is less attraction between the outer electron and the nucleus. The outer electron is lost more easily from a sodium atom than from a lithium atom.

The candidate remembers that the question involves a comparison between sodium and lithium.

The candidate ensures lithium is mentioned by name.

Examiner comment

The candidate has used a range of specialist terms correctly, answering in a logical sequence with clear explanations. The candidate uses their knowledge of the reactions of group 1 elements with water to explain the observations. These include why the reaction starts after the barrels enter the water, the corrosive nature of the clouds produced and the explosions. The candidate could have mentioned that the reaction is exothermic, so showing why the hydrogen ignited and the clouds were steamy. The explanation of the difference in reactivity of sodium and lithium is presented key idea by key idea, remembering that a comparison is needed.

C3 Analysis, ammonia and organic compounds

Chemical tests are important to forensic scientists, and to scientists working in the areas of health or the environment. In this section you will learn how to use a range of chemical tests, including flame tests and precipitation tests. These tests are used to detect and identify various elements and compounds. You will also learn how to calculate the concentrations of solutions, and how to use titration to find how much acid or alkali there is in a solution.

Industrial processes need energy and they produce waste substances. It is important for economic reasons, and for sustainable development, that both energy use and waste are reduced. The Haber process produces ammonia, which is needed for chemicals such as fertilisers, dyes and explosives. You will find out about the conditions needed for this process to work economically, and the reasons for choosing them.

Compounds that contain carbon, covalently bonded to other elements, are called organic compounds. Alcohols, carboxylic acids and esters are examples of such organic chemicals. You will learn about why they are important, how they are used and how these uses are related to their properties, structures and reactions.

Test yourself

1 Explain what happens in a precipitation reaction.

2 Describe how you could determine, using an experiment, the volume of acid needed to exactly neutralise an alkali.

3 Describe three factors that influence the rate of a reaction.

4 If the forward reaction in a reversible reaction is exothermic, what can we say about the reaction in the opposite direction?

5 (a) What is a covalent bond?

(b) What does the term *simple molecule* mean?

Objectives

By the end of this unit you should be able to:

- describe the expected results of flame tests and precipitation tests
- interpret and evaluate the results of chemical analyses to identify elements and compounds
- calculate chemical quantities in acid–alkali titrations
- describe the conditions used in the Haber process
- describe the effects of changing the temperature and pressure on a reversible reaction
- recognise alcohols, carboxylic acids and esters from their names and formulae
- describe the properties of alcohols, carboxylic acids and esters.

Testing for metal ions

Taking it further

Atoms emit light when their electrons are 'excited'. The energy from the Bunsen burner flame can 'promote' electrons into a higher energy level. When they return to their original energy level, these excited electrons emit energy as light. The different jumps from level to level produce different coloured light.

Science in action

The Bunsen burner was originally designed to investigate the flame colours produced by metal ions. Robert Bunsen and Gustav Kirchhoff discovered rubidium and caesium by studying minerals and mineral water. Both metals are named after the Latin words for the flame colours they produce: *rubidus* means dark red and *caesius* means grey-blue.

Identifying and analysing substances

Chemists have developed a range of simple chemical tests to analyse substances. These tests reveal the presence or absence of certain metal ions.

Flame tests

Fireworks burn with coloured flames because of the various metal compounds they contain. In the laboratory, different metal ions give off different coloured flames when solutions of these ions are heated in a Bunsen burner flame.

Table 1 The colours produced by different metal compounds.

Metal in the compound	Colour of flame
lithium	crimson
sodium	orange
potassium	lilac
calcium	red
barium	green

Practical

There are three common ways to carry out **flame tests**:

- spray the metal compound solution into a Bunsen burner flame
- use wooden splints soaked overnight in the metal compound solution
- use a loop of nichrome wire to hold a small volume of solution in the flame.

If a nichrome wire is used it must be cleaned between each flame test. The loop is dipped in concentrated hydrochloric acid, then held in the flame. This process is repeated until the wire no longer affects the flame colour. Traces of sodium ions are particularly difficult to remove.

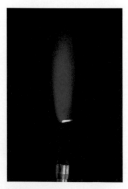

Lithium ions produce a crimson flame.

Sodium ions produce an orange flame.

Potassium ions produce a lilac flame.

Calcium ions produce a red flame.

Barium ions produce a green flame.

Hydroxide precipitates

Copper and iron form coloured compounds, like many other transition metals. Coloured hydroxide **precipitates** form when a few drops of sodium hydroxide solution are added to solutions of the metal ions. For example:

$$Cu^{2+}(aq) + OH^-(aq) \rightarrow Cu(OH)_2(s)$$

Table 2 The colours of some precipitates.

Metal ion	Colour of precipitate
copper(II), Cu^{2+}	blue
iron(II), Fe^{2+}	green
iron(III), Fe^{3+}	brown

Aluminium, calcium and magnesium ions also form hydroxide precipitates with sodium hydroxide solution. These are difficult to tell apart because they are all white. However, only the aluminium hydroxide precipitate dissolves to form a colourless solution when excess sodium hydroxide solution is added.

Questions

1. Describe the flame colours produced by lithium, sodium, potassium, calcium and barium compounds.

2. Describe the colours of the precipitates produced by adding sodium hydroxide solution to solutions of copper(II), iron(II), iron(III) and calcium ions.

3. The label on a bottle of powder reads 'IRON CHLORIDE (HARMFUL)'. Describe how you could carry out a simple laboratory test to determine whether it contains iron(II) chloride or iron(III) chloride.

4. Describe how you could carry out flame tests to distinguish between solutions of sodium chloride and potassium chloride.

5. Explain how you could use sodium hydroxide solution to distinguish between solutions of magnesium nitrate and aluminium nitrate.

6. A laboratory technician analyses some mineral water to detect the presence of calcium ions. Explain why carrying out a flame test would be preferable to using sodium hydroxide solution.

7. A student analyses some white crystals to detect the presence of potassium ions. **(a)** Describe how a flame test could be carried out using a nichrome wire loop. **(b)** Explain why testing with sodium hydroxide solution would not work.

8. A mixture of salts in solution is analysed using a flame test. The flame colour is orange. A white precipitate forms when a few drops of sodium hydroxide are added. **(a)** What does the flame test result mean? **(b)** What conclusions can be drawn from the sodium hydroxide test? **(c)** When more sodium hydroxide solution is added, the white precipitate dissolves. What conclusion can be drawn from this? **(d)** Write an ionic equation for the reaction that produced the white precipitate.

A green precipitate of iron(II) hydroxide. This gradually changes to brown as it oxidises to iron(III) hydroxide.

A brown precipitate of iron(III) hydroxide.

Copper(II) ions produce a blue precipitate of copper(II) hydroxide with sodium hydroxide solution.

Aluminium, calcium and magnesium ions produce a white precipitate.

Science in action

Flame tests are one of a range of analytical techniques called spectroscopy. Visible light is not the only way to analyse and identify substances. Infrared spectroscopy uses infrared light to analyse substances, particularly those containing carbon. Infrared light is absorbed by covalent bonds in the substance, and this absorbance can be detected. Covalent bonds between different atoms absorb infrared light of different frequencies, allowing the bonds present to be identified.

Testing for negatively charged ions

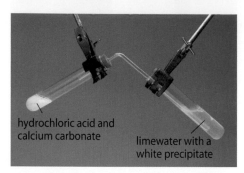

Dilute acids react with carbonates to produce carbon dioxide, which is detected using limewater.

hydrochloric acid and calcium carbonate

limewater with a white precipitate

Testing for carbonate ions

Compounds that contain carbonate ions, CO_3^{2-}, react with dilute acids to form carbon dioxide. For example, drops of hydrochloric acid added to a piece of calcium carbonate cause bubbling on the surface of the solid. Most carbonates are insoluble in water. However sodium carbonate *is* soluble. Adding dilute acid to sodium carbonate solution causes brief bubbling:

$$Na_2CO_3(aq) + 2HCl(aq) \rightarrow 2NaCl(aq) + H_2O(l) + CO_2(g)$$

sodium + hydrochloric → sodium + water + carbon
carbonate   acid   chloride   dioxide

Limewater, calcium hydroxide solution, is used in a **confirmatory test** for the production of carbon dioxide. A white precipitate of calcium carbonate forms when carbon dioxide is bubbled through limewater:

$$Ca(OH)_2(aq) + CO_2(g) \rightarrow CaCO_3(s) + H_2O(l)$$

calcium hydroxide + carbon dioxide → calcium carbonate + water

Testing for sulfate ions

Barium sulfate is insoluble in water, but barium chloride is not. This difference is the basis for a laboratory test for the presence of sulfate ions, SO_4^{2-}. A white precipitate of barium sulfate forms when barium chloride solution is added to solutions containing sulfate ions. For example:

$$Na_2SO_4(aq) + BaCl_2(aq) \rightarrow 2NaCl(aq) + BaSO_4(s)$$

sodium sulfate + barium chloride → sodium chloride + barium sulfate

Science in action

Soluble barium salts, such as barium chloride, are toxic. The barium ions are absorbed through the wall of the intestines into the bloodstream, and affect the nervous system. Rat poison may contain barium carbonate, which reacts with hydrochloric acid in the rat's stomach to produce barium chloride. Barium sulfate is insoluble, non-toxic, and opaque to X-rays. It is used in 'barium meals', given to patients with intestinal problems so that the outline of their intestines can be seen in medical X-ray photographs.

Practical

When testing for the presence of sulfate ions, the test solution is first acidified with a few drops of dilute hydrochloric acid. This is to prevent interference from other ions. A few drops of barium chloride solution are then added.

A positive result for sulfate ions.

Testing for halide ions

Halide ions are formed by the halogens, the elements in group 7 of the periodic table. Silver fluoride is soluble in water but the other silver halides are insoluble. Their insolubility is the basis for a laboratory test for the presence of chloride ions, bromide ions or iodide ions. A precipitate forms when silver nitrate solution is added to solutions containing these ions. For example:

$$NaCl(aq) + AgNO_3(aq) \rightarrow NaNO_3(aq) + AgCl(s)$$

sodium chloride + silver nitrate → sodium nitrate + silver chloride

Each halide has a different coloured precipitate, as shown in Table 1.

Table 1 Colours of the different silver halide precipitates.

Halide ion	Precipitate formed	Colour of precipitate
Chloride, Cl^-	silver chloride, AgCl	white
Bromide, Br^-	silver bromide, AgBr	cream
Iodide, I^-	silver iodide, AgI	yellow

Questions

1. Describe how carbonate ions are detected.

2. Describe the colours of the precipitates produced by adding silver nitrate solution to solutions containing chloride, bromide or iodide ions.

3. **(a)** Explain why solutions containing sulfate ions are acidified with dilute hydrochloric acid before testing with barium chloride solution. **(b)** Suggest why they are not acidified with sulfuric acid, $H_2SO_4(aq)$.

4. **(a)** Explain why solutions containing chloride ions are acidified with dilute nitric acid before testing with silver nitrate solution. **(b)** Suggest why they are not acidified with hydrochloric acid, HCl(aq).

5. Explain why fluoride ions cannot be detected using silver nitrate solution.

6. Limescale forms in kettles in hard water areas. It contains calcium carbonate. **(a)** Describe what you would see if dilute hydrochloric acid were added to limescale. **(b)** Write a balanced equation for the reaction. **(c)** Explain why a drop of limewater on a glass rod would turn cloudy white if held near the reaction mixture.

7. **(a)** Describe what you would see if barium chloride solution and potassium sulfate solution, $K_2SO_4(aq)$, were mixed together. **(b)** Write a balanced equation for the reaction.

8. A mixture of sodium salts in solution is analysed. Fizzing is observed when dilute hydrochloric acid is added. A white precipitate is produced when barium chloride solution is then added. In a separate test, a yellow precipitate forms when silver nitrate solution is added to the mixture, acidified with nitric acid.

   Explain which ions are present in the mixture. Write balanced equations for the reactions observed.

A*

Science in action

Silver salts are light sensitive. They decompose in light, forming silver crystals. This is the basis of traditional film-based photography. The camera film and photographic paper contain silver salts, including silver chloride and silver bromide.

Taking it further

Lead nitrate solution may be used instead of silver nitrate solution to test for the presence of halide ions. Lead chloride forms a white precipitate, lead bromide forms a cream-coloured precipitate and lead iodide forms a bright yellow precipitate.

Examiner feedback

You need to remember these three colours for the examination.

Practical

When testing for the presence of halide ions, the test solution is first acidified with a few drops of dilute nitric acid. This is to prevent interference from other ions. A few drops of silver nitrate solution are then added.

Silver halide precipitates.

You take a burette or pipette reading from the bottom of the meniscus.

Titrations for calculations

Titrations may be used to determin the volumes of acid and alkali needed to make a neutral salt solution (see lesson C2 6.2). They may also be used to determine the concentration of an acid or alkali, if the concentration of the other reactant is known. It is important that titrations are carried out accurately.

The **meniscus** is the curve in the surface of a liquid. The burette and pipette used in accurate titrations must be read from the bottom of the meniscus. The smallest graduations on a burette are 0.1 cm³ apart. The bottom of the meniscus may be on a line, or part-way between two lines. So the precision of a burette reading is $\pm$ 0.05 cm³. The acid is usually put in the burette.

The alkali is usually added to a conical flask. The standard volume for a titration is 25 cm³. A measuring cylinder can be used, but a 25 cm³ pipette is more accurate. The alkali is drawn up into the pipette using a pipette filler. A few drops of a suitable **indicator**, such as phenolphthalein, are also added to the conical flask. Phenolphthalein is pink when alkaline and colourless when acidic.

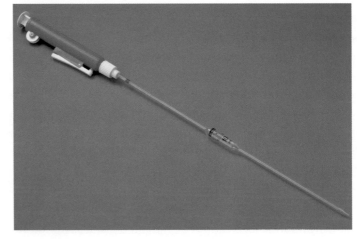

A pipette and pipette filler.

Acid is added from the burette to the alkali until the end-point is reached. With phenolphthalein, this is when the colour just changes from pink to colourless. The titre, the volume of acid needed to exactly neutralise the alkali, is the difference between the first and second readings on the burette. More precise results are obtained by repeating the titration.

Concentration calculations

The concentration of a dissolved substance can be measured in grams per dm³ (cubic decimetre). For example, suppose 4 g of sodium hydroxide is dissolved in 2 dm³ of water:

$$\text{concentration in g/dm}^3 = \frac{\text{mass in g}}{\text{volume in dm}^3}$$

so concentration of sodium hydroxide = 4 ÷ 2 = 2 g/dm³

Laboratory volumes are usually measured in cm³ rather than in dm³. You need to divide a volume in cm³ by 1000 to convert to dm³. For example:

100 cm³ = 100 ÷ 1000 = 0.100 dm³

Example 1

Calculate the concentration when 1.00 g of sodium hydroxide is dissolved in 50 cm³ of water.

Using the magic triangle, you can see that

$$\text{volume} = \text{mass} \div \text{concentration}$$

$$50 \text{ cm}^3 = 50 \div 1000 \text{ dm}^3 = 0.05 \text{ dm}^3$$

concentration = mass ÷ volume.

$$= 1.00 \div 0.05 = 20 \text{ g/dm}^3$$

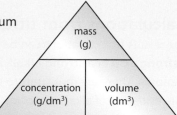

Figure 1 A 'magic triangle' for calculating concentration in g/dm³. Cover up the quantity you want to calculate. You can then see whether to multiply or divide the other two values to work out the answer.

Chemists often use concentrations measured in moles/dm³ instead of g/dm³. This is because these give information about the number of dissolved particles rather than their mass. For example, if 0.2 moles of sodium hydroxide are dissolved in 0.5 dm³ of water:

$$\text{concentration in g/dm}^3 = \frac{\text{amount in moles}}{\text{volume in dm}^3}$$

$$\text{so concentration} = 0.2 \div 0.5$$

$$= 0.4 \text{ moles/dm}^3$$

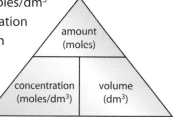

Figure 2 A 'magic triangle' for calculating concentration in moles/dm³.

Example 2

Calculate the concentration when 0.050 moles of sodium hydroxide are dissolved in 25 cm³ of water.

Volume = 25 ÷ 1000 = 0.025 dm³

Concentration = 0.050 ÷ 0.025 = 2 moles/dm³

Questions

1   Describe how to carry out a titration involving hydrochloric acid, sodium hydroxide solution and phenolphthalein indicator.

2   Calculate the concentration of a solution containing the following. **(a)** 10 g of sodium hydroxide in 2.5 dm³ of water. **(b)** 2.5 g of sodium hydroxide in 1.25 dm³ of water.

3   Convert the following volumes to dm³. **(a)** 500 cm³. **(b)** 250 cm³. **(c)** 10 cm³.

4   Convert the following volumes to cm³. **(a)** 2 dm³. **(b)** 0.5 dm³. **(c)** 0.025 dm³.

5   Calculate the concentration in moles/dm³ of a solution containing the following. **(a)** 0.2 moles of potassium hydroxide in 400 cm³ of water. **(b)** 0.01 moles of potassium hydroxide in 100 cm³ of water.

6   Calculate the mass of sodium hydroxide contained in 25 cm³ of 0.5 g/dm³ sodium hydroxide.

7   Calculate the volume of 0.1 mole/dm³ hydrochloric acid, in dm³, that contains 0.25 moles of hydrochloric acid.

8   Sulfuric acid H_2SO_4(aq) breaks down completely in solution to form hydrogen ions H^+(aq) and sulfate ions SO_4^{2-}(aq). **(a)** Write a balanced equation for this process. **(b)** Use your answer to part b to help you calculate the amount, in moles, of hydrogen ions produced when 0.25 moles of sulfuric acid dissolves in water. **(c)** Calculate the concentration of hydrogen ions formed, in moles/dm³, when 0.25 moles of sulfuric acid dissolve in 250 cm³ of water.

Using results from analyses

Learning objectives

- calculate chemical quantities from titrations
- interpret and evaluate the results of chemical tests to identify elements and compounds for forensic, health or environmental purposes.

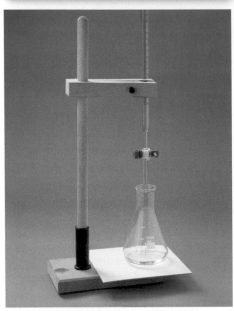

Apparatus set up for a titration.

Examiner feedback

Pass your answer through a 'reality check'. In the worked example, the volumes of the two reagents were almost the same, so the concentration of the acid should be close to the concentration of the alkali. In addition, since less acid was needed, it must be more concentrated than the alkali.

Science in action

A positive result shows the presence of a particular ion in solution. However, a negative result does not necessarily mean that a particular ion is not present. It may simply be present in a concentration that is too low to be detected. This is why sensitive instruments are needed in modern chemistry.

Calculations from titrations

The results from titrations can be used to determine the concentration of a **strong acid** or a **strong alkali**, if the concentration of the other reactant is known. There are four steps to the calculation. For example, where the concentration of alkali is known:

Step 1 Write down the balanced equation.

Step 2 Calculate the amount of alkali in moles from its concentration and volume.

Step 3 Check the balanced equation and your answer to Step 2 to see how many moles of acid will have reacted in the titration.

Step 4 Calculate the concentration of the acid using your answer to Step 3 and the volume of the acid.

Example 1

25.0 cm^3 of 0.10 mole/dm^3 sodium hydroxide was exactly neutralised by 20.0 cm^3 of hydrochloric acid. Calculate the concentration of the acid.

Step 1 $NaOH(aq) + HCl(aq) \rightarrow NaCl(aq) + H_2O(l)$

Step 2 Volume of $NaOH(aq) = 25.0 \div 1000 = 0.025 \text{ dm}^3$

Amount of $NaOH = 0.1 \times 0.025 = 0.0025$ moles

Step 3 One mole of NaOH reacts with one mole of HCl, so 0.0025 moles of NaOH react with 0.0025 moles of HCl.

Step 4 Volume of $HCl(aq) = 20.0 \div 1000 = 0.020 \text{ dm}^3$

Concentration of $HCl(aq) = 0.0025 \div 0.020 = 0.125 \text{ mole/dm}^3$

Interpreting chemical tests

Metal ions in compounds can be detected using flame tests and sodium hydroxide solution (see lesson C2 4.1). Carbonate ions can be detected using hydrochloric acid, sulfate ions using barium chloride solution in the presence of dilute hydrochloric acid, and halide ions using silver nitrate solution in the presence of dilute nitric acid (see lesson C2 4.2).

Scientists use chemical tests to analyse compounds.

It is possible to identify an ionic compound using two or more of these tests. This is useful for forensic scientists, who may need to analyse a substance to determine whether or not it contains a particular element or compound. The tests are also useful for environmental scientists and scientists working in the health industry.

Example 2

A white solid containing one ionic compound dissolved to form a colourless solution. This produced a crimson flame in a flame test. It did not produce any bubbles when dilute hydrochloric acid was added. A white precipitate formed when barium chloride solution was added. Identify the compound.

The crimson flame showed the presence of lithium ions, Li^+. The lack of bubbles showed that carbonate ions were not present. The white precipitate showed that sulfate ions, SO_4^{2-}, were present. The compound must be lithium sulfate, Li_2SO_4.

Six ionic compounds. The blue crystals might be a copper compound, but analytical tests are needed to identify all six confidently.

Route to A*

Sulfuric acid ionises in solution to form hydrogen ions, H^+, and sulfate ions, SO_4^{2-}. It cannot be used to acidify a test solution before adding aqueous barium chloride solution, because it would form a white precipitate of barium sulfate. Similarly, hydrochloric acid ionises to form hydrogen ions and chloride ions Cl^-. It cannot be used to acidify a test solution before adding aqueous silver nitrate, because it would form a white precipitate of silver chloride. Hydrochloric acid is used for barium chloride tests, and nitric acid for silver nitrate tests. The chloride ions and nitrate ions these acids contain are already present in the barium chloride or silver nitrate solutions.

Questions

1. $25.0\ cm^3$ of 0.20 mole/dm^3 sodium hydroxide was exactly neutralised by $20.0\ cm^3$ of nitric acid. Calculate the concentration of the acid.

2. A solution of an ionic compound produces a green flame in a flame test. It did not produce any bubbles when dilute nitric acid was added. A white precipitate formed when silver nitrate solution was added. Identify the compound.

3. $25.0\ cm^3$ of 0.1 mole/dm^3 hydrochloric acid was exactly neutralised by $20.0\ cm^3$ of potassium hydroxide solution. Calculate the concentration of the alkali.

4. A white precipitate formed when sodium hydroxide solution was added to a sample from a crime scene. **(a)** Explain why the sample is unlikely to contain copper. **(b)** Describe how you could determine whether the sample contains aluminium ions rather than calcium ions or magnesium ions.

5. Bones contain calcium compounds. Calcium can be identified in compounds either by using flame tests or by using sodium hydroxide solution. Which analytical test is more useful, and why?

6. Car battery acid is concentrated sulfuric acid. $25.0\ cm^3$ of 0.10 mole/dm^3 sodium hydroxide solution was exactly neutralised by $10.0\ cm^3$ of diluted battery acid. Calculate the concentration of the diluted battery acid.

7. 'Low salt' table salt contains potassium chloride in addition to sodium chloride. Describe how you would distinguish between solutions of sodium chloride, potassium chloride and potassium bromide.

8. Four bottles have lost their labels. They contain aluminium sulfate, iron(II) sulfate, magnesium chloride or sodium carbonate. Describe how you would identify the contents of each bottle.

Taking it further

'Confirmatory tests' are used to obtain more evidence to support a conclusion from a laboratory test. One such test involves using limewater to confirm that the gas given off when dilute hydrochloric acid is added to a carbonate is carbon dioxide.

Ammonia is used in confirmatory tests for the silver nitrate tests for halide ions. Dilute ammonia solution is added to the silver halide precipitates Silver chloride redissolves, but silver bromide and silver iodide do not. Concentrated ammonia is then added to the two remaining precipitates. Silver bromide redissolves, but silver iodide does not. These tests can also reveal if a mixture of halide ions is present.

Equilibria

The reversible reaction of anhydrous copper sulfate with water.

Reversible reactions

Some chemical reactions are **reversible**. This means that both the forward and reverse reactions can take place. For example, when blue hydrated copper sulfate crystals are heated, they form white anhydrous copper sulfate and water. If water is added to white anhydrous copper sulfate, then blue hydrated copper sulfate can be re-formed:

$$CuSO_4.5H_2O \rightleftharpoons CuSO_4 + 5H_2O$$

hydrated copper sulfate (blue)   anhydrous copper sulfate (white)   water

Forward and reverse reactions

It is useful to think about reversible reactions in terms of the forward (left to right) and reverse (right to left) reactions.

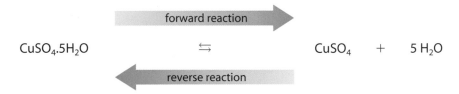

forward reaction

$$CuSO_4.5H_2O \rightleftharpoons CuSO_4 + 5H_2O$$

reverse reaction

The state of equilibrium

The reaction in which ammonium chloride decomposes into ammonia and hydrogen chloride (see lesson C2 3.4) is also reversible:

$$NH_4Cl \rightleftharpoons NH_3 + HCl$$

ammonium chloride   ammonia   hydrogen chloride

As the white ammonium chloride is heated at the bottom of the tube it breaks apart into ammonia and hydrogen, which react together to re-form ammonium chloride in the cooler parts of the tube.

When ammonium chloride is heated in an open boiling tube, some of the ammonia and hydrogen chloride formed react to re-form ammonium chloride, but some escapes into the atmosphere. This is an **open system**, because substances can enter or leave.

However, if the tube is sealed, the system is **closed** because substances cannot enter or leave. In a closed system, the ammonia and hydrogen chloride cannot escape. They will react to re-form ammonium chloride. After a time, the rate at which the ammonium chloride decomposes equals the rate at which the ammonia and hydrogen chloride recombine. When the rate of the forward reaction equals the rate of the reverse reaction the system is in a state of **equilibrium**.

When a system is at equilibrium, both the forward and reverse reaction are taking place at the same rate. This means that, although both reactions are happening, the amount of each substance in the equilibrium remains the same. The symbol $\rightleftharpoons$ is used to show a reaction is at equilibrium.

$$NH_4Cl \rightleftharpoons NH_3 + HCl$$

| ammonium chloride | | ammonia | | hydrogen chloride |

A useful analogy of a system at equilibrium is somebody running on an escalator. Once the person is running at the same speed as the escalator is moving, a state of equilibrium has been reached.

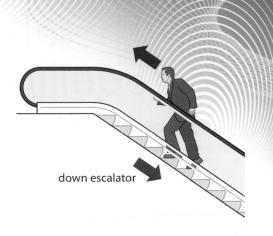

Figure 1 The runner is in equilibrium: he runs upwards at the same rate as the escalator moves downwards, so he stays in one place.

The position of the equilibrium

A system is at equilibrium when the rates of the forward and reverse reactions are the same. However, the position of the equilibrium can vary. Using the escalator analogy, the person could be at equilibrium anywhere on the escalator: near the top, in the middle or at the bottom.

With a chemical equilibrium, if the equilibrium position lies to the left, it means that at equilibrium there are more of the chemicals on the left of the equation (the reactants) than of those on the right. For example, the reaction between nitrogen and hydrogen to form ammonia reaches a state of equilibrium that usually lies to the left. This means that there is more nitrogen and hydrogen than ammonia in the equilibrium mixture.

$$N_2 + 3H_2 \rightleftharpoons 2NH_3$$

| nitrogen | | hydrogen | | ammonia |

If an equilibrium lies to the right, it means that there are more of the chemicals on the right of the equation (the products) than those on the left in the equilibrium mixture.

Examiner feedback

It is important to distinguish reversible reactions and the state of equilibrium. Reversible reactions can take place in both directions. A state of equilibrium is reached when both reactions are taking place at the same time and at the same rate.

Questions

1 What does the symbol '$\rightleftharpoons$' mean in an equation?

2 Give two examples of reversible reactions.

3 What does the symbol '$\rightleftharpoons$' mean in an equation?

4 What is the difference between an open system and a closed system?

5 Describe what is happening when a reaction reaches a state of equilibrium.

6 Look at the equation for the following equilibrium:

carbon monoxide + hydrogen $\rightleftharpoons$ methanol

(a) Write a word equation for the forward reaction.
(b) Write a word equation for the reverse reaction.

7 When calcium carbonate is heated, it breaks down to form calcium oxide and carbon dioxide:

$$CaCO_3(s) \rightleftharpoons CaO(s) + CO_2(g)$$

If this is done in an open system all the calcium carbonate will decompose. However, if the system is closed, it reaches a state of equilibrium. Explain why equilibrium is not reached if the system is open.

8 The reaction between hydrogen and iodine to form hydrogen iodide reaches a state of equilibrium in a closed system. Under normal conditions, this equilibrium lies to the right:

$$H_2 + I_2 \rightleftharpoons 2HI$$

Describe what is happening in the equilibrium mixture and what is meant by the equilibrium lying to the right.

The effect of conditions on equilibria

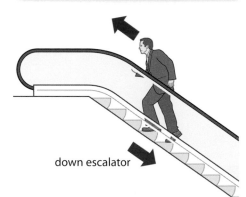

down escalator

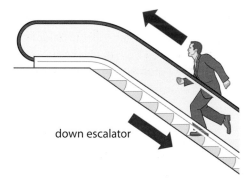

down escalator

Figure 1 A new equilibrium position is established after the speed of the escalator is increased.

Changing the position of an equilibrium

The position of an equilibrium is affected by changes in conditions such as temperature and pressure. If the conditions are changed, the position of the equilibrium moves to oppose the change.

In lesson C3 5.1, we described an equilibrium as being like a person running up a down escalator. Changing the reaction conditions is like changing the speed of the escalator. If the speed is increased, the person will have to run faster to oppose the change. A new equilibrium will be established in a different position. This new equilibrium position lies further down the escalator.

The fact that changing conditions affects the position of an equilibrium is useful. It means that we can change the conditions to make more of the substances we want to make.

The effect of temperature on the position of an equilibrium

In an equilibrium, the reaction in one direction is exothermic and the reaction in the other direction is endothermic. This is due to the law of conservation of energy. Exothermic reactions release energy to the surroundings, which get hotter. Endothermic reactions take in energy from the surroundings, which get colder.

For example, the decomposition of dinitrogen tetroxide to form nitrogen dioxide is endothermic (+58 kJ/mole). The reverse reaction, to form dinitrogen tetroxide from nitrogen dioxide, is exothermic (−58 kJ/mole).

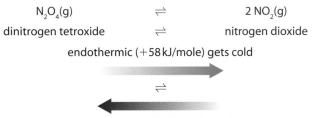

$$N_2O_4(g) \rightleftharpoons 2\,NO_2(g)$$

dinitrogen tetroxide $\rightleftharpoons$ nitrogen dioxide

endothermic (+58 kJ/mole) gets cold

exothermic (−58 kJ/mole) gets hot

If the temperature is raised, the position of the equilibrium will move to oppose the change, i.e. to take in energy. Endothermic reactions get cold, so the position of the equilibrium moves in the direction of the endothermic reaction. This means that the yield of the chemicals made in the endothermic reaction increases and the yield of those made in the exothermic reaction decreases.

If the temperature is lowered, the position of the equilibrium will move to oppose the change, to release energy. Exothermic reactions get hot, so the position of the equilibrium moves in the direction of the exothermic reaction. This means that the yield of the chemicals made in the exothermic reaction increases and the yield of those made in the endothermic reaction decreases.

In the dinitrogen tetroxide equilibrium, raising the temperature will move the position of the equilibrium further to the right to favour the endothermic reaction.

This makes more brown NO_2 and less colourless N_2O_4, so the equilibrium mixture gets darker. If the temperature is lowered, the position of the equilibrium will move further to the left to favour the exothermic reaction. The equilibrium mixture gets paler in colour as there is more colourless N_2O_4 and less brown NO_2.

The effect of pressure on the position of an equilibrium

In an equilibrium involving gases, changes in pressure will affect the position of the equilibrium. The more gas particles there are in the mixture, the greater the pressure. In the dinitrogen tetroxide equilibrium, there is one gas molecule on the left for every two gas molecules on the right. This means that pressure increases with the forward reaction and decreases with the reverse reaction.

If the pressure of a system at equilibrium is increased, the equilibrium position will change to lower the pressure. The equilibrium position moves towards the side with fewest gas molecules.

If the pressure of a system at equilibrium is decreased, the equilibrium position will move to increase the pressure. The equilibrium position moves towards the side with most gas molecules.

In the dinitrogen tetroxide equilibrium, if the pressure is increased, the equilibrium position moves to the left: the side with fewer molecules and so lower pressure. This means there will be more colourless N_2O_4 and less brown NO_2 in the mixture, which will become paler. If the pressure is decreased, the equilibrium moves to the right, where there are more gas molecules to increase the pressure. There will be more brown NO_2 and less colourless N_2O_4, so the mixture will become darker.

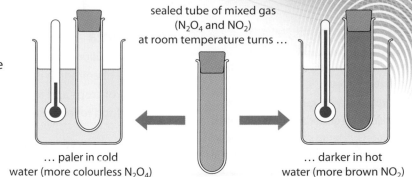

sealed tube of mixed gas
(N_2O_4 and NO_2)
at room temperature turns …

… paler in cold
water (more colourless N_2O_4)

… darker in hot
water (more brown NO_2)

Figure 2 The effect of temperature on the equilibrium between dinitrogen tetroxide (colourless) and nitrogen dioxide (brown).

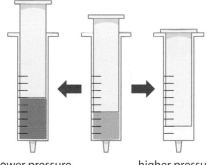

lower pressure
(more brown NO_2)

higher pressure
(less brown NO_2)

Figure 3 The effect of pressure on the equilibrium between dinitrogen tetroxide and nitrogen dioxide.

Questions

1. Describe the energy change in an exothermic reaction.

2. Describe the energy change in an endothermic reaction.

3. In an equilibrium, the energy change for the forward reaction is –56 kJ/mole. What is the energy change for the reverse reaction?

4. What does it mean if the position of an equilibrium moves to the right?

5. How does the number of particles of a gas in a container affect the pressure of the gas?

6. The reaction between hydrogen and iodine to form hydrogen iodide reaches a state of equilibrium. The forward reaction is exothermic.

    $$H_2(g) + I_2(g) \rightleftharpoons 2HI(g)$$

    (a) What will happen to the position of the equilibrium if the temperature is increased?
    (b) Explain your answer.

7. Ethene reacts with steam to form ethanol in an equilibrium reaction.

    $$CH_2{=}CH_2(g) + H_2O(g) \rightleftharpoons CH_3CH_2OH(g)$$

    (a) What will happen to the position of the equilibrium if the pressure is increased? **(b)** Explain your answer.

8. Methanol can be made by the reaction of carbon monoxide and hydrogen in an equilibrium reaction. The forward reaction is exothermic.

    $$CO(g) + 2H_2(g) \rightleftharpoons CH_3OH(g)$$

    Explain how temperature and pressure will affect the position of the equilibrium, and how conditions can be altered to give a high yield of methanol.

The Haber process

Learning objectives

- describe how ammonia is made in the Haber process
- explain why the conditions are chosen in industrial equilibria, such as the Haber process, in terms of equilibrium position, energy requirements, reaction rate and environmental impact.

Ammonia

Over one million tonnes of ammonia, NH_3, are made in the UK every year. The main use of ammonia is to make fertilisers (see lesson C2 6.5). Fertilisers increase the yield of crops and are very important in making sure that we can grow enough food to feed everyone on the planet.

Making ammonia

Ammonia is made by the reaction of nitrogen with hydrogen.

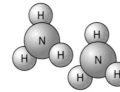

The nitrogen is obtained by fractional distillation of liquefied air, which is 78% nitrogen. The hydrogen is often made from natural gas, methane (CH_4), or other sources, such as oil.

The reaction between hydrogen and nitrogen to make ammonia is reversible and slow. This produces a very low yield of ammonia. Fritz Haber, a German chemist, came up with a way of producing large amounts of ammonia from this reaction by changing the conditions to move the equilibrium position of the reaction further towards the right and to speed it up.

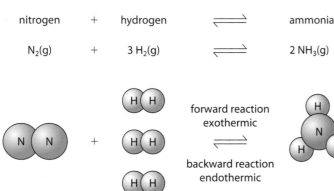

Figure 1 Making ammonia.

In the Haber process, hydrogen and nitrogen are passed over an iron catalyst at a pressure of 200 atmospheres and a temperature of 450 °C. These are the optimum conditions for making ammonia.

Fritz Haber received the Nobel Prize for Chemistry in 1918 for developing the Haber process.

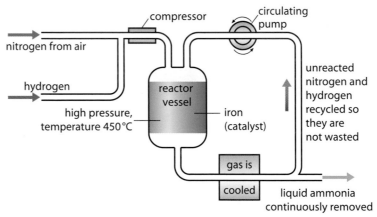

Figure 2 The Haber process.

The forward reaction is exothermic and so the lower the temperature, the higher the yield of ammonia. There are more gas molecules on the left-hand side of the equilibrium, so the higher the pressure, the higher the yield of ammonia.

The temperature of 450 °C that is used is a compromise. The forward reaction is exothermic and, as Figure 3 shows, lower temperatures give a higher yield of ammonia. However, this reaction is slow and higher temperatures are needed to produce ammonia at a good rate; 450 °C is used as a compromise to strike a balance between the yield and the rate of reaction.

The pressure of 200 atmospheres is also a compromise. As there are fewer gas molecules on the right-hand side of the equation, higher pressure gives more ammonia. However, high pressure is very expensive: large amounts of energy are needed to create the high pressure, and the apparatus needed to withstand high pressures is very expensive. Two hundred atmospheres is used as a compromise between yield and cost.

Iron is used as a catalyst in this process. The catalyst does not change the yield of ammonia, but it increases the rate of reaction.

Recycling

After the reaction, the equilibrium mixture is cooled. Ammonia has a higher boiling point than hydrogen and nitrogen, and so will liquefy first as the mixture is cooled. The liquid ammonia is then removed from the mixture and the left-over nitrogen and hydrogen is recycled to make more ammonia.

Ammonia production requires massive amounts of energy. The Haber process provides the optimum conditions to ensure that energy is used efficiently. Efficient use of energy saves money and has less environmental impact.

Other industrial processes

Many industrial processes involve equilibrium reactions. As with the Haber process, the reaction conditions chosen in each case are a compromise between getting a good yield, having the reaction happen quickly, keeping the energy input down and minimising the impact on the environment. Different conditions are needed for different industrial processes.

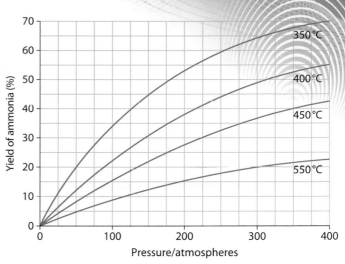

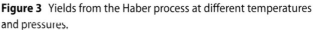

Figure 3 Yields from the Haber process at different temperatures and pressures.

Examiner feedback

When explaining why conditions are used, always make a comparative statement. For example, in the case of the Haber process, 'a lower temperature gives a higher yield', rather than 'a low temperature gives a high yield'.

Also, when explaining why conditions are a compromise, always state what the compromise is between. For example, for temperature in the Haber process, the compromise is between yield and reaction rate.

Questions

1 What is the chemical formula of ammonia?

2 What is the main use of ammonia?

3 **(a)** Name the two elements needed to make ammonia. **(b)** Where do these two elements come from? **(c)** Give two reasons why it is hard to make ammonia by reaction of these two elements.

4 **(a)** Name the catalyst used in the Haber process. **(b)** What effect does this catalyst have on the rate of reaction? **(c)** What effect does this catalyst have on the yield of ammonia?

5 **(a)** How is the ammonia removed from the equilibrium mixture in the Haber process? **(b)** What happens to the left-over nitrogen and hydrogen?

6 Ammonia is made in the Haber process at a temperature of 450 °C and a pressure of 200 atmospheres. **(a)** Explain why this temperature is a compromise condition. **(b)** Explain why this pressure is a compromise condition.

7 What factors must be taken into consideration when deciding on the conditions used for an industrial process that involves an equilibrium reaction?

8 Hydrogen for the Haber process can be made from the reaction of methane with steam, which is slow and reaches a state of equilibrium. The forward reaction is endothermic.

$$CH_4(g) \ + \ H_2O(g) \ \rightleftharpoons \ CO(g) \ + \ 3\,H_2(g)$$

Discuss what conditions may be used to make hydrogen in this process, taking into consideration a range of factors.

Assess yourself questions

1 Some chemical reactions are reversible. In a closed system, reversible reactions reach a state of equilibrium. For example, the reaction between ammonia and hydrogen chloride to form ammonium chloride reaches a state of equilibrium in a closed system:

$$NH_3(g) + HCl(g) \rightleftharpoons NH_4Cl(s)$$

(a) What is a closed system? *(1 mark)*

(b) How does the rate of the forward reaction compare with the reverse reaction at equilibrium? *(1 mark)*

(c) What happens to the quantities of each of the substances at equilibrium? *(1 mark)*

2 Methanol (CH_3OH) is an alcohol and is used as a solvent, a fuel and an antifreeze. It is made from carbon monoxide and hydrogen in a reaction that reaches a state of equilibrium. The forward reaction is exothermic.

$$CO(g) + 2H_2(g) \rightleftharpoons CH_3OH(g)$$

(a) Will the yield of methanol be greater at high or low temperature? Explain your answer. *(2 marks)*

(b) Will the yield of methanol be greater at high or low pressure? Explain your answer. *(2 marks)*

3 Use this list of equilibria to answer the questions that follow.

Reaction 1 $2A(g) + B(g) \rightleftharpoons 2P(g)$
forward reaction is exothermic

Reaction 2 $2C(g) \rightleftharpoons P(g)$
forward reaction is endothermic

Reaction 3 $F(g) + G(g) \rightleftharpoons 2P(g)$
forward reaction is exothermic

Reaction 4 $2H(g) + I(g) \rightleftharpoons 3P(g) + Q(g)$
forward reaction is exothermic

(a) In which reaction(s), if any, will there be a higher yield of P at higher temperatures? *(1 mark)*

(b) In which reaction(s), if any, will there be a higher yield of P at higher pressures? *(1 mark)*

(c) In which reaction(s), if any, will increasing the pressure have no effect on the yield of P? *(1 mark)*

(d) In which reaction(s), if any, will the highest yield of P be at higher temperatures and higher pressures? *(1 mark)*

(e) In which reaction(s), if any, will the highest yield of P be at lower temperatures and higher pressures? *(1 mark)*

(f) In which reaction(s), if any, will the highest yield of P be at higher temperatures and lower pressures? *(1 mark)*

(g) In which reaction(s), if any, will the highest yield of P be at lower temperatures and lower pressures? *(1 mark)*

4 Over one million tonnes of sulfuric acid is made every year in the UK. Sulfur trioxide is used to make sulfuric acid. Sulfur trioxide is made from sulfur dioxide in the Contact process:

$$2SO_2(g) + O_2(g) \rightleftharpoons 2SO_3(g)$$

This reaction is slow and reversible. The process is carried out at a temperature of 450 °C and pressure of two atmospheres and with a catalyst of vanadium oxide, V_2O_5.

(a) The forward reaction is exothermic. Explain what this means. *(1 mark)*

(b) A higher yield of sulfur trioxide would be formed at lower temperatures. Explain why lower temperatures are not used. *(1 mark)*

(c) A higher yield of sulfur trioxide would be formed at higher pressures. Explain why. *(1 mark)*

(d) Explain why higher pressures are not used. *(3 marks)*

(e) What is a catalyst? *(2 marks)*

(f) What effect does the catalyst have on the yield of sulfur trioxide? *(1 mark)*

(g) Why is a catalyst used in this process? *(1 mark)*

5 Ammonia is made in the Haber process by the reaction of nitrogen and hydrogen:

$$N_2(g) + 3H_2(g) \rightleftharpoons 2NH_3(g) + energy$$

The graphs below show how the yield of ammonia changes with temperature and pressure.

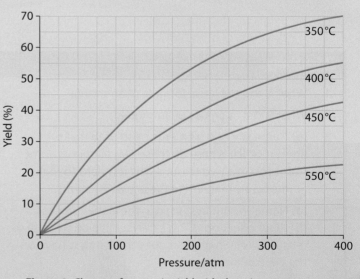

Figure 1 Change of ammonia yield with changing temperature and pressure.

(a) What is the source of nitrogen for this reaction? *(1 mark)*

(b) What is the source of hydrogen for this reaction? *(1 mark)*

(c) How do changes in temperature affect the yield of ammonia? Explain why. *(2 marks)*

(d) The temperature used is 450 °C. Explain why this is a compromise temperature. *(3 marks)*

(e) How do changes in pressure affect the yield of ammonia? Explain why. *(2 marks)*

(f) The pressure used is 200 atmospheres. Explain why this is a compromise pressure. *(3 marks)*

(g) Name the catalyst used in the Haber process. *(1 mark)*

(h) Explain why the catalyst is used. *(1 mark)*

(i) How is the ammonia removed from the equilibrium mixture? Explain why this works. *(3 marks)*

(j) What happens to the left-over hydrogen and nitrogen? *(1 mark)*

6 Compound X is formed in an equilibrium reaction. The graphs below show how the yield of X varies with temperature and pressure.

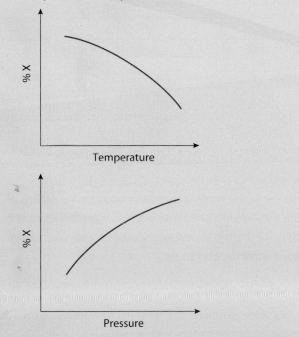

Figure 2 Effect of temperature and pressure on yield of X.

(a) Is the reaction that forms X exothermic or endothermic? Explain your answer. *(2 marks)*

(b) Does the reaction that forms X give an increase or decrease in the number of gas molecules? Explain your answer. *(2 marks)*

7 Cobalt chloride paper is used as a test for water. The equilibrium shown below is used to indicate the presence of water. The paper contains the cobalt compounds shown below:

$$CoCl_2.6H_2O \rightleftharpoons CoCl_2.2H_2O + 4H_2O$$
$$\text{pink} \qquad\qquad \text{blue}$$

(a) If the paper is dried in an oven, what colour will the paper be? Explain your answer. *(2 marks)*

(b) If the dried paper comes into contact with water, what colour will the paper be? Explain your answer. *(2 marks)*

8 Gas C is formed in the following equilibrium reaction:

$$A(g) + B(g) \rightleftharpoons C(g)$$

Table 1 How the yield of gas C changes as the temperature is varied.

Temperature/°C	20	100	200	300	400	500
Yield of C (%)	45	60	70	78	83	85

(a) Plot a graph to show how the percentage yield of C varies with temperature. Draw a line of best fit. *(4 marks)*

(b) Is the reaction to form C endothermic or exothermic? Explain your answer. *(2 marks)*

(c) A company that manufactures C chooses to carry out the reaction at 200 °C. Why does it not use a higher temperature that would give a higher yield? *(3 marks)*

9 The following equilibrium exists in a solution of cobalt ions with chloride ions. The forward reaction is endothermic:

$$Co^{2+}(aq) + 4Cl^-(aq) \rightleftharpoons CoCl_4^{2-}(aq)$$
$$\text{pink} \qquad\qquad\qquad\qquad \text{blue}$$

Describe how changes in temperature will affect the colour of the equilibrium mixture. Explain your answer. *(3 marks)*

10 Gas T is formed in an equilibrium reaction. The following table shows how the yield of T varies with pressure.

Pressure/atm	10	50	100	150	200	250
Yield of T (%)	65	45	30	23	16	14

(a) Plot a graph to show how the percentage yield of T varies with pressure. Draw a line of best fit. *(4 marks)*

(b) Is there an increase or decrease in the number of gas molecules in the reaction that forms T? Explain your answer. *(2 marks)*

Alcohols

Examiner feedback

When drawing the structure of alcohols, show all the bonds and remember to draw all the H atoms. Some students leave off the H atoms and lose marks.

Alcoholic drinks contain ethanol.

Sodium reacting with ethanol.

What are alcohols?

The word **alcohol** is usually associated with alcoholic drinks. These drinks contain an alcohol called ethanol, but there are many other alcohols besides ethanol.

Alcohols are **organic compounds**. Organic compounds are compounds containing carbon, although a few carbon compounds, such as carbon monoxide and carbon dioxide, are not classed as organic compounds.

All alcohols contain the **functional group** –OH. A functional group is an atom or group of atoms responsible for most of the chemical reactions of a compound.

The many different alcohols together form a **homologous series.** This is a family of compounds with the same general formula and similar chemical properties. For alcohols, the general formula is $C_nH_{2n+1}OH$. Methanol, CH_3OH, is the simplest alcohol. Next is ethanol, CH_3CH_2OH, then propanol, $CH_3CH_2CH_2OH$. Each alcohol in the series has one more CH_2 group than the one before. There is also a gradual change in physical properties across the series; for example, the boiling points rise as the molecules get larger. Some examples of alcohol molecules are shown in the table.

Table 1 The names, structures and formulae of the first three alcohols.

Name	methanol	ethanol	propanol
Structure	H | H—C—O—H | H	H H | | H—C—C—O—H | | H H	H H H | | | H—C—C—C—O—H | | | H H H
Formula	CH_3OH	CH_3CH_2OH (C_2H_5OH)	$CH_3CH_2CH_2OH$ (C_3H_7OH)

The names of all the alcohols end in '–ol'. The first part of the name indicates how many carbon atoms there are in the molecule and is linked to the names of alkanes (see lesson C1 4.1). The alkane with one carbon atom is methane, so the alcohol with one carbon atom is methanol. Similarly, the alkane with two carbon atoms is ethane, so the alcohol is ethanol, and so on.

Properties and uses of alcohols

● The simplest alcohols are colourless liquids. They dissolve in water to form neutral solutions. For example, alcoholic drinks such as wine and beer contain ethanol, and other substances, dissolved in water.

● Alcohols are good solvents. They dissolve some substances that water does not. For example, ethanol is often used as a solvent for medicines, perfume and cosmetics.

● All alcohols react with sodium and produce hydrogen gas. The reactions are similar to that of water with sodium, but less vigorous.

● Alcohols also burn very well in air in a combustion reaction. When they burn they form carbon dioxide and water:

$$2CH_3OH + 3O_2 \rightarrow 2CO_2 + 4H_2O$$

methanol + oxygen → carbon dioxide + water

Ethanol is harmful. In small amounts it causes drunkenness, while in larger amounts it is **toxic** and causes death. Other alcohols are more dangerous: they cause death even in small amounts.

Renewable fuels

The fact that alcohols burn well means that they can be used as fuels. Petrol in the UK now contains 5% ethanol made from renewable crops such as maize or sugar. Ethanol made in this way is from a renewable source and is also **carbon neutral**. In Brazil, many cars run on ethanol. Methylated spirits, which contain alcohols, are also used as a fuel. One use of methylated spirits is in camping stoves.

Some camping stoves use methylated spirits as fuel.

Burning is an example of an oxidation reaction. Alcohols can be partially oxidised by air to form compounds called carboxylic acids, which are discussed in the next lesson.

Questions

1 Which of the following compounds are alcohols?
   propanol, cyclohexane, cyclohexanol, ethene, methane, methanol
2 Alcohols are a type of organic compound. What are organic compounds?
3 Give three uses of ethanol.
4 The alcohols are a *homologous series* containing the *functional group* –OH. Explain the meaning of the terms in italics.
5 Butanol is an alcohol with four carbon atoms in a chain. **(a)** Draw the structure of butanol. **(b)** Give the chemical formula of butanol. **(c)** Identify the functional group in butanol.
6 **(a)** What happens if the alcohol propanol is added to sodium? **(b)** What happens if the alcohol propanol is added to water?
7 **(a)** Write a balanced equation for the combustion of ethanol. **(b)** To what use is this reaction put?
8 Ethanol is used as a fuel and can be made from crops. Give some advantages of using ethanol made in this way as a fuel over fuels made from crude oil.

Carboxylic acids

What are carboxylic acids?

Carboxylic acids are a homologous series of organic compounds that contain the functional group –COOH. Some examples of carboxylic acid molecules are shown in Table 1.

Table 1 The names, structures and formulae of the first three carboxylic acids.

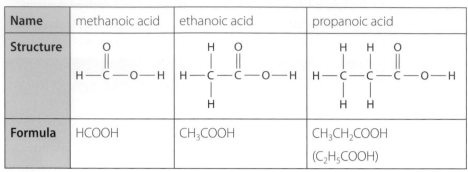

Name	methanoic acid	ethanoic acid	propanoic acid
Structure			
Formula	HCOOH	CH$_3$COOH	CH$_3$CH$_2$COOH (C$_2$H$_5$COOH)

Making carboxylic acids from alcohols

When alcohols are burned they form carbon dioxide and water. However, if alcohols are mildly oxidised, they form carboxylic acids. This can be done with mild **oxidising agents**, which are chemicals that will gently oxidise another substance. One example is acidified potassium dichromate.

Alcohols are also oxidised to carboxylic acids by oxygen in air in the presence of microbes. Ethanol, for example, is oxidised to ethanoic acid. If a bottle of wine is left to stand in air, the microbes oxidise the ethanol in the wine to ethanoic acid. This gives the wine a sour taste; it 'goes off'.

Properties and uses of carboxylic acids

There are many useful carboxylic acids. Vinegar is a dilute solution of ethanoic acid in water. The ethanoic acid gives the vinegar its sharp, sour taste. Vinegar is used as a preservative in several foods, such as pickled onions, where the ethanoic acid kills bacteria that would make the food go off. Wine vinegar is made by leaving wine to react with air over time. The word vinegar comes from the French *vin aigre*, meaning 'sour wine'.

Another carboxylic acid is the pain-killing drug aspirin. Vitamin C, also called ascorbic acid, is a carboxylic acid often found in fruit and vegetables.

All of these contain carboxylic acids.

Carboxylic acids are **weak acids**. Acids are substances that dissolve in and react with water to form hydrogen ions (see lesson C2 6.1). In a weak acid, only a small fraction of the molecules break apart into ions. As there are fewer hydrogen ions in an aqueous solution of a weak acid compared with a strong acid of the same concentration, the pH of the weak acid is higher. In strong acids, all the molecules break into ions. Common solutions of carboxylic acids typically have a pH between 2 and 4. They react like other acids, but do so more slowly than strong acids of the same concentration. One example is the reaction of carboxylic acids with carbonates, such as sodium carbonate or calcium carbonate, where they react producing bubbles of carbon dioxide.

Some carboxylic acids have unpleasant tastes or smells. Butanoic acid is responsible for the smell of rancid butter and is present in human sweat. Methanoic acid is found in ant stings.

Some ants can spray out a jet of methanoic acid up to a metre long.

Carboxylic acids also react with alcohols to make **esters**. These compounds will be studied in lesson C1 6.3.

Questions

1   Which of the following compounds are carboxylic acids?
    propanol, ethene, methanoic acid, hexanoic acid, ethanol
2   What is the functional group in carboxylic acids?
3   **(a)** What is vinegar? **(b)** Which of the following would be the most likely pH of some vinegar: 1, 4, 7 or 11?
4   **(a)** Name the alcohol from which propanoic acid can be made.
    (b) Name the type of reaction taking place. **(c)** Describe two ways of converting the alcohol into propanoic acid.
5   Butanoic acid is a carboxylic acid with four carbon atoms in a chain.
    (a) Draw the structure of butanoic acid. **(b)** Give the chemical formula of butanoic acid.
6   **(a)** What would happen if methanoic acid was added to water? **(b)** What would happen if methanoic acid was added to a solution of sodium carbonate?
7   Wine vinegar, which contains ethanoic acid, is made from wine, which contains ethanol. **(a)** Draw the structures of ethanol and ethanoic acid. **(b)** Identify the functional groups of ethanol and ethanoic acid. **(c)** Explain how the ethanol is converted to ethanoic acid.
8   Carboxylic acids are organic compounds. Explain what they are and give some ways in which carboxylic acids are useful and some in which they are not.

Examiner feedback

Some students state that 'in weak acids only a small fraction of the ions are ionised'. This is wrong as ions cannot be ionised. It is that a small fraction of the molecules are ionised.

Taking it further

As the hydrogen ions in a solution of a weak acid are used up in reactions, more acid molecules break apart to form more hydrogen ions for reaction. In time, all the acid molecules can ionise and react. However, the reaction is slower than that of a strong acid because there is never a high concentration of hydrogen ions.

Science in action

The sting from nettles is caused by methanoic acid. Dock leaves are often found near nettles and are often rubbed onto nettle stings to ease the pain. However, research suggests that this is just a placebo effect and the dock leaf has no effect on the sting. It is simply that the person who has been stung believes that the dock leaf will reduce the pain.

Taking it further

Fatty acids are carboxylic acids containing a long carbon chain. For example, the fatty acid stearic acid (octadecanoic acid) is $CH_3(CH_2)_{16}COOH$.

Esters

Route to A* A*

The first part of the name of an ester comes from the alcohol, and the second part from the carboxylic acid. For example, the name of the ester methyl ethanoate comes from the alcohol methanol and the carboxylic acid ethanoic acid.

What are esters?

Esters are organic compounds that contain the functional group –COO–. Some examples of ester molecules are shown in Table 1.

Table 1 The names, structures and formulae of three esters.

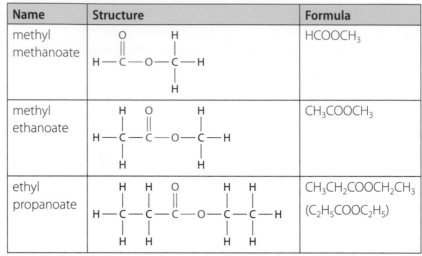

Name	Structure	Formula
methyl methanoate		$HCOOCH_3$
methyl ethanoate		CH_3COOCH_3
ethyl propanoate		$CH_3CH_2COOCH_2CH_3$ ($C_2H_5COOC_2H_5$)

Making esters

Esters can be made by the reaction of alcohols with carboxylic acids. The reaction is slow and so is carried out in the presence of a concentrated acid, such as sulfuric acid, which acts as a catalyst:

$$\text{carboxylic acid} + \text{alcohol} \xrightarrow{\text{catalyst}} \text{ester} + \text{water}$$

For example:

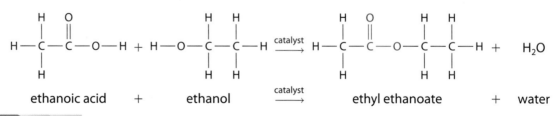

ethanoic acid   +   ethanol $\xrightarrow{\text{catalyst}}$ ethyl ethanoate   +   water

Examiner feedback

For the exam you will need to know how esters are made and some of their uses. However, you do not need to be able to name esters or balance equations for making esters.

Table 2 Some examples of esters along with the alcohols and carboxylic acids from which they are formed.

Alcohol	Carboxylic acid	Ester
methanol	ethanoic acid	methyl ethanoate
propanol	ethanoic acid	propyl ethanoate
ethanol	methanoic acid	ethyl methanoate

Properties and uses of esters

Esters all have very distinctive smells. They are responsible for the smells and flavours of many fruits. Some esters found in fruits are shown in Table 3. Esters can either be extracted from fruits or made artificially from alcohols and carboxylic acids. They can then be used as flavourings in foods.

The tastes and smells of most fruits comes from esters

Table 3 The structures of esters found in apples, pears and pineapples.

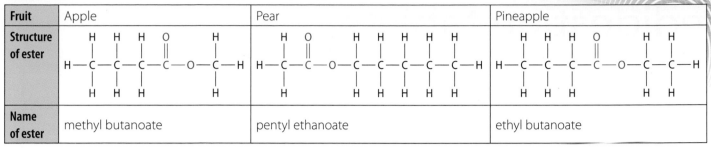

Fruit	Apple	Pear	Pineapple
Structure of ester	methyl butanoate structure	pentyl ethanoate structure	ethyl butanoate structure
Name of ester	methyl butanoate	pentyl ethanoate	ethyl butanoate

Esters are **volatile** liquids. This means that they have low boiling points and readily evaporate, so they can reach the scent receptors in your nose. They also smell pleasant, so they are often used as perfumes.

Questions

1. Which of the following compounds are esters?

   methanol, methyl ethanoate, propanoic acid, butane, propyl methanoate

2. Which of the following is the correct structure of the ester propyl ethanoate?

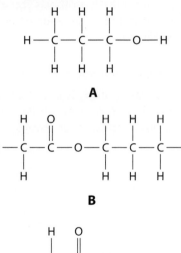

   A

   B

   C

3. Give two uses of esters.

4. Esters are volatile liquids. What are volatile liquids?

5. Why is it important that esters are volatile for their uses in perfumes and fragrances?

6. The ester, pentyl ethanoate, is responsible for the taste and smell of pears. It can be made from the reaction of pentanol with ethanoic acid or it can be extracted from pears. This ester can then be used to give foods a pear flavour. A good example is pear-drop sweets. It is cheaper to make pentyl ethanoate than it is to extract it from pears. **(a)** Write a word equation for the formation of pentyl ethanoate by reaction of pentanol with ethanol acid. **(b)** What else would be needed for this reaction and why? **(c)** Most food manufacturers make pentyl ethanoate to use as a flavouring rather than extracting it from pears. Why might they do this?

7. An ester found in oranges is octyl ethanoate. It has the formula $CH_3COO(CH_2)_7CH_3$. Draw the structure of this ester showing all the bonds.

8. Ethyl ethanoate is an ester used as the solvent in some glues. The structure of ethyl ethanoate is shown.

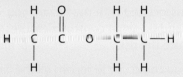

   ethyl ethanoate

   It can be made by the reaction of ethanol with ethanoic acid in the presence of concentrated sulfuric acid. Draw the structures of ethanol and ethanoic acid. Identify the functional group in the ethyl ethanoate and state the role of the concentrated sulfuric acid in this reaction.

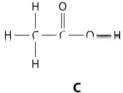

ISA practice: effectiveness of indigestion tablets

A company called Indicure, which makes indigestion tablets, carried out some tests to compare the effectiveness of different amounts of the active ingredient of the tablets. Indicure works by neutralising the excess hydrochloric acid in the stomach.

Section 1

1   Write a hypothesis about the effectiveness of different amounts of the active ingredient and its effect on excess stomach acid. Use information from your knowledge of the neutralisation to explain why you made this hypothesis.
(3 marks)

2   Describe how you are going to do your investigation.
You should include:
- the equipment that you are going to use
- how you will use the equipment
- the measurements that you are going to make
- how you will make it a fair test.

You are provided with the active ingredient in a powdered form.

You may include a labelled diagram to help you to explain your method.

In this question you will be assessed on using good English, organising information clearly and using specialist terms where appropriate.
(6 marks)

3   Think about the possible hazards in your investigation.
   (a) Describe one hazard that you think may be present in your investigation.
   (1 mark)
   (b) Identify the risk associated with the hazard you have described, and say what control measures you could use to reduce the risk.
   (2 marks)

4   Design a table that you could use to record all the data you would obtain during the planned investigation.
(2 marks)

Total for Section 1: 14 marks

Section 2

A group of students, Study Group 1, carried out an investigation into the hypothesis. They used several different masses of the active ingredient, and measured how much dilute hydrochloric acid was needed to neutralise the active ingredient in each case. Figure 1 shows their results.

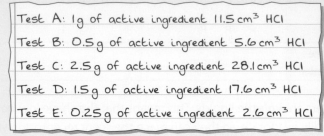

Test A: 1g of active ingredient 11.5 cm³ HCl
Test B: 0.5g of active ingredient 5.6 cm³ HCl
Test C: 2.5g of active ingredient 28.1 cm³ HCl
Test D: 1.5g of active ingredient 17.6 cm³ HCl
Test E: 0.25g of active ingredient 2.6 cm³ HCl

Figure 1 Study Group 1's results.

5   **(a)** Plot a graph of these results.   *(4 marks)*
   (b) What conclusion can you make from the investigation about a link between the amount of active ingredient and the amount of acid needed to neutralise it? You should use any pattern that you can see in the results to support your conclusion.
   (3 marks)
   (c) Do the results support the hypothesis you wrote in answer to question 1? Explain your answer. You should quote some figures from the data in your explanation.
   (3 marks)

Here are the results from three other study groups.

A second group of students, Study Group 2, carried out a similar investigation. They used 1 mole per dm³ hydrochloric acid. Figure 2 shows their results.

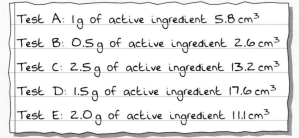

Test A: 1g of active ingredient 5.8 cm³
Test B: 0.5g of active ingredient 2.6 cm³
Test C: 2.5g of active ingredient 13.2 cm³
Test D: 1.5g of active ingredient 17.6 cm³
Test E: 2.0g of active ingredient 11.1 cm³

Figure 2 Study Group 2's results.

Study Group 3 was a third group of students. They decided that they wanted to know what happened to the stomach pH. They used a constant volume of hydrochloric acid and added different amounts of active ingredient. Table 1 shows their results, and Figure 3 is a plot of the mean data.

Table 1 Study Group 3's results.

Mass of active ingrediant/g	pH of solution after 10 min			
	Trial 1	Trial 2	Trial 3	Mean
0.25	2.6	3.3	3.1	3.0
0.5	3.8	3.4	3.3	3.5
1.0	4.9	4.6	4.3	4.6
1.5	5.7	5.6	6.1	5.8
2.0	6.6	6.4	6.5	6.5
2.5	6.3	6.6	6.6	6.5

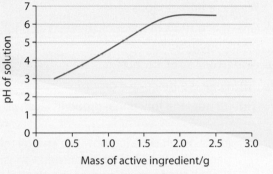

Figure 3 Graph showing the results of Study Group 3.

Study Group 4: a group of scientists working for Indicure repeated the work of Study Group 3, but this time they measured the pH every minute for 10 minutes. They were interested in which combination of mass of active ingredient and time would give the quickest relief to indigestion sufferers. They knew that an indigestion-free stomach has a pH of 3.5. Their results are shown in Table 2. Figure 4 is a graph of this data.

Table 2 Results from Study Group 4.

Time/ min	pH of solution with each mass of active ingrediant				
	0.25 g	0.5 g	1.0 g	1.5 g	2.0 g
0	1.0	1.0	1.0	1.0	1.0
1	1.2	1.4	1.5	1.9	2.3
2	1.4	1.7	1.9	2.8	3.4
3	1.6	2.0	2.4	3.5	4.3
4	1.8	2.3	2.9	4.2	5.3
5	2.0	2.5	3.3	4.8	6.3
6	2.2	2.8	3.7	5.4	6.5
7	2.4	3.1	4.1	5.8	6.5
8	2.6	3.3	4.5	5.9	6.5
9	2.8	3.5	4.6	5.9	6.5
10	3.0	3.5	4.7	5.9	6.5

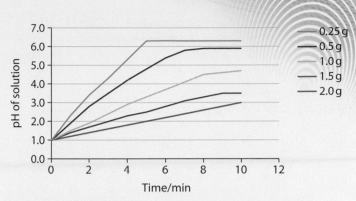

Figure 4 Graph showing the results of Study Group 4.

6 (a) Draw a sketch graph of the results from Study Group 2. *(3 marks)*

(b) Look at the results from Study Groups 2 and 3. Does the data support the conclusion you reached about the investigation in question 5(b)? Give reasons for your answer. *(3 marks)*

(c) The data contain only a limited amount of information. What other information or data would you need in order to be more certain whether or not the hypothesis is correct? Explain the reason for your answer. *(3 marks)*

(d) Look at Study Group 4's results. Compare them with the data from Study Group 1. Explain how far the data support or do not support your answer to question 5(b). You should use examples from the results from Study Group 4 and from Study Group 1. *(3 marks)*

7 (a) Compare the results of Study Group 1 with Study Group 2. Do you think that the results for Study Group 1 are *reproducible*? Explain the reason for your answer. *(3 marks)*

(b) Explain how Study Group 1 could use results from other groups in the class to obtain a more *accurate* answer. *(3 marks)*

8 Suggest how ideas from the original investigation and the other studies could be used by Indicure to decide how much of the active ingredient should be used to return a stomach with indigestion to a pH of 3.5. *(3 marks)*

Total for Section 2: 31 marks

Total for the ISA: 45 marks

Assess yourself questions

1 The structures of four compounds, A, B, C and D, are shown. Decide whether each one is an alcohol, a carboxylic acid or an ester. *(4 marks)*

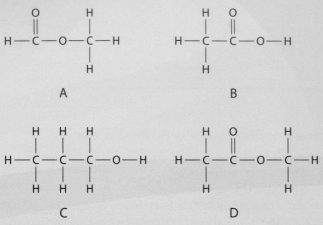

Figure 1 Structures of unidentified alcohols, carboxylic acids and esters.

2 (a) The names of some organic compounds are given below. Match the names to the compounds.

ethyl ethanoate, methanol, methanoic acid

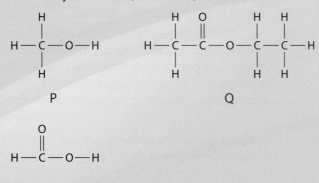

Figure 2 The structures of ethyl ethanoate, methanol and methanoic acid.

(3 marks)

(b) Which one of these compounds would be used:

(i) as a flavouring? *(1 mark)*

(ii) as a fuel? *(1 mark)*

3 The labels have come off three bottles of organic compounds. One bottle contains methanol, one bottle contains propanoic acid and the third bottle contains propyl ethanoate. Plan a series of chemical tests to identify which compound is which.

In this question you will be assessed on using good English, organising information clearly and using specialist terms where appropriate. *(6 marks)*

4 Table 1 gives some information about some alcohols.

Table 1 Data on some alcohols.

Alcohol	Formula	Relative formula mass/M_r	Energy released when burned / kJ/g
methanol	CH_3OH	32	22.7
ethanol	C_2H_5OH	46	29.7
propanol	C_3H_7OH	60	33.6
butanol	C_4H_9OH	74	36.1
pentanol	$C_5H_{11}OH$	88	

(a) Alcohols are a homologous series of compounds. What is a homologous series? *(2 marks)*

(b) Give the formula of hexanol, the next alcohol in the homologous series. *(1 mark)*

(c) Give two uses of alcohols. *(2 marks)*

(d) (i) What is a functional group? *(1 mark)*

(ii) Identify the functional group in alcohols. *(1 mark)*

(e) (i) Plot a graph showing how the energy released per gram of alcohol when burned varies with the relative formula mass of the alcohols. Leave space for pentanol on your graph. Draw a line of best fit. *(4 marks)*

(ii) Describe the relationship between the energy released per gram and the relative formula mass. *(1 mark)*

(iii) Use your graph to predict the energy released per gram when pentanol is burned. *(1 mark)*

5 Many cars use petrol as the fuel. Petrol sold in the UK is 95% petroleum from crude oil and 5% is ethanol made from fermentation of crops. In some parts of the world cars run on fuel made of a 90%–10% mixture of petroleum and ethanol, or 85%–15%, or even pure ethanol. It is estimated that one acre (just over 4000 m²) of farmland is needed to fuel a car running on pure ethanol. The cost of production of petroleum depends on the price of crude oil. Some data about the costs of production for petroleum and ethanol are shown in Table 2.

Table 2 Costs of production for petroleum and ethanol.

	2000	2002	2004	2006	2008	2010
Cost of production of petroleum/ pence per litre	0.12	0.09	0.15	0.26	0.36	0.31
Cost of production of ethanol/pence per litre	0.16	0.14	0.21	0.31	0.34	0.29

(a) Plot a single graph to show how the costs of ethanol and petroleum have varied between 2000 and 2010. Draw a line of best fit for each fuel. *(5 marks)*

(b) Give some social and economic advantages and disadvantages of using ethanol as a fuel in cars, either as a mixture with petroleum or on its own. *(4 marks)*

6 The ester, pentyl butanoate, is shown below. It is found in strawberries and is also used to flavour foods with the taste of strawberries.

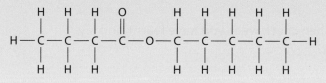

Figure 3 The ester pentyl butanoate.

(a) Identify the functional group in this ester. *(1 mark)*

(b) Name two types of compound that can be used to make this ester. *(2 marks)*

(c) Give some social and economic reasons why some people would prefer to extract this ester from strawberries to use as a flavouring in other foods, and why some people would prefer to use this ester made artificially. *(2 marks)*

7 Alcoholic drinks, for example wine, contain ethanol. If a bottle of wine is left open to the air for a few days it turns sour. Wine vinegar can be made from wine that has gone off in this way.

(a) Draw the structure of ethanol. *(1 mark)*

(b) Draw the structure of and name the molecule formed from the ethanol as the wine goes off. *(2 marks)*

(c) What type of reaction has taken place? *(1 mark)*

(d) What is vinegar? *(2 marks)*

8 The boiling points of some esters are shown in Table 3 at the top of the next column.

(a) Esters are volatile liquids. Explain what volatile means in this context. *(2 marks)*

(b) How does the fact that esters are volatile allow us to smell them? *(2 marks)*

(c) **(i)** Plot a graph showing how the boiling point of the esters varies with the relative formula mass. Draw a line of best fit. *(4 marks)*

(ii) Describe the relationship between the boiling point of the esters and the relative formula mass. *(1 mark)*

(iii) Use your graph to predict the boiling point of butyl ethanoate. *(1 mark)*

Table 3 Boiling points of esters.

Ester	Formula	Relative formula mass/M_r	Boiling point/°C
methyl ethanoate	CH_3COOCH_3	74	57
ethyl ethanoate	$CH_3COOC_2H_5$	88	77
propyl ethanoate	$CH_3COOC_3H_7$	102	102
butyl ethanoate	$CH_3COOC_4H_9$	116	
pentyl ethanoate	$CH_3COOC_5H_{11}$	130	149

9 Look at the following reaction sequence and identify the substances labelled A to D. *(4 marks)*

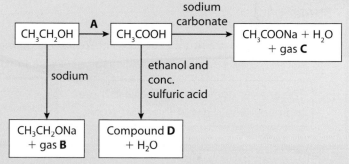

Figure 4 A reaction sequence.

10 Propanoic acid is a weak acid with the formula CH_3CH_2COOH. Hydrochloric acid is a strong acid with formula HCl.

(a) What is a strong acid? *(1 mark)*

(b) What is a weak acid? *(1 mark)*

(c) Propanoic acid and hydrochloric acid both react with sodium carbonate.

(i) Describe a similarity about the reactions of propanoic acid and hydrochloric acid with sodium carbonate. *(1 mark)*

(ii) Describe a difference about the reactions of propanoic acid and hydrochloric acid with sodium carbonate. *(1 mark)*

(d) Describe another test that could be used to distinguish solutions of propanoic acid and hydrochloric acid of the same concentration, and give the results of this test. *(2 marks)*

(e) State a reaction that propanoic acid would undergo that hydrochloric acid would not. *(1 mark)*

Here are three students' answers to the following question:

Ammonia is made by the reaction of hydrogen with nitrogen. This is a slow and reversible reaction.

$$N_2(g) + 3\,H_2(g) \rightleftharpoons 2\,NH_3(g) + energy$$

The reaction is carried out at a temperature of 450 °C, a pressure of 200 atmospheres and with an iron catalyst. Explain in detail why these conditions are used and why they are compromise conditions.

In this question you will be assessed on using good English, organising information clearly and using specialist terms where appropriate.
(6 marks)

Read the answers together with the examiner comments. Then check what you have learnt and try putting it into practice in any further questions you answer.

B Grade answer

Student 1

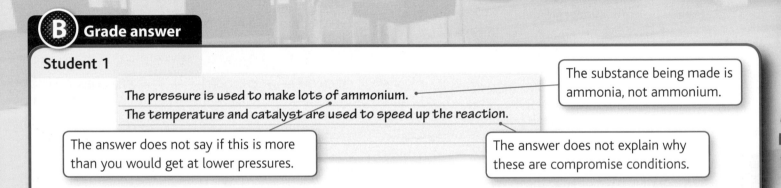

The pressure is used to make lots of ammonium.

The temperature and catalyst are used to speed up the reaction.

> The substance being made is ammonia, not ammonium.

> The answer does not say if this is more than you would get at lower pressures.

> The answer does not explain why these are compromise conditions.

Examiner comment

Apart from the use of the name ammonium rather than ammonia, everything the student has written is correct but the answer would miss many of the marks. In terms of pressure, the student has not indicated whether more ammonia is made at a higher or a lower pressure. Although it states that 'lots of ammonia' is made at pressures, it is not clear enough that this means more is made at higher pressures. The answer also does not explain why high pressures give lots of ammonia and why the pressure used is a compromise.

In terms of temperature, the student has indicated that it is used to speed up the reaction, but it is not clear whether the rate is increased by higher or lower temperature. From this answer, it may be that it is faster at lower temperatures. The answer does not explain how or why the temperature affects the yield and why this temperature is a compromise.

The student has explained why the catalyst is used, i.e. to speed up the reaction.

- In questions that ask for several things, ensure that each part is answered. You could highlight each thing you are asked to do and tick them off as you cover them.
- If asked to explain how something affects something else, make sure that you explain the effect as well as just describing what the effect is.
- When answering a longer question, it is helpful to structure your answer into paragraphs or sections.
- The use of good English is important, with the use of proper sentences with good spelling and grammar.

 A **Grade answer**

Student 2

> Two hundred atmospheres pressure is used because the higher the pressure, the more ammonia is made. However, high pressure is very expensive so the pressure used is a compromise between cost and yield. A temperature of 450 °C is used because the higher the temperature, the faster the reaction. However, the lower the temperature, the higher the yield, so this temperature is a compromise between rate and yield. The catalyst is used to speed up the reaction.

The answer does not mention why higher pressure and lower temperature give higher yield.

This answer shows how changes in pressure and temperature affect the yield.

This answer explains why both the pressure and temperature are compromise conditions.

Examiner comment

The answer clearly shows how changes in pressure affect the yield and cost, and why the pressure used is a compromise between cost and yield. However, it does not explain why changes in pressure affect the yield.

It clearly shows how changes in temperature affect the yield and reaction rate, and why the temperature used is a compromise between the rate and yield. However, it does not explain why changes in temperature affect the yield.

The answer explains clearly why the catalyst is used.

A* **Grade answer**

Student 3

> Pressure: The higher the pressure the higher the yield, because there are fewer gas molecules on the right of the equation than on the left. However, high pressure is very expensive due to the cost of apparatus that can withstand high pressure and the energy needed to establish the pressure. The pressure used is a compromise between the cost and the yield.
>
> Temperature: The lower the temperature the higher the yield because the forward reaction is exothermic. However, this reaction is slow and would be too slow at low temperatures. The higher the temperature, the faster the reaction; the temperature used is a compromise between the rate and the yield.
>
> Catalyst: The reaction is slow so the catalyst is used to speed up the reaction. The addition of the catalyst does not affect the yield.

The answer explains why higher pressure gives a higher yield.

The answer explains why lower temperature gives a higher yield.

Examiner comment

This answer has covered everything that was asked. For both temperature and pressure the student has clearly explained how and why the yield is affected as well as explaining why each of the conditions is a compromise. The student also clearly explains why the catalyst is used. The way the student has laid out the answer to give separate sections on pressure, temperature and the catalyst is also helpful.

Examination-style questions

1   The following questions relate to the electronic structure of the elements, their position in the Periodic Table and their chemical properties.

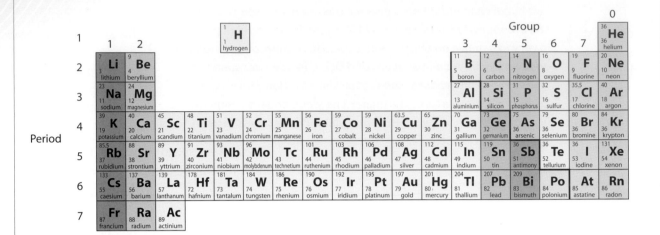

(a)   For the purposes of this question an unknown element has been given the symbol X. The properties of X are given in the table below.

Appearance	Silver-grey metal
Melting point	High melting point, 1890.0 °C
Density	6.1 g/cm^3.
Thermal conductivity	Good
Electrical conductivity	Good
Chloride	It forms two crystalline salts with chlorine: green XCl_2 and red XCl_3
Carbonate	Its carbonate XCO_3 is insoluble in water
Oxide	It forms an oxide X_2O_5
Uses	X is used as an additive in steels for added strength; the oxide is used as a catalyst in the production of sulfuric acid

To which section of the Periodic Table does X belong?

Give five pieces of evidence from the description above that help you to place element X in this section, explaining their significance.   *(5 marks)*

(b)   Elements in Group 4 of the Periodic Table show a more complex range of properties than those seen in groups 1 or 7, yet the principles that explain these properties are fundamentally the same.

Describe the trend seen in Group 4 and use your knowledge of the underlying principles to explain fully the range of chemical properties exhibited.

In this question you will be assessed on using good English, organising information clearly and using specialist terms where appropriate.   *(6 marks)*

2 Rain water contains dissolved carbon dioxide. Rain water falling on rocks containing calcium carbonate or calcium sulfate becomes 'hard' by gaining calcium ions (Ca^{2+}), along with hydrogencarbonate (HCO_3^-) or sulfate (SO_4^{2-}) ions.

(a) Describe the chemical reactions that lead to calcium ions and hydrogencarbonate ions entering the rain water, giving a balanced chemical equation for the main reaction. *(3 marks)*

(b) What simple chemical method can be used to remove calcium ions from water that contains calcium ions and hydrogencarbonate ions and/or sulfate ions?

Give a balanced chemical equation for this reaction – a simple ionic equation will suffice.

(3 marks)

3 **(a)** Here is a comparison table for some fuels and their properties.

Fuel	Energy produced by combustion / kJ/g	Molecules of waste gases produced	Comments
Hydrogen	143	Water only	Expensive to produce (often by the electrolysis of water) and difficult to store in the liquid form
Natural gas	54	Two water molecules for every carbon dioxide molecule	Natural fossil fuel that needs relatively little refining
Oil	44	Approximately one water molecule for every carbon dioxide molecule	Natural fossil fuel that must be refined before it can be used; spills can pollute
Coal	24	Carbon dioxide only	Natural fossil fuel that can be mined and easily stored

Discuss the relative advantages and disadvantages of these fuels in terms of ease of use, energy content and effect on the environment, explaining your reasoning as fully as possible.

In this question you will be assessed on using good English, organising information clearly and using specialist terms where appropriate. *(6 marks)*

(b) Ethanol, produced by the fermentation of sugar, can be used as a biofuel. The equation for its combustion is:

$$C_2H_5OH + 3O_2 \rightarrow 2CO_2 + 3H_2O$$

(i) Sketch an energy diagram for this reaction, clearly labelling the reactant and product energy levels. Use your diagram to help explain why this reaction is exothermic. *(3 marks)*

(ii) What bonds are formed during the reaction shown in the equation? How many of each? Use the bond energy table below to calculate the total energy (per mole) given out when these bonds form. *(4 marks)*

Bond	Bond energy / kJ/mole
O − H	464
C − H	413
C − C	347
C = O	805
O = O	498

4 Ammonia is a very important compound. Over 1 million tonnes are made each year in the UK. It is made from hydrogen and nitrogen in a reversible reaction that reaches a state of equilibrium in a closed system. Energy is released by the forward reaction:

$$N_2(g) + 3H_2(g) \rightleftharpoons 2NH_3(g)$$

(a) Where does the nitrogen for this process come from? *(1 mark)*

(b) How does the rate of the forward reaction compare to the reverse reaction at equilibrium? *(1 mark)*

(c) Iron is often used as a catalyst in this reaction. Why is a catalyst used? *(1 mark)*

(d) How is the ammonia removed from the reaction mixture in this process? *(1 mark)*

(e) The graph below shows how changes in temperature and pressure affect the yield of ammonia.

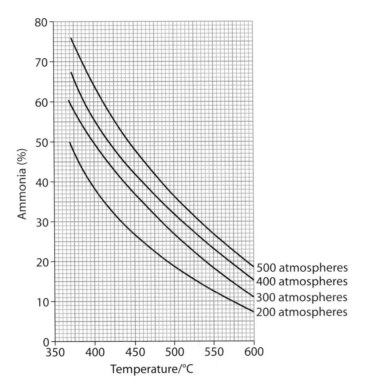

Typical conditions that are actually used are a temperature of 450 °C and a pressure of 200 atmospheres. Explain why these conditions are used and described as compromise conditions.

In this question you will be assessed on using good English, organising information clearly and using specialist terms where appropriate. *(6 marks)*

5   (a) Ethanoic acid is a carboxylic acid. It is a weak acid. Hydrochloric acid is a strong acid.

        (i) Explain the difference between a strong acid and a weak acid.   *(2 marks)*

        (ii) The difference between the strong acid hydrochloric acid and the weak acid ethanoic acid can
              be shown using universal indicator. Describe and give the results of a chemical test that would
              also show the difference between the strong and weak acids.   *(3 marks)*

    (b) A glass of wine was left to stand for a few days. It turned sour due to the formation of ethanoic acid.

        (i) Explain why ethanoic acid formed.   *(2 marks)*

        (ii) A student carried out a titration to find the concentration of ethanoic acid (CH_3COOH) in the
              wine She found that 25.0 cm^3 of the wine reacted with 22.1 cm^3 of 0.8 mole/dm^3 sodium
              hydroxide solution. Ethanoic acid reacts with sodium hydroxide in a 1 : 1 molar ratio.

              Calculate the concentration of ethanoic acid in the wine in mole/dm^3.   *(2 marks)*

        (iii) Calculate the concentration of ethanoic acid (CH_3COOH) in the wine in g/dm^3. (relative atomic
               masses: H = 1, C = 12, O = 16)   *(1 mark)*

    (c) Ethyl ethanoate is an ester that is used as the solvent in nail varnish. It can be made from the
         reaction between ethanol and ethanoic acid. Identify which of the substances A to D are

        (i) ethyl ethanoate

        (ii) ethanol

        (iii) ethanoic acid.   *(3 marks)*

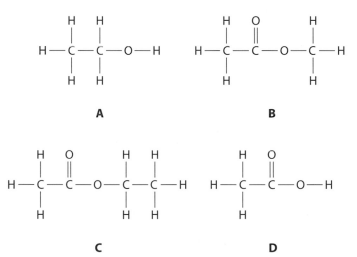

A   B

C   D

    (d) Give another use of esters.   *(1 mark)*

6   Samples of four compounds have been placed in beakers. However, the beakers were not labelled and it
    is not certain which compound is in which beaker.

    The four compounds are:

    • sodium sulfate

    • sodium chloride

    • potassium chloride

    • sodium carbonate.

    Plan a series of chemical tests to show which beaker contains which compound. Describe which tests
    you would perform and how the results would show which compound was which.

    *In this question you will be assessed on using good English, organising information clearly and using
    specialist terms where appropriate.*   *(6 marks)*

Glossary

A

accurate – An accurate measurement is close to the true value.

acid – A compound with a pH of less than 7, which turns litmus paper red and releases hydrogen ions in solution.

acid rain – Rain that has a pH of less than 5.6 because of dissolved pollutants, usually sulfur dioxide.

acidic – Having the properties of an acid.

activated carbon – Charcoal that has been broken up into very fine granules to increase its surface area so that it is able to adsorb other substances.

activation energy – The minimum amount of energy particles must have to react.

aggregate – A material such as gravel, crushed rock or ash, made up of coarse particles. It is used in building, in particular for making concrete.

air – A mixture of gases, mainly nitrogen and oxygen, that covers the Earth's surface.

alcohol – An organic compound from a homologous series containing the functional group –OH; for example ethanol.

algae (singular alga) – A group of organisms that photosynthesise but, unlike plants, do not have leaves or roots. These organisms include seaweeds and some single-celled organisms.

alkali – A base that dissolves in water to form hydroxide ions and give a solution of pH greater than 7.

alkali metal – A metal in group 1, such as lithium, sodium or potassium. They have characteristic flame colours and react with water to give hydrogen and a hydroxide.

alkaline – Having the properties of an alkali.

alkane – A hydrocarbon that has only C–H and single C–C bonds, with the general formula C_nH_{2n+2}.

alkene – A hydrocarbon with one or more C=C double bonds.

alloy – A mixture of metals blended to have specific properties.

atmosphere – The layer of gases surrounding the Earth.

atom – A particle with no overall electric charge, made up of a combination of protons, neutrons and electrons; an atom is the smallest particle of an element that retains the properties of that element.

atomic number – The number of protons in the nucleus of atoms of a particular element.

B

balanced equation – A chemical equation with the same numbers and types of atoms on both sides of the arrow.

base – A compound, usually a metal oxide or metal hydroxide, that reacts with an acid to neutralise it.

baseline – A baseline measurement is one taken before an experiment starts to establish the usual value.

bauxite – An ore containing a high concentration of aluminium oxide.

biodegradable – Able to be broken down easily into simpler chemicals by natural, biological processes.

biodiesel – A renewable fuel made from vegetable oil that can be used in place of diesel from crude oil.

biofuel – A fuel made from living sources such as plant matter.

biogas – A fuel made by the anaerobic fermentation of waste food or sewage by microbes, or when waste wood is heated in a limited supply of air; a mixture of mostly methane and carbon dioxide.

bioleaching – The process of extracting metals from their ores using bacteria or other microorganisms.

blast furnace – A furnace in which ore containing iron oxide is heated with limestone and coke to produce iron.

bond energy – The amount of energy needed to break a chemical bond.

brittle – Easily cracked or broken.

burette – A long, thin, calibrated glass tube with a tap at the base, used to measure volumes of liquid accurately, for example in a titration.

by-product – A substance produced in a chemical process that is not the main product being made.

C

calibrated – Marked with a scale, so that readings can be taken.

calorimeter – An apparatus for measuring the energy change during a reaction.

calorimetry – Experimental technique for measuring the energy change in a reaction.

carbon-neutral – Having no overall carbon dioxide emissions; usually refers to a fuel that releases as much carbon dioxide when it is produced and burned as was absorbed by the plants from which it is made as they grew.

carboxylic acid – An organic compound from the homologous series containing the functional group –COOH; for example ethanoic acid, CH_3COOH.

catalyst – A chemical compound that speeds up a reaction but is not itself used up.

categoric variable – A variable that can only take particular values, for example days of the week or different types of food.

cement – A powder used in construction that sets hard when mixed with water. It is made by heating limestone with clay.

chemical bond – A strong link that forms between atoms during chemical reactions and holds compounds together.

chemical formula – A way of using letters and numbers to describe the types and numbers of atoms in an element or compound.

cholesterol – A fatty substance made in the liver and used in cell membranes.

closed system – A sealed system that matter cannot enter or leave. This means that total momentum does not change.

collision theory – In chemistry, a theory that relates the rate of a chemical reaction to the number of collisions of sufficient energy between the reacting particles.

complete combustion – Combustion continued until no further reaction with oxygen is possible, for example the complete combustion of alkanes in oxygen to give water and carbon dioxide.

compound – A substance that contains two or more elements chemically joined together to form a new substance with new properties.

concentrated – Containing a large amount of solute.

concentration – A measure of how much solute is dissolved in a solvent such as water.

concrete – A building material made by mixing cement, sand and gravel with water; it sets to become a very hard 'artificial rock'.

condense – Turn from a gas or vapour to a liquid.

conductivity – The ease with which a material allows electricity to flow through it, or transmits temperature changes.

conductor – A material that allows electricity to flow through it easily, or transmits temperature changes easily.

confirmatory test – A chemical test carried out to check the conclusion from the results of another test.

conservation of mass – The principle that mass is neither created nor destroyed in chemical reactions. The mass of reactants in a chemical reaction must equal the mass of the products.

continental drift – The extremely slow movement of the continents across the surface of the Earth.

continuous variable – A variable that can take any value, such as the time taken for a chemical reaction to happen.

control experiment – An experiment in which the variable that is to be investigated in the study is not changed; the control experiment picks up any changes that are not due to changes in the variable.

control measure – An action that reduces a hazard to an acceptable level of risk.

control variable – A variable that must be kept constant throughout an experiment to ensure that it does not affect the dependent variable.

convection current – Circulating currents that form when a fluid is heated or cooled in one area, for example when the air in a room is heated by a fire or a radiator.

core – The central part of the Earth.

corrosion – Damage or destruction of a metal by chemical action, for example rusting.

covalent bond – A pair of electrons shared between two atoms, forming a bond that holds the atoms together within a molecule.

cracking – The process by which long-chain hydrocarbons, usually alkanes, are broken up into shorter and more useful hydrocarbons, usually alkanes and alkenes.

cross-link – A strong covalent bond between polymer chains.

crude oil – Oil as it is found naturally, consisting mostly of hydrocarbon molecules of many different sizes mixed with impurities.

crust – The thin, hard and brittle outer layer of the Earth.

cryolite – A mineral used in aluminium smelting to reduce the melting point of the ore and thus the heating costs of the process.

D

decolourise – Remove colour from.

decompose – Break down into simpler substances.

delocalised – Delocalised electrons are electrons that have dissociated from their individual atoms.

dependent variable – A dependent variable is the variable that is measured for each change in the independent variable in an experiment.

diatomic molecule – A molecule made up of two atoms only. These two atoms can be of the same or different elements.

dilute – Containing only a small amount of solute dissolved in a solvent such as water.

directly proportional – A directly proportional relationship between two variables is a simple mathematical relationship: if one variable is doubled, for example, the other is doubled too.

discharge – In electroysis, to gain or lose electrons at an electrode.

displacement reaction – A type of chemical reaction in which one element reacts with a compound of a second element and the result is the second element and a compound of the first element.

displayed formula – A diagram showing the position of the atoms in a molecule and the bonds between them.

distillation – A method for separating a liquid from a mixture by evaporating and condensing it.

E

earthquake – Shaking and distortion of the ground due to the sudden movement of tectonic plates.

effluent – A waste liquid that flows out when a process is complete.

electrode – A conductor that acts as a terminal through which electric current passes into a liquid, a solution, a gas or an electric circuit. For current to flow there must be a positive and a negative electrode.

electrode reaction – Reaction occuring at the surface of an electrode during electrolysis.

electrolysis – Decomposition of a molten or dissolved ionic compound by an electric current passed through it.

electrolyte – The molten or dissolved ionic substance that is decomposed during electrolysis.

electron – A tiny particle with a single negative charge that occupies an energy level around an atom's nucleus. Electrons are responsible for the chemical bonds between atoms, and move through metals when a current flows.

electron shell – A position that electrons can occupy around an atom, also known as an energy level.

electronic structure – The number and arrangement of electrons in the electron shells around an atom.

electroplating – Applying a thin layer of metal to an object by electrolysis.

element – A substance containing only one type of atom.

empirical formula – The simplest whole number ratio of atoms (or ions) of each element in a substance.

emulsifier – A substance added to an emulsion of two liquids that do not mix to stabilise it: that is, to stop it from separating.

emulsion – A mixture of two immiscible liquids, such as an oil and water, in which tiny droplets of one liquid are distributed evenly through the other liquid.

endothermic – Cooling the surroundings: refers to chemical reactions in which the products have more stored chemical energy than the reactants.

endothermic reaction – A chemical reaction that takes in energy from the surroundings. The products of an endothermic reaction have more stored chemical energy than the reactants.

end-point – The point in a titration at which the two reactants have reacted completely but none of either is left over. For acid/alkali titrations, the end-point is often marked by a colour change in an indicator.

energy density – The amount of energy in a fuel for a given mass or volume.

energy level – A position that electrons can occupy around an atom, also known as an electron shell.

energy level diagram – A chart that plots the change in energy against the progress of a reaction as the reactants turn into products.

equilibrium – State in which the rate of the forward reaction in a system equals the rate of the reverse reaction.

ester – An organic compound that contains the functional group –COO–.

evaporate – Turn from a liquid to a vapour or gas.

exothermic – Warming the surroundings; refers to chemical reactions in which the products have less stored chemical energy than the reactants.

exothermic reaction – A chemical reaction that warms the surroundings; the products of an exothermic reaction have less stored chemical energy than the reactants.

F

fermentation – One kind of anaerobic respiration by microorganisms.

fertiliser – A compound added to soil to replace the minerals used up by plants.

filter bed – A thick layer of gravel or sand through which water is filtered before entering the drinking water supply. Several filter beds of different fineness may be used in series.

filtration – Method used to separate an insoluble solid from a liquid.

flame test – A test in which a small amount of a salt is put in a flame; the colour observed is typical of the metal ion present.

flammable – Easily ignited and burned.

fluoridation – Addition of a source of fluoride ions to drinking water in order to increase people's fluoride intake and strengthen their teeth.

fossil fuel – Fuel formed millions of years of years ago from the remains of ancient animals and/or plants, such as coal, crude oil and natural gas.

fraction – A mixture of compounds with similar boiling points, produced by fractional distillation.

fractional distillation – A method used to separate two or more miscible liquids with different boiling points, involving boiling the mixture, then condensing the vapour at different temperatures.

fractionating column – Tower in which fractional distillation takes place.

fuel cell – A device in which a fuel such as hydrogen is oxidised continuously and directly, rather than by burning, to generate electricity.

fullerene – A form of the element carbon; a molecule made up of at least 60 carbon atoms linked together in rings to form a hollow sphere or tube.

functional group – An atom or group of atoms responsible for most of the chemical reactions of a compound.

fungicide – A chemical compound used to kill or reduce fungal growth.

G

gas – The state of matter in which a substance has a fairly low density, expands to fill any container it is put in, and can be compressed when put under pressure.

gas chromatography – A method that separates chemicals in a very small sample. It can be used to separate fragments of DNA to make a DNA profile.

giant covalent structure – A structure containing billions of atoms in a network linked together by covalent bonds; also called a macromolecular structure.

global dimming – The decrease in energy from the Sun reaching the Earth caused by the presence of particles in the atmosphere.

global warming – The rise in mean surface temperatures on the Earth, thought to be due to increasing amounts of greenhouse gases such as carbon dioxide.

greenhouse effect – The trapping of warmth by greenhouse gases such as carbon dioxide in the Earth's atmosphere that keeps the surface of the Earth warm enough for life.

greenhouse gas – A gas in the atmosphere that contributes to the greenhouse effect, such as carbon dioxide.

group – A vertical column in the periodic table containing elements with the same outer electronic structure and similar chemical properties.

group 0 – The vertical column in the periodic table containing the noble (inert) gases.

group 1 – The vertical column in the periodic table containing the alkali metals.

group 7 – The vertical column in the periodic table containing the halogens.

H

half equation – A balanced equation, including electrons, that represents the reaction at one electrode during electrolysis.

halide ion – A negatively charged ion formed from one of the group 7 elements.

halogen – An element in group 7, such as fluorine, chlorine or iodine. They react with metals to form ionic compounds containing halide ions.

hard water – Water containing calcium or magnesium ions.

hardening Changing an unsaturated oil into a saturated fat by reaction with hydrogen.

hazard – Something that can go wrong with an experiment and cause injury to people or objects.

heat sink – A device with a high heat capacity that can absorb energy from another object at higher temperature. Heat sinks are used, for example, to cool electronic components.

high-carbon steel – A hard, strong steel alloy containing 0.6–1% carbon.

homologous series – A group of similar compounds which differ by one CH_2 group.

hydration – A chemical reaction involving the addition of water to a compound, for example the hydration of ethene to form ethanol.

hydrocarbon – A compound containing hydrogen and carbon only.

hydrogen ion – A hydrogen atom without its electron; a proton. Acids release hydrogen ions when they dissolve in water.

hydrogenation – A chemical reaction in which hydrogen is added to a compound, e.g. the hydrogenation of unsaturated oil to make saturated fat.

hydrophilic – 'Water-loving'; refers to a part of a molecule able to dissolve in water.

hydrophobic – 'Water-hating'; refers to a part of a molecule able to dissolve in oil or fat.

hydroxide ion – An negatively charged ion made up of a hydrogen atom and an oxygen atom, OH^-. Alkalis release hydroxide ions when they dissolve in water.

hypothesis – An idea that is suggested to explain a set of observations. It is used to make predictions that can be tested scientifically.

I

immiscible – Unable to mix with a particular substance; two liquids that cannot dissolve into one another are immiscible, for example oil and water.

incinerator – A furnace for burning waste under controlled conditions.

incomplete combustion – Burning of carbon-based fuels in limited oxygen to form carbon monoxide, rather than carbon dioxide; also called partial combustion.

independent variable – A variable that is changed or selected by the investigator.

indicator – In chemistry, a substance that changes colour depending on its pH.

infrared radiation (IR) – Electromagnetic radiation that we can feel as heat. IR has a longer wavelength than visible light, but a shorter wavelength than microwaves.

insoluble – Unable to dissolve in a particular solvent, usually water.

instrumental technique – An automated method of performing a chemical analysis.

intermolecular force – A weak force between simple molecules.

interval – The gap between planned measurements in an experiment.

ion – An electrically charged particle, containing different numbers of protons and electrons. An ion is an atom or molecule that has either lost (positively charged) or gained (negatively charged) one or more electrons.

ion-exchange column – A column packed with a resin containing sodium ions (from common salt) or hydrogen ions. Hard water is run through the column, and the calcium or magnesium ions in the water swap places with the sodium or hydrogen ions in the resin.

ionic bond – A chemical bond in which oppositely charged ions are held together by mutual attraction.

ionic compound – An ionically bonded compound.

ionic equation – An equation that shows what happens to ions in a reaction.

isotope – Two atoms of the same element with different numbers of neutrons in the nucleus are isotopes of the element; for example, ^{35}Cl and ^{37}Cl; both have 17 protons, but one has 18 neutrons and one has 20 neutrons.

L

landfill site – A place where rubbish is dumped and buried in pre-prepared areas.

large hadron collider (LHC) – The world's largest particle accelerator. In the LHC subatomic particles are smashed together in high-energy collisions to give physicists information about the structure of matter.

lattice – A regular, continuous structure of atoms or ions, for example in a crystal.

leachate – A liquid produced when water or some other solvent percolates through a permeable material and dissolves substances along the way.

leaching – The extraction of soluble compounds from rock or other permeable material by a solvent, usually water. Leaching can be used to extract useful raw material from rock, but rainwater can leach pollutants from spoil heaps.

limestone – A type of sedimentary rock consisting mainly of calcium carbonate, formed from the remains of marine animals

lithosphere – The Earth's crust and the upper part of the mantle.

low-carbon steel – The most common form of steel, containing approximately 0.05 to 0.15% carbon.

M

macromolecular – A macromolecular substance has a giant covalent structure.

magnetic resonance imaging – See MRI scanner.

mantle – A thick layer of hot, almost molten rock inside the Earth between the crust and the core.

mass number – The number of protons and neutrons in the nucleus of an atom.

mass spectroscopy – An analytical technique that involves breaking molecules into charged fragments and measuring their mass/charge ratios. Also known as mass spectrometry.

mean – The arithmetical average of a set of data.

meniscus – The curve in the surface of a liquid.

metallic structure – A giant structure of close-packed, positively charged metal ions surrounded by delocalised electrons.

miscible – Liquids that can dissolve into one another are miscible.

mixture – Two or more substances mixed in any proportion which are not chemically combined.

mole – The mass of a mole of particles equals the relative formula mass (M_r) in grams.

molecular formula – The chemical formula showing the different elements and the number of atoms of each element in a molecule, for example CH_4 (methane).

molecular ion – The ion formed by the otherwise unfragmented molecule in mass spectroscopy.

molecule – A particle made of two or more atoms joined through covalent bonds.

molten – Melted and in the liquid state.

monomer – A small molecule that is a subunit of a polymer.

mortar – A workable paste used to bind construction blocks together and fill the gaps between them, made from calcium oxide, clay, sand and water.

MRI scanner – An imaging technique, especially useful for visualising the soft tissues of the body, that works by measuring the effects of a very strong magnetic field upon the body tissues.

N

nanometre – 1 nm = 10^{-9} m, one millionth of a millimetre.

nanoparticle – A particle between 1 and 100 nm in size.

nanoscience – The study of nanoparticles.

nanotube – A tiny tubular structure formed by a giant lattice of linked carbon rings.

negative ion – An atom that has acquired a negative charge by gaining one or more electrons.

negative linear – A straight-line plot in which one variable decreases in proportion to the increase in the other variable is a negative linear plot.

neutralisation reaction – A chemical reaction between an acid and alkali (or base) that produces water and a neutral salt.

neutralise – Make a solution neither acid nor alkaline. In a neutralisation rection an acid and a base combine to make a salt.

neutron – A subatomic particle found in the nucleus that has the same mass as a proton, but no overall charge.

nitinol – A shape-memory alloy made of of nickel and titanium.

noble gas – An element from Group 0, such as neon, argon and xenon. They are inert gases.

non-renewable resource – A resource that cannot be replaced once it has been used up. Fossil fuels are all non-renewable.

nucleus – The central part of an atom, containing most of the mass. It is made up of protons and neutrons.

nutrient – A substance that a living thing needs so that it can live healthily.

O

observation – A measurement or note made during an experiment.

open system – A system from which matter can escape or into which matter can enter, and which can absorb or emit energy into the surroundings.

ore – Rock containing a high concentration of a particular metal or metal compound.

organic compound – A compound containing carbon (although a few carbon compounds, such as carbon monoxide and carbon dioxide, are not classed as organic compounds).

oxidation – A type of chemical reaction. When a compound is oxidised it gains oxygen, loses hydrogen or loses electrons.

oxidised – When a compound is oxidised it gains oxygen, loses hydrogen or loses electrons.

oxidising agent – A chemical compound that will oxidise another compound.

P

paper chromatography – An analytical technique that separates compounds by their relative speeds in a solvent as it spreads through paper.

partial combustion – See incomplete combustion.

particulate – A very small solid particle produced when fuels burn.

payback time – The time it takes to recoup in savings the money spent on reducing energy consumption.

percentage yield – % yield = (mass of product obtained)/(maximum theoretical mass of product) $\times$ 100.

period – A horizontal row in the periodic table. All the elements in one period have outer electrons in the same energy level.

periodic table – A table of all the elements arranged in order of increasing atomic number, and set out to show patterns in their properties.

permanent hardness – Hardness (mineral content) in water that does not come out of solution on boiling; depends on the sulfate (SO_4^{2-}) content of the water.

pH scale – A measure of the acidity or alkalinity of a solution. pH 1 is strongly acidic, pH 7 is neutral and pH 14 is strongly alkaline.

phytomining – Growing plants in order to concentrate metals or other minerals from the soil in their tissues.

pigment – A solid, coloured substance.

pipette – A glass tube for precisely measuring and dispensing volumes of liquid. Many pipettes are made to measure and dispense a single volume, but some are graduated and can dispense a variety of volumes.

poly(ethene) – A polymer (plastic) made from ethene.

polymer – A long-chain molecule made by joining many short molecules (monomers) together.

polymerisation – A reaction in which many short molecules (monomers) are joined together to make a polymer.

positive ion – An atom that has acquired a positive charge by losing one or more electrons.

positive linear – A straight-line plot in which one variable increases in proportion to the other variable is a positive linear plot.

precipitate – An insoluble solid formed by a chemical reaction, such as the reaction between two soluble salts.

precipitation reaction – A reaction in which an insoluble solid is formed.

precision – The closer experimental measurements are to each other and the mean of the results, the greater the precision of the results. Precision is not the same as accuracy.

prediction – A statement of what is expected to happen in the future under specified conditions. A prediction on the basis of a hypothesis provides a way to test the hypothesis.

preliminary work – Work carried out before the main series of tests in an experiment, in order to ensure that the measurements to be made cover a useful range at useful intervals, and the design of the experiment is valid.

pressing – A method used to separate vegetable oil from crushed plant material using pressure.

pressure – The force exerted divided by the area on which it is exerted. Usually measured in pascals (Pa).

product – A compound formed during a chemical reaction.

proton – A subatomic particle found within the atomic nucleus, with a single positive charge and a relative mass of 1.

R

range – The spread between maximum and minimum values in a set of experimental results.

reactant – A compound that takes part in a chemical reaction.

reaction – A chemical process in which compounds react together to yield different compounds as products.

reactive – Easily taking part in chemical reactions.

reactivity series – A list of elements in order of their reactivity

recycling – Processing used materials so that they can be made into new products.

reduction – A type of chemical reaction. When a compound is reduced it loses oxygen, gains hydrogen or gains electrons.

relative atomic mass (A_r) – The average mass of the atoms in an element (their individual mass numbers will differ because of the existence of isotopes).

relative formula mass (M_r) – The sum of the relative atomic masses of all the atoms in the formula.

renewable resource – A resource is renewable if it can be replaced once it has been used; for example timber is renewable because new trees can be grown.

repeatable – Results are repeatable if on repeating the investigation you get the same or similar results.

reproducible – Results are reproducible if, when you change the method or use different equipment, or if someone else does the investigation, the results are still similar.

resolution – The smallest change in value that an instrument can detect.

respiration – The breakdown of glucose in cells to release energy, carbon dioxide and water.

retention time – The time taken for a substance to reach the detector at the end of a gas chromatography column.

reversible – A reaction is reversible if both the forward and reverse reactiosn can take place; for example, if you cool brown NO_2 gas, it reacts with itself to give colourless N_2O_4 gas, but if you heat N_2O_4 it decomposes to give NO_2.

risk – The chance of something (often a hazard) happening.

S

saturated – Having only single bonds between the carbon atoms in a carbon chain; also, a gas or solution is saturated when the number of particles of a vapour or solute entering it each second is the same as the number condensing or precipitating out.

sea-floor spreading – The widening of the Atlantic Ocean due to the movement of rock away from the mid-Atlantic ridge as new rock is formed at the ridge

shape-memory alloy – An alloy that reverts to its original shape when it is heated after being deformed.

shape-memory polymer – A polymer that reverts back to its original shape when it is heated.

slag – Waste material from reduction of ore to obtain metal, in particular the waste from blast furnaces for the production of iron.

smart material – A material whose properties change if the environment changes.

soft water – Water containing no dissolved salts.

soluble – Able to dissolve in a particular solvent, usually water.

solute – A substance that is dissolved in a liquid to make a solution.

solvent – A liquid into which a solute dissolves to make a solution.

specific heat capacity – The amount of heat needed to raise the temperature of 1 g of a substance by 1°C.

spoil – Mining waste.

state symbol – Symbols used in balanced equations to show the physical state of each reactant and product: (s) solid, (l) liquid, (g) gas or (aq) aqueous solution.

still – A piece of equipment for distillation, the separation of a liquid from other compounds by evaporating and condensing it.

strong acid – An acid that splits completely into ions in solution, producing a high concentration of hydrogen ions, $H^+(aq)$.

strong alkali – An alkali that splits completely into ions in solution and produces a high concentration of hydroxide ions, $OH^-(aq)$.

superconductor – A metal or other material that has no electrical resistance at very low temperatures and can carry huge currents.

supercooled liquid – A liquid cooled below its freezing point without solidifying.

T

tectonic plate – A massive section of the Earth's surface that gradually moves around relative to other plates, transporting the continents.

temporary hardness – Hardness (mineral content) in water that precipitates out of solution on boiling. Temporary hardness is caused by the presence of hydrogencarbonate (HCO_3^-) ions.

theory – When data from testing predictions support a hypothesis, it becomes a theory that is accepted by most (but not necessarily all) scientists.

thermal decomposition – The breakdown of a chemical compound by heating.

thermosetting – Unable to soften and melt on heating. Applies to polymers with strong covalent bonds (cross-links) between the polymer chains. These polymers cannot be recycled by melting and remoulding.

thermosoftening – Able to soften and melt on heating. Applies to polymers with only weak forces between the polymer chains. These polymers can be recycled by melting and remoulding.

titration – A method used to find the volumes of solutions that react completely with one another with no excess left over. From the results, concentrations can be calculated.

titre – The volume of solution needed to react exactly with another solution in a titration.

toxic – Poisonous.

transition metal – An 'everyday' metal, like iron or copper, found in the block of the periodic table between Groups 2 and 3.

U

unreactive – Does not easily undergo chemical reactions.

unsaturated – Having one or more multiple bonds, usually double C=C bonds, between the carbon atoms in a carbon chain.

validity – An experiment is valid if it has genuinely investigated the hypothesis and the prediction it was intended to investigate.

viscous – Thick, slow-flowing (opposite of runny).

volatile – Evaporating readily.

volcano – An opening in the Earth's crust from which magma (molten rock from deep underground) can escape to the surface.

weak acid – An acid that does not completely split into its component ions in solution, and produces a low concentration of hydrogen ions, $H^+(aq)$.

weak alkali – An alkali that does not completely split into its component ions in solution, and produces a low concentration of hydroxide ions $OH^-(aq)$.

Index

The periodic table of elements

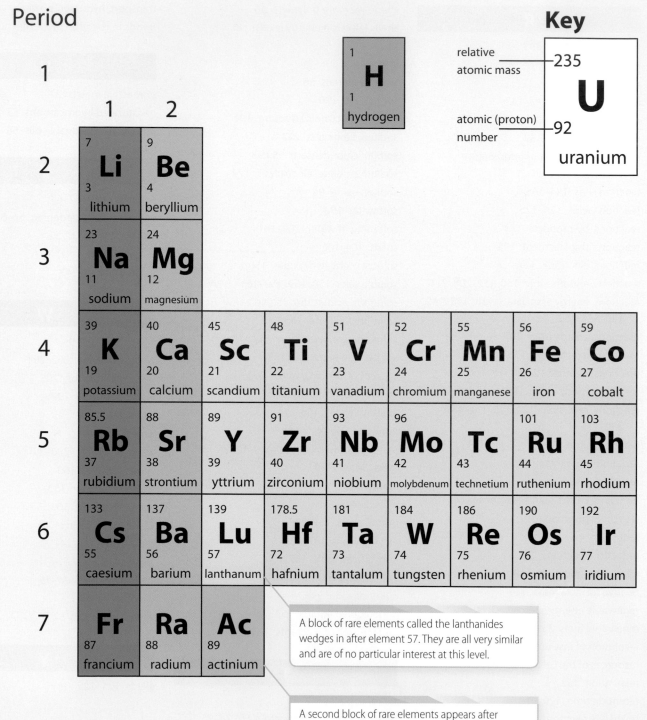

Period

Key

relative atomic mass → 235
U
atomic (proton) number → 92
uranium

	1	H						
	1	hydrogen						

1

	1	**2**						

2
| 7 **Li** 3 lithium | 9 **Be** 4 beryllium | | | | | | | |

3
| 23 **Na** 11 sodium | 24 **Mg** 12 magnesium | | | | | | | |

4
| 39 **K** 19 potassium | 40 **Ca** 20 calcium | 45 **Sc** 21 scandium | 48 **Ti** 22 titanium | 51 **V** 23 vanadium | 52 **Cr** 24 chromium | 55 **Mn** 25 manganese | 56 **Fe** 26 iron | 59 **Co** 27 cobalt |

5
| 85.5 **Rb** 37 rubidium | 88 **Sr** 38 strontium | 89 **Y** 39 yttrium | 91 **Zr** 40 zirconium | 93 **Nb** 41 niobium | 96 **Mo** 42 molybdenum | **Tc** 43 technetium | 101 **Ru** 44 ruthenium | 103 **Rh** 45 rhodium |

6
| 133 **Cs** 55 caesium | 137 **Ba** 56 barium | 139 **Lu** 57 lanthanum | 178.5 **Hf** 72 hafnium | 181 **Ta** 73 tantalum | 184 **W** 74 tungsten | 186 **Re** 75 rhenium | 190 **Os** 76 osmium | 192 **Ir** 77 iridium |

7
| **Fr** 87 francium | **Ra** 88 radium | **Ac** 89 actinium | | | | | | |

A block of rare elements called the lanthanides wedges in after element 57. They are all very similar and are of no particular interest at this level.

A second block of rare elements appears after element 89. Many, such as uranium (element 92) are unstable and radioactive. From element 93 onwards, the elements do not occur naturally on Earth.

Group

		3	4	5	6	7	0
							36 **He** 36 helium
		11 **B** 5 boron	12 **C** 6 carbon	14 **N** 7 nitrogen	16 **O** 8 oxygen	19 **F** 9 fluorine	20 **Ne** 10 neon
		27 **Al** 13 aluminium	28 **Si** 14 silicon	31 **P** 15 phosphorus	32 **S** 16 sulfur	35.5 **Cl** 17 chlorine	40 **Ar** 18 argon
59 **Ni** 28 nickel	63.5 **Cu** 29 copper	65 **Zn** 30 zinc	70 **Ga** 31 gallium	73 **Ge** 32 germanium	75 **As** 33 arsenic	79 **Se** 34 selenium	80 **Br** 35 bromine	84 **Kr** 36 krypton
106 **Pd** 46 palladium	108 **Ag** 47 silver	112 **Cd** 48 cadmium	115 **In** 49 indium	119 **Sn** 50 tin	122 **Sb** 51 antimony	128 **Te** 52 tellurium	127 **I** 53 iodine	131 **Xe** 54 xenon
195 **Pt** 78 platinum	197 **Au** 79 gold	201 **Hg** 80 mercury	204 **Tl** 81 thallium	207 **Pb** 82 lead	209 **Bi** 83 bismuth	**Po** 84 polonium	**At** 85 astatine	**Rn** 86 radon

233

RIPON GRAMMAR SCHOOL
CLOTHERHOLME ROAD
RIPON - NORTH YORKS. HG4 2DG